Jesuit Contribution to Science

Agustín Udías

Jesuit Contribution to Science

A History

Springer

Agustín Udías
Department of Geophysics and Meteorology
Complutense University of Madrid
Madrid, Spain

ISBN 978-3-319-08364-3 ISBN 978-3-319-08365-0 (eBook)
DOI 10.1007/978-3-319-08365-0
Springer Cham Heidelberg New York Dordrecht London

Library of Congress Control Number: 2014947464

Printed on acid-free paper

Springer is part of Springer Science+Business Media (www.springer.com)

Preface

The presence of the Jesuits in different fields of the natural sciences is an interesting phenomenon that has attracted academic and general attention and can be found in the literature. The Jesuits are popularly known as religious persons who are involved in scientific work, and they appear as such in some science fiction novels. A few years after its founding in 1540 by Saint Ignatius of Loyola, the Society of Jesus undertook its educational endeavor as the key instrument of its apostolic work. From the beginning, as a novelty for the time, a special attention was given in the first colleges to the teaching of mathematics and astronomy. This coincided with the origins of modern science, and Jesuit professors were in contact with many of its key figures, such as Galileo, Kepler, Huygens, and Newton. Jesuit missionaries introduced European mathematics and astronomy to China and India; made the first maps of the unknown regions of America, Asia, and Africa; and brought to Europe the first news about the geography, animals, and plants of those lands. The scientific work of the Jesuits during the sixteenth to eighteenth centuries until the suppression of the order in 1773 is the subject of a large number of studies. Less known is the Jesuit scientific contribution during the nineteenth and twentieth centuries, especially in the fields of astronomy, geophysics, and meteorology. Although a good number of books and articles have been published about partial aspects of the history of Jesuit scientists, especially for the first period, there is no complete history that covers both periods in a unified form. This is the aim of the present book.

Among other things, this book tries to answer three main questions: what moved the Jesuits to give an important place to mathematics and astronomy in their first colleges in the sixteenth and seventeenth centuries, why did they after their restoration in the nineteenth century establish a worldwide network of observatories, and what happened after the 1970s that caused the closing of the observatories and the decrease of Jesuit presence in science. Given the wide extension of the material this book tries to cover, I have had to select the persons and contributions that I consider most relevant, without attempting to study them at the depth that would be possible. It may be intriguing to some that a religious order of the Catholic Church has had a continuous contribution to make to science still maintained today. This book may help to answer these questions by giving an overall image, as complete as possible, of the history of that contribution. Moreover, I try to find a relation between the scientific work of the Jesuits and their spirituality. Thus, this book is directed to all those interested in the scientific work of the Jesuits and who are curious about the reasons behind it.

The book begins with a discussion on Christopher Clavius, Professor of Mathematics in the Roman College between 1567 and 1595, the initiator of this tradition (Chap. 1). It was in his time that relations between the Jesuits and Galileo began, an important factor in the history of science. The first five chapters cover the Jesuits' scientific contributions especially in mathematics, astronomy, and physics, before the suppression of the Society in 1773. Contributions of professors of mathematics, astronomy and physics are presented in Chap. 2. Two key figures of this period are Kircher and Boscovich, both discussed in Chap. 3. Scientific work progressed at the almost 600 Jesuit colleges in Europe where 32 astronomical observatories were installed, the first in several countries. Chapter 4 presents the extraordinary adventure of the Jesuit astronomers in China, starting with the arrival to Beijing of Matteo Ricci in 1601. For 150 years, the Jesuits were the directors of the Imperial Astronomical Observatory and held the dignity of mandarins. Less known is the work of Jesuit astronomers in India and their failed journey to Indochina. Chapter 5 deals with Jesuit naturalists, geographers, and explorers. The discovery of America gave the Jesuits the possibility to study and to make known in Europe the characteristics and novelties of the geography and fauna and flora of those hitherto unknown lands. The Jesuits explored those new lands from Canada to Patagonia, and they were the first Europeans to navigate the great rivers, the Mississippi, the Amazon, and the Orinoco. Their interest in geography led to an impressive work of cartography, preparing the first maps of many regions of America, China, India, Tibet, and Ethiopia. This work was interrupted by the suppression of the order by Pope Clement XIV in 1773.

After their restoration in 1814, the Jesuits resumed their scientific work. To this period, the book dedicates its last five chapters. In this period, the Jesuits' scientific work changed with respect to the earlier periods and had a somewhat apologetic character aimed against those who attacked the Church at that time in opposition to science. The Jesuits' scientific work was an answer to such attacks. From 1825, the Jesuits established a new network of about 70 observatories throughout the world, as presented in Chap. 6. Many of those installed in Africa, Asia, and Central and South America were the first such scientific institutions in those countries. The contribution to the study of terrestrial magnetism (Chap. 6), tropical hurricanes (Chap. 7), and earthquakes (Chap. 8) was especially important. Exceptional among the most relevant figures, presented in Chap. 9, are Secchi in astrophysics, Perry in geophysics and astronomy, Macelwane in seismology, and Teilhard de Chardin in geology and paleontology, who was also famous for his influential thought on the relation between evolution and Christian faith. At the present time, the Jesuits manage 133 universities and more than 300 secondary schools. In Chap. 10, I deal with modern Jesuit scientists in universities and schools, where they combine teaching and research, and there is special mention of those dedicated to biology.

After this overview of the five centuries of the Jesuit scientific tradition, unique in the Catholic Church, one may still have questions about the reasons behind it. In the Epilogue, I try to answer this question by linking dedication to science with the specific Jesuit spirituality. Saint Ignatius insisted that the Jesuits strive to find God in all things. In this spirit, the Jesuits endeavor in their apostolic work to enter the field of scientific research as a mission territory. Thus, patient work in observatories and laboratories is for the Jesuits as appropriate as preaching or pastoral work in parish churches. For them, science as knowledge and as an instrument for the good of mankind is also a means for the propagation of the Christian faith. Following this long tradition, the Jesuits have trodden an uncommon path to the frontiers where the Christian message has not yet reached.

Given the large amount of material to be covered, broad brushstrokes are needed to present this interesting history. Many items and persons that may have deserved mention had unfortunately to be left out. But I have made an effort to present at least the most representative items. The Jesuits and their work are presented in the context of the history of science. In particular, I delve into the relation of the Jesuits with important scientists of their time. In order to highlight Jesuit scientists biographical information (dates of birth and death) is only given for them. The bibliography, though quite large, has been restricted to the most important works. Except for a few of them, books by Jesuit scientists referred to in the text are not included in the bibliography. I have made abundant use of the *Diccionario Histórico de la Compañía de Jesús* edited by O'Neill and Domínguez and of the *Bibliothèque de la Compagnie de Jésus* by Sommervogel, verifying in them all the Jesuits mentioned, although not quoting them explicitly in each case. The book discusses a total of 363 Jesuit scientists: 50 mathematicians; 45 physicists; 110 astronomers; 70 geophysicists, geologists, and meteorologists; 4 chemists; 21 biologists; 39 naturalists, geographers, and cartographers; and 24 explorers (Appendix 1). In fact this is only a small sample of the large number of Jesuits dedicated to science since the foundation of the Society of Jesus. The *Dictionary of Scientific Biography*, edited by Gillespie, included 29 of the Jesuits mentioned here (Appendix 2). These may be considered the most representative.

Finally, I want to thank all those who have contributed to this work, first of all Joseph Munitiz S.J. of Campion Hall (Oxford), who patiently corrected the English text. I acknowledge the help from librarians and archivists of Jesuit libraries and archives, especially those of Rome, Paris, Cologne, and London, and from many Jesuit observatories and universities that made available their publications and documents. I also thank the many Jesuit companions who have encouraged me in my work.

Madrid, Spain Agustín Udías

Contents

Chapter 1
Clavius and Mathematics in the *Collegio Romano*

1.1 Mathematics in the First Jesuit Colleges

The history of Jesuit involvement in the sciences begins with the founding of the first Jesuit schools, educational work that continues to the present day.[1] A few years after its founding in 1540 by Saint Ignatius of Loyola (1491–1556), the Society of Jesus began to get involved in the problem of education, starting with the training of its own members who had not had sufficient studies. In 1544 there were seven colleges or residences for Jesuit students near the universities of Paris, Louvain, Cologne, Padua, Alcalá de Henares, Valencia and Coimbra. A few years later, Jesuits began to establish their own institutions, they themselves taking charge of teaching and accepting non-Jesuit students. Soon most of the Jesuit schools were for lay students. In 1548 in Messina (Sicily) the Jesuits established what is considered the first Jesuit school with the characteristics of all future schools: exclusively Jesuit professors, lay students, fixed income which allowed free access and a program which included humanities, philosophy and theology. In 1546 at the request and with the support of Francisco de Borja (1510–1572), at that time Duke of Gandía (he later became a Jesuit and was elected in 1565 Superior General of the Order), a college was established in Gandía (Valencia, Spain). Pope Paul III gave it university status (*Studium generale*) and it became the first Jesuit university, though in fact it was always a very modest institution. The establishment of colleges and universities was for Saint Ignatius a new orientation of the recently created religious order, which

[1] Biographic and bibliographic data are taken from Charles E. O'Neill and José M. Domínguez (eds.), *Diccionario histórico de la Compañía de Jesús (DHCJ)*. Madrid: Universidad Pontificia Comillas, vols. 1–4 (2001) and Charles Sommervogel, *Bibliothèque de la Compagnie de Jésus. Bibliographie*. Brussels, vols. 1–10 (1890–1909); Charles C. Gillispie (ed.), *Dictionary of Scientific Biography*, vols. 1–16. New York: Charles Scribner and Sons (1970–1980). Two recent general summaries about Jesuits in science are: Jaime Fernández and Eduardo di Vita, *Presença dos jesuitas no mundo científico*. São Paulo (Brazil): Edições Loyola (2004) and François Euvé, *Mathématique, astronomie, biologie et soin des âmes. Les jésuites et les sciences*. Brussels: Lessius (2012).

A. Udías, *Jesuit Contribution to Science: A History*,
DOI 10.1007/978-3-319-08365-0_1

he had not envisaged in the beginning. However, colleges became, already during Saint Ignatius' life, the most important instrument of the Order's apostolic work. At his death in 1556 the Society of Jesus already had 35 colleges in different countries of Europe and one in India. In this way 10 years after its founding the Society had made educational work the key element of its apostolic work.[2] The need and the social demand for schools in the rapidly growing cities of sixteenth and seventeenth century Europe explains the rapid expansion of Jesuit schools which provided a new style and pedagogical practice. This method of apostolic work fitted well a new religious order that had been founded by ten Masters of Arts (*Magistri Artium*) of the University of Paris. In 1551 Jesuits founded a modest college in Rome which soon grew, and in 1556 Pope Paul IV gave it university status with chairs of philosophy and theology. Twenty years later in 1576 this was confirmed by Pope Paul V. In 1584, with the generous support of Pope Gregory XIII, a magnificent building was inaugurated and the school began to be known as the Roman College (*Collegio Romano*) (Fig. 1.1); since 1873, it has been named the Gregorian University, in memory of its patron. Saint Ignatius himself gave great importance to this school which became the model to be followed by all Jesuit colleges.[3]

The great success of the first Jesuit schools was due in part to a new orientation in teaching based on the method followed in the university of Paris, the *modus parisiensis*, and the use of programs adapted to the times. The Jesuits placed emphasis on the order and discipline to be followed in their teaching and on the ordered distribution of classes according to the needs and capacities of the students. Exercises by the students, such as repetitions and disputations, were an important practice which complemented professorial lectures. On some occasions, schools held special academic acts or public disputations (*disputationes*) before invited

Fig. 1.1 *Collegio Romano*, Rome (Arch. Romanum S.I.)

[2] John W. O'Malley, *The first Jesuits*. Harvard: Harvard University Press (1993) (Chap. 6, The schools), 200–242.

[3] Ricardo García Villoslada, *Storia del Collegio Romano, dal suo inizio alla soppressione della Compagnia de Gesù (1773)*. Rome: Universitàt Gregoriana (Analecta Gregoriana vol. LXVI) (1954); Sabino Maffeo, Il Collegio Romano e l'insegnamento de la scienze. In: Aldo Altamore and Sabino Maffeo (eds.), *Angelo Secchi. L'avventura scientifica del Collegio Romano*. Foligno: Quater (2012), 15–42.

audiences, which served to give publicity to the academic progress of the students and served as an effective tool of propaganda. They also helped to gain economic support for the schools and thus secure their finances with fixed rents that allowed their free character. Jesuit schools came to satisfy a need which was not being met in those times and was a clear answer to a true social demand; this explains their rapid spread through the main towns of Europe. Jesuits took charge also of universities with full courses of philosophy and theology. To obtain university status, schools needed a Papal or Royal decree which granted the right to award academic degrees. The Roman College had university rank from 1584 and among the first Jesuit universities were those of Gandía, Messina, Coimbra, Palermo, Vienna and Prague. The network of Jesuit schools and universities spread rapidly through Europe so that in the eighteenth century there were about 625 of them. Jesuits established also colleges and universities in America, India and the Philippines.

Programs in Jesuit schools covered three cycles, grammar (humanities), philosophy and theology, but not all the schools had all three. Most of them resembled today's secondary schools, with programs in humanities and some elements of philosophy and theology. Only in those considered as major colleges were there complete programs of philosophy and theology. Each cycle lasted, in general, three years, though in some cases philosophy was reduced to two years and, for Jesuit students, theology was extended to four years. The first cycle comprised the humanities with emphasis on the mastery of languages, especially, with the study of the Latin and Greek classics. The three years dedicated to philosophy were called: logic, physics and metaphysics, though programs varied in different schools. Teaching corresponded to Aristotle's philosophy, so that in physics the subject matter was that of his texts on natural philosophy: *Physica*, *De Coelo*, *De generatione et corruptione*, *De anima* and *Meteorologica*. To the philosophy cycle was added also the teaching of mathematics which included, besides pure mathematics, that is, geometry, arithmetic and algebra, other disciplines of applied mathematics such as astronomy, optics, mechanics and music.

The first official mention about teaching mathematics in Jesuit schools is found in the *Constitutions,* written by Saint Ignatius, part IV, where when treating about the subjects to be taught in the Jesuit universities he says: "there must be taught logic, physics, metaphysics and moral theology and also mathematics, with the due moderation for the end that is intended", and with reference to the professors he says: "There should be some professors who as public lecturers read philosophy or mathematics".[4] Jerónimo Nadal (1507–1580), a collaborator of Saint Ignatius, had proposed from as early as 1548, for the programs of the College of Messina, classes of mathematics and astronomy with specific mention of geometry, arithmetic, astronomy (*sphaera*), and the theory of the astrolabe and the planets. Nadal in 1552 proposed some of the books to be used in the classes of mathematics which included, besides the classic works of geometry and astronomy of Euclid and Ptolemy, books by modern authors such as Jordanus Nemorarius, (mathematician), Johann Müller (Regiomontanus, an astronomer famous for his astronomical tables),

[4] Saint Ignatius of Loyola, *Obras Completas. Edición Manual.* Madrid: Biblioteca de Autores Cristianos, (1991). Constituciones (Constitutions), Part IV, Chap. 12, C (451) and C (458).

Oronce Finé (cosmography) and Erazmus Ciolek Witelo (optics).[5] In the Roman College, the first professor of mathematics was the Spaniard Baltasar Torres (1481–1561), who had spent several years in Sicily before joining the Jesuits and knew there the renowned astronomer Francesco Maurolico, professor at the university of Messina. Torres in 1553 proposed a comprehensive program for the students of philosophy including geometry, arithmetic and astronomy. Torres besides the classic and medieval texts used also those of contemporary authors like Francesco Maurolico, Pedro Nunes and Jacques Pelletier. He can be considered the initiator of the brilliant career of the professors of mathematics of the Roman College.

The years when Jesuits established their first schools coincided with the beginning of the so-called scientific revolution and the origin of modern science. In 1543 Nicolaus Copernicus had published his book, *De revolutionibus orbium celestium*, in which he proposed the heliocentric system, which brought a radical change to astronomy and cosmology. Between 1572 and 1600 Tycho Brahe had carried out in his observatory in Denmark astronomical observations which helped the establishment of the new astronomy. His studies of the observation of a nova star in 1572 and of a comet in 1577 were decisive for the dismantling of the medieval view of the celestial spheres and established the orbital motion of planets. He held to the central position of the Earth, but all the other planets rotated around the Sun, a theory that would be held later by Jesuit astronomers. In 1595 Johannes Kepler published his first book defending Copernicus' system and between 1609 and 1619, using Brahe's observations, discovered the three laws of planetary motion. These astronomical innovations were accompanied by a new orientation of science which united the appeal to observations and experiments with mathematical formulation. This new approach to the knowledge of natural phenomena received the name of 'new science', a title given by Niccolò Tartaglia to his book, *Nuova scienza*, published in 1537 and to which William Gilbert contributed in 1600 with his work about magnetism. These developments could not be absent from the teaching in the new Jesuit schools. But since the philosophy professors, who explained what was then called *physica*, had to abide by the Aristotelian doctrine as already established, the new ideas were promoted thanks to their mathematical formulation by the teaching of the mathematics professors. Hence, it is through the teaching of mathematics that in Jesuit colleges a door was opened to the introduction of modern science.

1.2 Christopher Clavius, the Beginning of a Tradition

The beginning of the scientific tradition of the Society of Jesus is tied to the figure of Christopher Clavius (1537–1612), born in Bamberg, Germany (Fig. 1.2).[6] The exact German version of his family name is uncertain; it could have been Klau or

[5] Ladislaus Lukács, *Monumenta Paedagogica Societatis Iesu*, Vols. I–VII. Rome: Institutum Historicum Societatis Iesu, (1986), Vol. I, 25–26.

[6] Charles Naux, Le père Christophore Clavius (1537–1612). Sa vie et son œuvre; *Revue des Questions Scientifiques*, 154, (1983) 55–67, 181–193, 325–347; Eberhard Knobloch, Sur la vie et

Fig. 1.2 Christopher Clavius (1537–1612), founder of the Jesuit scientific tradition (Udías, 2003)

Schlüssel, for it appears always in a Latinized form. His love for his birth place is clear as he always added it to his name in the titles of his books (*Clavius Bambergensis*). Nor is anything known with certainty about his preliminary studies, prior to his being received into the Society by Saint Ignatius himself in Rome in 1555. It is presumed that he studied at the Jesuit College in Bamberg. In 1556 he began his philosophical studies in Coimbra, and it was there that he observed the solar eclipse of 1559, the first indication of his astronomical interests.

He returned to Rome in 1560 to finish his course in philosophy and started those in theology; he was ordained to the priesthood in 1564. It seems that his mathematical interests began very early, since in 1567, he replaced Torres in the chair of mathematics and held that post until 1595. In the final years of his life, up to his death in 1612, he remained active, supervising editions of his books and following recent astronomical observations and theories, such as those of Copernicus, Tycho Brahe, Galileo and Kepler. Clavius was fundamentally a great teacher, and Jesuit mathematicians and astronomers who came after him always considered him to be one who initiated the tradition of science, and in particular the interest for mathematics in the Society. Between 1570 and 1612, Clavius produced an extensive series of publications, not less than 23 books. Twelve of these are text books of arithmetic, geometry and

---

l'oeuvre de Christophore Clavius (1538–1612). *Revue d'Histoire des Sciences*, 41 (1989), 331–356; James M. Lattis, *Between Copernicus and Galileo. Christopher Clavius and the collapse of Ptolemaic cosmology*. Chicago: The University of Chicago Press, (1994).

algebra; three are commentaries on the classic geometrical works of the Greek mathematicians Euclid and Theodosius, and also on the astronomy of Johannes Sacrobosco, while six are devoted to defending the reform of the Gregorian calendar. A year before his death in 1611, he published his complete works (*Opera Mathematica*) in five volumes. His literary output covered the whole realm of mathematical disciplines, as they were then known, and his textbooks continued in use for many years.

Among Clavius's works, special mention should be made of his edition with commentary of the 15 books of Euclid (Books 14 and 15, although attributed to Euclid, are by later Greek authors): *Euclidis elementarum libri XV (*1574). This book won for Clavius the title of "Euclid of our age", and went into three editions while the author was still alive, and another in 1691. He used the Latin translation of Federico Comandino with some corrections. In the prologue to this work, Clavius stresses the importance of geometry for an understanding of nature, since the world as a whole is the result of geometry. With this remark, Clavius came close to the trends of the "new science", which required knowledge of mathematics for the description of natural phenomena. In his *Euclidis* Clavius presents not only Euclid's proofs, but adds long commentaries with new material taken mainly from Proclus (a Greek mathematician of the fifth century), but also taken from modern authors and original ones by himself. In some cases he presents several proofs for the same theorem in order to show that there is not a single way to arrive to the same conclusion. Among the problems to which Clavius paid special attention were those concerning parallel lines and polygons inscribed and circumscribed to a circle, where he clarifies Euclid's propositions. Perhaps his major innovation was the addition of numerical solutions to geometrical proofs. This approach was further developed in his *Geometria Practica.* Greek mathematicians had separated geometrical theory from its applications, Clavius, however, put together theory and practice in a comprehensive view of geometry. In some cases his numerical solutions to geometric problems were close to the developments that René Descartes made later in his analytic geometry. For many years, the *Commentarius* became the geometry text-book in most of the colleges run by the Jesuits; it illustrates the emphasis placed on geometry in the teaching of mathematics in these colleges.

Clavius published another work of practical character on arithmetic entitled *Arithtmetica Practica*, which was especially successful, appearing in 25 editions by 1738 (10 in Latin and 15 in Italian). This shows what an excellent teacher and popularizer of arithmetic Clavius was. Its success came partly from the book's clarity, and partly from the need, felt at that time for a work that would facilitate calculations in the growing trade and industry. Among the innovations introduced by Clavius in this work, was the placing of a dot at the end of numbers to separate them, and above numbers to differentiate thousands in large numbers. The present-day use of the dot in decimal notation derives from this. The book includes also an original algorithm to perform division similar to the modern one. Clavius completed his mathematical treatises with a text-book on algebra (*Algebra*, 1608). Algebra, although originated by Diophantus of Alexandria in the fourth century, had been developed by the Arabic mathematician Al-Khwarizmi and his followers since the eighth century and was introduced into Europe with the Latin translations of their books on the thirteenth century. Gerolamo Cardano in 1545 and François Viète in 1591 gave a new impulse to algebra. Clavius used in his *Algebra* the works of the Portuguese Pedro Nunes, whom he could have known during his stay in Coimbra. In this work, of less importance than his geometry, Clavius introduced some new developments and innovations in the notation and added a collection of 70 problems solved by algebra. He used different signs for the unknowns and their powers and a dot between unknowns and numerical factors. Although Clavius' *Algebra* was soon superseded by Descartes' analytical geometry, it was a step in this direction. Marin Mersenne, Descartes' friend and confident, would advise reading Clavius' Algebra, for he said that there one could always find interesting questions.

The principal astronomical work of Clavius was his commentary on the *Tractatus de sphaera* (Treaty on the sphere) written by the Englishman Johannes Sacrobosco (John Holywood), a professor in Paris in the thirteenth century. This book is a simplified presentation of the astronomy of Ptolemy which gained great popularity and became the most used textbook throughout the Middle Ages. Ptolemy had completed his great synthesis of astronomy in Alexandria in the second century, collecting in one work all ancient Greek astronomical learning. His treatise was known later by its Arabic title, *Almagest*, which entered Europe by its Latin translation from Arabic in the twelfth century. In his own work, *In sphaeram Joannis de Sacrobosco commentaries* (Commentary on the Sphere of John de Sacrobosco), published in 1570 but edited six times (the last in 1611) before his death, Clavius sets out his view of what was then contemporary astronomy. His commentaries are really much more extensive than the original text of Sacrobosco. He presents a geocentric Ptolemaic astronomy, insisting on the physical reality of the celestial spheres, both the eccentric and the epicycles, around the Earth which occupies the center of the universe. Following the astronomer of Bologna, Giovanni Antonio Magini, Clavius proposed the existence of 12 celestial spheres in which the heavenly bodies are placed with their epicycles: seven for the five planets, the Sun and the Moon, one for the fixed stars three more to explain certain motions of the stars, and finally the sphere of the empyrean heaven, inhabited by God, angels and saints. In this respect, Clavius maintained the traditional view of an immoveable Earth placed at the center of the universe, in accordance with both the physics of Aristotle and a literal interpretation of the Bible, which mentions in some places the movement of the Sun and the stability of the Earth. However, in successive editions Clavius gradually introduced more recent astronomical findings. In the 1581 edition he pays tribute to the astronomical work of Copernicus, whom he calls "the outstanding renovator of astronomy in our times whom posterity will celebrate and admire like a second Ptolemy", even if he never accepted his system.

Although Clavius recognized the great value of Copernicus as a source of observations, tables and calculations, he considered the theory of Ptolemy preferable to what he called the "inventions" of Copernicus. Thus he writes in the introduction to this subject that many observations and the common opinion of philosophers disprove the heliocentric system. He presented many observations and phenomena which prove that the Earth is at the center of the universe. Although the most important arguments were those taken from the Aristotelian physics, he also referred to the texts of the Bible. However, he never used these texts against Copernicus, or suggested that his system was dangerous for the Christian faith, as some others did. In the last edition of 1611, Clavius mentions discoveries such as those of the novae stars seen in 1570, 1600 and 1604 and also the comet of 1577 among others by Tycho Brahe; all of which indicated that the celestial spheres were not incorruptible, contrary to the Aristotelian doctrine. He also mentions those made by Galileo with the telescope in 1609 and 1610 of the phases of Venus and the satellites of Jupiter which showed that not everything was in orbit around the Earth. In the light of these observations Clavius commented that the celestial orbits needed to be reformed. It is not clear what he meant, by this comment but it does not seem to indicate a possible acceptance of the systems of Copernicus or Tycho Brahe, but simply a modification of the Ptolemaic system. Clavius treated also some practical aspects of observational astronomy in another book *Astrolabium* (Astrolabe, 1593) where he makes a clear presentation of the resolution of spherical triangles and the stereographic projection.

For many years, it had been obvious that the Julian Calendar, decreed for the whole Roman Empire by Julius Caesar in 45 B.C. and observed in Western Christendom, was inexact and in need of reform; the date of the Spring equinox had moved further away from the 21st of March and this had an effect on the date for the celebration of Easter. Clavius was such a well-known figure that Pope Gregory XIII invited him to join the commission appointed for the reform of the calendar. The accepted reform had been originally proposed by the already deceased Luigi Giglio. When the reform was proclaimed by the Pope in 1582, it involved eliminating 10 days (4–15 October) in the calendar of that year.[7] Clavius was the person asked to explain and defend the reform, and he did so in a long work, *Romani calendarii a Gregorio XIII restituti explicatio,* (An explanation of the Roman calendar restored by Gregory XIII) published in 1595. This was followed by five more volumes, requested by Clement VIII, in defense of the reform which had come under attack, especially from Protestant circles. He dedicated two volumes to refute the objections presented by the French philologist and humanist Juste Scaliger and the mathematician and teacher of Kepler, Michael Maestlin. Scaliger had proposed a different reform of the Julian Calendar whose errors Clavius showed. The role Clavius played in the reform of the calendar increased his fame even more, especially outside the Jesuit world. Today, one can see on the tomb of Gregory XIII in St. Peter's Basilica a relief that show a cleric offering the Pope a book about the reform of the calendar; the person shown is thought to be Clavius, and if this is the case, he is the only Jesuit, apart from St. Ignatius, shown in the Basilica.

[7] G. V. Coyne, M. A. Hoskin and O. Pedersen (eds.), *Gregorian Reform of the Calendar.* Proceedings of the Vatican conference to commemorate its 400th anniversary, 1582–1982. Vatican: Specola Vaticana (1983); Carmelo Oñate Guillén, Christopher Clavius y el calendario gregoriano. *Letras de Deusto*, 89 (2000), 55–70.

1.3 Clavius and Mathematics in the *Ratio studiorum*

The rapid spread of Jesuit colleges made it necessary to establish a common regulatory system to be followed. St. Ignatius himself had already in the fourth part of the Constitutions, established the basis of the order of studies. Nadal in 1548 wrote a set of rules for the college of Messina which included the teaching of mathematics. Diego de Ledesma (1524–1575), professor of theology in the Roman College, wrote in 1564–1565 more complete set of rules, *De ratione et ordine studiorum Collegii Romani* (About the order and distribution of studies of the Roman College), on the teaching of humanities and omitted all mention of mathematics.[8] However, Baltasar Torres, the first professor of mathematics in this college, had already in 1557–1560 inserted rules about the teaching of mathematics.[9] For the first year of philosophy, he proposed the first six books of Euclid, practical arithmetic, astronomy (*sphaera*) and geography; and for the second year, the astrolabe, theory of planets, perspective geometry, sun dials and the ecclesiastical calendar. For each subject, he assigned three or four months, and for more interested students proposed private lessons about more advanced subjects such as the book on spherical geometry of Theodosius and Menelaus and the astronomy of Francesco Maurolico. Basing themselves on these teaching experiences, Jesuits began to write in 1580 the definitive regulation of teaching for all schools known as the *Ratio atque Institutio Studiorum Societatis Iesu* (Order and distribution of studies of the Society of Jesus). There are three versions, the first of 1586, the second more complete of 1591 and the third and final of 1599.[10]

Clavius first started to intervene in the composition of the *Ratio studiorum* in 1580 with proposals for the teaching of mathematics.[11] His attitude, which might be described as belligerently in favor of this discipline, can be seen in various of his works written between 1580 and 1593: *Ordo servandus in addiscendis disciplinis mathematicis* (Order to be followed in teaching mathematical disciplines), *Modus quo disciplinae mathematicae in scholis Societatis possent promoveri* (Ways to promote mathematical disciplines in the schools of the Society) and *De Mathematica instructio* (Instruction about mathematics).[12] In the first work (*Ordo*) Clavius presents three programs of mathematics for the three year period of philosophical studies. The first is more complete, the second more brief for those who do not intend a complete training in mathematics, and the third even shorter to be taught in only two years. The

---

[8] *Mon. Paed. S. I.*, vol. II, 519–62.

[9] *Mon. Paed. S. I.*, vol. II, 434–435.

[10] *Mon. Paed. S. I.*, vol. V, 41–156 (text of 1586), 231–349 (text of 1591), 357–454 (text of 1599); Carmen Labrador, M. Bertrán-Quera, A. Diez Escanciano and J. Martínez de la Escalera, *La Ratio Studiorum de los jesuitas*. Madrid: Universidad Pontificia Comillas (1986).

[11] Antonella Romano, *La contre-réforme mathématique. Constitution et diffusion d'une culture mathématique jésuite à la Renaissance*. Rome: École Française de Rome (1999), 94–134.

[12] *Mon. Paed. S. I.*, vol. VII, 110–115 (*Ordo*), 115–117 (*Modus*), 117–118 (*Instructio*); Romano, *La contra-réforme*, 94–132.

first shows a complete coverage of the mathematics of his times. Geometry was the main subject with Euclid's complete *Elements*, the spherical geometry of Theodosius and Menelaus, arithmetic, algebra and trigonometry, astronomy, theory of planets, astrolabe, geography, sun dials, mechanics and speculative music. Clavius made reference to the classical Greek authors, medieval astronomers and mathematicians like Georg Peurbach, renowned for his theory of the planets and the mathematician Jordanus Nemorarius and also to contemporary authors like Francesco Maurolico and Jacques Pelletier.

In the *Modus* Clavius put forward a series of practical suggestions to increase the role of teaching mathematics. He asked professors to be well informed about the subjects and to teach with authority, so that the students will have a high appreciation for mathematics. He proposed the celebration of disputations or public acts with presentations by students, similar to those required for philosophy, and also that questions of mathematics be included in the final exams of philosophy and that professors of mathematics take part as members of the exam board. Clavius suggested that professors of mathematics should participate in the public acts of philosophy so that students could see the connection between mathematics and philosophy. Finally he demanded that professors of philosophy forbear from saying that mathematics is not a science, nor talk or write against the usefulness of mathematics, as it seems that some sometimes did. All these recommendations were aimed at giving mathematics equal importance to that given to natural philosophy in the teaching of philosophy. Such an attitude implied an innovation in the programs for philosophy and allowed the entry there of modern science which was just beginning to gain ground with mathematical formulations and an experimental foundation. Clavius emphasizes above all the need for mathematics when dealing with physics, since "through their ignorance of mathematics some professors have been guilty of many and egregious errors". He claimed that "without mathematics natural philosophy is maimed". He was aware that the natural philosophy of his times, tied down by presuppositions base on Aristotelian doctrine, could make progress only with the help of mathematics. He saw clearly that advances in science were linked to the application of mathematics to the study of natural phenomena, as was already being done by the pioneers of modern science. In the third work (*Instructio*) he proposed to prepare for teaching those students with particular interest and attitude for mathematics with special more advanced classes after they have completed the philosophical studies. In consequence, Clavius felt that Jesuit students should not remain isolated from the developments of modern science. He also mentioned the need for adequate knowledge of mathematics in order to be able to take part in the meetings with important persons, where questions about mathematics were often discussed. Christoph Scheiner, one of his disciples, commented about the growing social interest in mathematics in his times when he wrote to Paul Godin, professor of mathematics in Graz that "it is evident that mathematics is the net with which we can catch noblemen and magnates and bring them to God's service".[13]

[13] Steven J. Harris, Les chaires de mathématiques, In Luce Giard, *Les jesuites a la Renaissance*. Paris : Presses Universitaires de France (1995) 253–254.

This indicates that for Jesuits teaching of mathematics, like any other activity, was finally oriented toward an apostolic end.

In the text of the first version of the *Ratio* in 1586, there is a chapter with the title of *De mathematicis* where the influence of Clavius is very obvious.[14] There is a recommendation to study mathematics, as "without them, our entire academic program will be sadly depleted", and mention is made of the practical utility of mathematics for society in general and the Church. It is stipulated that in the second year of philosophy, the *Elements* of Euclid, should be taught, along with geography and astronomy and Clavius is asked to write a program for these subjects. The lack of good professors of mathematics is acknowledged and it is proposed that a special three-year program should be established to help train a small group of young Jesuits (those with the right talents) from different provinces. It is hoped that out of this "academy" (*academia rerum mathematicarum*) will emerge outstanding mathematicians, able to spread knowledge of their specialty in all the provinces of the Society (provinces are called the regional divisions of the Society). The text of 1591 repeats these recommendations, but the duration of the especial training in mathematics for the members of the academy is reduced to 6 months after the studies of philosophy and Clavius is asked again to write a program for these studies.[15] In the definitive version of the *Ratio* in 1599, the reference to mathematical studies is shorter and repeats what was proposed in the previous version.[16] There is no mention of the "mathematical academy" and Clavius's name is not explicitly mentioned. However, further specialization for a small group of students on a private basis is envisaged. The comments that had been reaching Rome on the earlier drafts had stressed the difficulty in some provinces to provide teachers of mathematics and questioned the usefulness of such studies. Clavius had to compromise and make do with much less than he would have wished.

During the years that he was teaching at the Roman College, Clavius established a school of mathematics that served as a model for all the colleges of the Society. The *Ratio* established that in all major colleges, that is, those with full philosophical studies, there must be at least one professor of mathematics, but this was not always fulfilled. At the beginning of the eighteenth century, out of approximately 620 colleges run by the Society in Europe only 95 were equipped with chairs in mathematics. While in France, Italy and Central Europe there was a relatively high number of stable chairs of mathematics in other countries this was not the case. For example in Spain out of 126 schools only the Imperial College of Madrid had a stable chair of mathematics during many years.[17] In practice the programs vary widely among the different schools. In the first period Clavius' influence was very strong with an emphasis on the teaching of geometry. With time, colleges in different countries

[14] *Mon. Paed. S. I.*, vol. V, 109–110 (*De mathematicis*).

[15] *Mon. Paed. S. I.*, vol. V, 236 (*De mathematicis*), 384–385 (*Regulae professoris mathematicae*).

[16] *Mon. Paed. S. I.*, vol. V, 362, 402 (*Regulae professoris mathematicae*).

[17] Agustín Udías, Profesores de matemáticas en los colegios de la Compañía en España. *Archivum Historicum S. I.*, 79 (2010), 3–27.

took their own paths, for example, in France there soon grew an interest for algebra, and later, differential and integral calculus were introduced.

1.4 The Debate About the Certainty of Mathematics

An important problem concerning mathematics – one already raised by medieval philosophers – was its relation to natural philosophy. Aristotle had made it clear that it does not belong to natural philosophy, since it deals only with the quantifiable aspects of things and not with things themselves. For him mathematical knowledge, unlike that of physics, is based on the imagination and not on the senses. In general, scholastic philosophers followed this doctrine. Therefore, for them the abstraction of mathematical knowledge placed it outside the truth of science, which was defined as the correspondence between knowledge and real things. Consequently, they did not consider mathematics, unlike physics, as a true science in the strict Aristotelian sense. They extended this reasoning to all those sciences in which mathematics was applied to natural phenomena, the so-called "middle sciences" (*scientiae mediae*), like astronomy, optics and mechanics. In Clavius' time, Alessandro Piccolomini, an Italian humanist and philosopher, had a great influence with his work *Commentarium de certitudine mathematicarum disciplinarum* (Commentary on the certainty of mathematical disciplines, 1547); there he agreed with Aristotle about the epistemological inferiority of mathematics with respect to logic and physics. Against this position the mathematician Francesco Barozzi, in a book published in 1560 on certitude in mathematics, argued in favor of the truly scientific character of mathematics using arguments from the Greek mathematician, Proclus.

This debate was also current between Jesuits: with professors of philosophy on the one side and those of mathematics on the other, with firm convictions on each side.[18] Benito Perera (1535–1610), who taught philosophy between 1558 and 1567 and theology until 1597, was prominent among the professors of philosophy in the Roman College who were against considering mathematics as a true science. His position was clear: mathematics is not science because it does not have true demonstrations and because it abstracts from consideration of "being" and "goodness". He held that according to Aristotle science is knowledge acquired through causes and this is not the case in mathematics. Many Jesuit professors of philosophy agreed with this view. Among them were the professors of the influential College of Coimbra (Portugal), who between 1592 and 1606 published a complete course of natural philosophy in five volumes under the title of *Commentarii Collegii Conimbricensis* (The commentaries of the College of Coimbra). These textbooks were re-edited

[18] Romano, *La contre-réforme*, 134–162; Ugo Baldini, *Legem impone subactis. Studi su filosofia e scienza dei gesuiti in Italia, 1540–1632*. Rome: Bulzoni (1992), 27–56; Machado Mota, Bernardo, A cultura matemática no contexto Jesuita anterior ao século XVIII, *Broteria* 170 (2010) 27–42.

several times and were used in many Jesuit colleges, so that their teaching had a great influence.

Clavius took part in this debate and presented his ideas in the introduction to the first volume of the *Opera mathematica*, which had the title: *In disciplinas mathematicas prolegomena* (Introductory considerations about the mathematical disciplines).[19] He based his opinions on the prologue of Proclus to the books of Euclid, which is influenced by the ideas of the neoplatonic philosopher Plotinus, and also on the considerations of Barozzi. Clavius begins by commenting that there are many and different interpretations of the doctrine of Aristotle, but, on the contrary, the demonstrations of Euclid's geometry always inspire the same certitude and truth. This indicates the superiority of geometry over natural philosophy. According to Clavius, if we rank sciences according to their degree of certitude, mathematics should occupy the first place. Its certainty is superior to the uncertainties of natural philosophy. This consideration has its clearest example in geometry, but it can be applied as well to all applications of mathematics in the explanation of natural phenomena, since the universe itself in its totality has been created according to geometry. In this way, geometry provides for Clavius knowledge of the structure and working of the natural world. Later, Galileo would use the same argument when he said that the book of nature is written in the language of mathematics and its characters are triangles, circles and other geometrical figures.[20] Clavius was in this way a pioneer in the view that mathematical sciences belong really to the natural sciences. He even asserts that mathematics is not only useful but necessary in metaphysics, since, quoting Plato, it directs the mind to the contemplation of divine things. He also uses the testimony of the Holy Fathers of the Church, such as Augustine, Jerome, Gregory of Nyssa and Basil. For example, Clavius quotes St. Augustine's remark that mathematical disciplines can be considered helpful to interpret Sacred Scripture. Clavius' continual reference to geometry as the principal mathematical discipline led to the proposal that the *Elements* of Euclid should begin all programs of mathematics.

Antonio Posevino (1533–1611), in the volume dedicated to mathematics in his bibliographical 18-volume work, *Bibliotheca selecta* (1593), supports Clavius' position regarding mathematics. Posevino defended the role of mathematics and its certitude, adopting an eclectic position that combined the doctrines of Plato and Aristotle. Plato's teaching is more favorable to mathematics than that of Aristotle, since he holds that mathematical relations constitute the ultimate essence of things. For this reason, there was some abandonment by the first authors of modern science of Aristotelian doctrine and its replacement by Neoplatonic ideas. Posevino's work, which had a widespread circulation among Jesuits, gave a certain official character to Clavius' position regarding mathematics. Giuseppe Biancani (1566–1624),

[19] Christopher Clavius, *Operum Mathematicorum Tomus Primus*. Mainz: Antoniu Hierat (1611), 3–9.

[20] Stillman Drake, *Discoveries and Opinions of Galileo*. Garden City, N.Y., Doubleday Anchor Books (1957) (The Assayer) 237–238.

a former student of Clavius, took up the subject again in an appendix of his work, *Aristotelis loca mathematica* (1615). There he defended the certitude of mathematical demonstrations against the opinion of those Jesuit professors of philosophy who, following the Aristotelian doctrine, defended Perera's position against mathematics. The Scottish Jesuit, Hugh Semple (1589–1654), a professor of mathematics at the *Colegio Imperial* of Madrid and later rector of the Scots College, took over Biancani's ideas and insisted in his work *De Mathematicis Disciplinis* (1635) that mathematics is true science. Vicenzo Figliucci (1566–1622), a professor at the College of Naples, took a more extreme position, pretending to develop a "theology of mathematics". Following Neoplatonic ideas, he proposed that mathematics has been put by God in the human mind, so that through it man can understand the world of nature.

With regard to problems concerning the relation between mathematics and philosophy, Jesuits were conditioned by the decree of the Fifth General Congregation (1593–1594), presided over by the Superior General Claudio Acquaviva (1543–1615); this stipulated that in philosophy Aristotelian doctrine had to be followed and that different doctrines were not to be taught. This norm had been proposed already by Saint Ignatius in the fourth part of the *Constitutions*. The requirement for Jesuits to hold uniformity in doctrine was another element to be considered when it came to introducing new ideas. Acquaviva in a letter of 1611 insisted on this regulation which was repeated in General Congregations 16 (1731) and 17 (1751).[21] This controversy about the certitude of mathematics is important, because mathematical formulation is precisely one of the characteristics of modern science, together with reliance upon observations and experiments. Clavius' insistence on the importance of teaching mathematics, which was finally accepted in the *Ratio* as normative for all Jesuit schools, put them in line with the new developments of science. Consequently, we will find the Jesuit contributions to modern science coming from the professors of mathematics, not from the professors of natural philosophy, who remained stuck for a long time in Aristotelian physics. Jesuit professors of mathematics often crossed the boundary between mathematics and natural philosophy to validate their conclusions about the nature of things. For example, astronomy was for them not only a mathematical instrument to predict the position of the heavenly bodies, but could provide knowledge about their nature.

1.5 Disciples and Successors

During his more than 40 years of teaching in the Roman College, Clavius had a great number of students, who later spread out to different countries of Europe and even to the faraway lands of America and Asia; some of them excelled in mathematics,

[21] *Epistolae selectae Praepositorum Generalium ad Superiores Societatis*. Rome: Polyglottis Vaticanis (1911), 207–209 and 209–214.

astronomy and other sciences. Most of those who continued working in mathematics belonged to the small group of promising students who attended the Academy of Mathematics.[22] The Academy began in 1594 and continued active to the death of Clavius in 1612. These students, known as the *mathematici*, followed a more advanced program than the normal students and received a more complete training. Clavius' textbooks correspond to the classes in the Academy, not to those given to the ordinary students, who received a more reduced training. In the Academy emphasis was put on geometry and astronomy, as already seen in Clavius' textbooks. The Academy was only for those Jesuit students preparing to teach mathematics or for those going to mission countries, especially India and China, where knowledge of these sciences were considered necessary. Their number was small, probably no more than ten each year. The studies lasted during the 3 years of philosophy, but not all completed them. The Academy had a very select library, separate from the general library of the college. It contained mathematical works from classic and modern authors; many of the modern books carried dedications from their authors. Good libraries were a tradition in all Jesuit colleges. At Clavius' death, this library was considered as one of the best in Europe. The Academy and its library were also a point of attraction for scientists visiting Rome, who could not dispense with such a visit. The Academy continued after Clavius' death with his successors, but its importance soon declined. There is no further mention of it after 1615, and it practically disappeared towards 1636.

Students from Clavius' Academy were his immediate successors at the Roman College and many became professors of mathematics in different countries of Europe: among them can be mentioned Giovanni Paolo Lembo, Odo van Maelcote and Nicola Zucchi (Rome), Giuseppe Biancani (Parma), Paul Guldin (Graz and Vienna) and Grégoire de Saint Vincent (Antwerp and Gante), about whom more will be said later. Among the missionaries, mention should be made above all of Matteo Ricci, student in Rome between 1575 and 1577, who used his knowledge of mathematics and astronomy to gain access to the learned circles of China and to the Imperial Court itself. The future director of the astronomical observatory in Beijing, Adam Schall, was a student in Rome in the German College (*Collegium Germanicum*) in 1611, although Clavius was no longer teaching then. The first generation of Jesuit professors of mathematics considered themselves as Clavius' disciples, either directly or through his books. They mentioned him with reverence as "our Clavius" and some spoke of him as an "oracle".

The first successor of Clavius at the Roman College was the German Christoph Grienberger (1564–1636). He began teaching mathematics in the college of Vienna and came to Rome in 1591. He stood in for Clavius in 1595, on the occasion of Clavius's journey to Naples, and later remained as professor until 1633. Besides his teaching, Grienberger was the official reviewer and censor of the mathematical

[22]Ugo Baldini, The Academy of Mathematics of the Collegio Romano from 1553 to 1612. In Mordechai Feingold (ed.). *Jesuit science and the Republic of Letters*. Cambridge (Mass.): The MIT Press (2003), 47–98.

works written by Jesuits. In astronomy he was mainly interested in practical and observational aspects of this science. He is credited with being among the first to use the equatorial mounting of telescopes with two axes, one in the direction of the Earth's axis, which allows a better observation of the heavenly objects. Grienberger organized a program with telescopic observations to collect a new catalogue of stars, which he never published. He was the first to organize public disputations or contests (*disputationes*) about mathematical subjects presented by students. These were an important practice in Jesuit colleges. A man of great modesty he was very reluctant to publish. Among his few publications is a collection of practical problems of mathematics presented in such public disputations. In view of the new telescopic observations in astronomy, like those of Galileo, Grienberger thought that it was necessary to give greater freedom of thought to mathematicians and philosophers with respect to strict Aristotelian doctrine.

Another Jesuit who taught mathematics during the lifetime of Clavius was Odo van Maelcote (1572–1615), professor between 1605 and 1610. Van Maelcote made very detailed observations of the nova star of 1604, and showed that this was a true star in the heavens and not, as the Aristotelian philosophers thought, some kind of atmospheric phenomenon. Orazio Grassi (1583–1636), professor between 1616 and 1628, was an excellent architect, who drew up the plans for the Church of Saint Ignatius (part of the Roman College). His observations and publication about the comet of 1618 led to a bitter controversy with Galileo, as will be seen below. Among other professors, some of whom were only professors during 1 or 2 years, mention should be made of the following. Nicola Zucchi (1586–1670) wrote about mechanics and magnetism and made observations of the surface of Jupiter and Mars. He was an accomplished maker of telescopes, one of which he gave as a gift to Kepler. In 1616 he experimented on an early reflection telescope with an ocular lens and a concave mirror. The reflecting telescope was perfected by Isaac Newton in 1668. Gabriele Beati (1607–1673), professor between 1638 and 1661, was the author of a popular textbook of astronomy, *Sphaera triplex, artificialis, elementaris ac coelestis* (Triple sphere, artificial, elementary and celestial, 1662). In this work Beati refers to the Copernican system as one that explains in the easiest and simplest way the astronomical observations, but cannot be accepted for physical reasons. He decided to accept the geo-heliocentric system of Tycho Brahe, which as will be seen, was the favorite system of Jesuit astronomers after 1620. Giles François de Gottignies (1630–1689), professor between 1661 and 1685, was an observational astronomer who made numerous observations, such as those of the spots on Jupiter and of the comets of 1664, 1665 and 1668. Orazio Borgondio (1675–1741), professor between 1712 and 1740, dedicated his time above all to the design and building of astronomical instruments (*Constructionum astronomicarum theoría et praxis,* Theory and practice of the construction of astronomical instruments, 1724). Two outstanding professor were Athanasius Kircher and Roger Boscovich, about whom more will be said in detail in Chap. 3. The last professor before the suppression of the Society was Giuseppe Asclepi (1706–1776), between 1760 and 1773. He studied the motion of several comets and made observations of the transit of Venus across the Sun in 1761.

1.6 The Jesuits and Galileo

There is no doubt that Galileo Galilei occupies a privileged central position in the development of modern science. It is important to see his relations with Clavius and the Jesuit mathematicians, who in those years were also embarking on the new sea of modern science.[23] The first contact of Galileo with Clavius took place in 1587 during a visit of the former to Rome. Galileo, 23 years old, was then giving public and private instructions at Siena and Florence. He gave Clavius a copy of a work about the center of gravity of solids. From this first contact, a correspondence and friendship began between the two, in spite of their almost 30 year difference in age, which lasted till the death of Clavius. Between 1589 and 1596 Galileo used Clavius's book *In Sphaeram* in his classes of astronomy in Pisa and Padua, when he still expounded the Ptolemaic system. In 1610 Galileo, already a convinced Copernican, published his work *Sidereus Nuncius* (Starry Messenger), where he presented his new discoveries with the telescope using this instrument for the first time for astronomical observations. Among his discoveries were the phases of Venus, the satellites of Jupiter, the existence of mountains and valleys on the Moon's surface, and the observation of stars not visible with the naked eye. These observations were for him clear evidence of the validity and reality of the Copernican system. In 1611, during a new visit to Rome, Galileo was received in the Roman College, where an academic assembly was organized in his honor. Odo van Maelcote, gave a welcome address in Galileo's honor in which he refers to him as "the one that with all merit deserves to be held among the most outstanding astronomers of our time", praise which was not well received by some professors of philosophy. Galileo returned very satisfied from this journey; he wrote that in Rome everybody seemed to have been well disposed toward him, especially the Jesuit fathers. Biancani was someone who also had good relations with Galileo. Biancani's sympathy to Galileo and his work was the motive behind some problems he had with his superiors.

The same year, 1611, Cardinal Roberto Bellarmino (1542–1621) asked the mathematicians of the Roman College for their opinion about Galileo's discoveries with the telescope. Bellarmino had joined the Society of Jesus in 1560 and had been professor in Louvain and at the Roman College up to 1599, when he was appointed Cardinal. From 1605 he belonged to the Roman Curia as a member of the Congregations of the Holy Office (Inquisition) and of the Index. He played an important role in the ecclesiastical process against Galileo. The answer to Bellarmino's request about Galileo's discoveries was signed by Clavius, Grienberger, Maelcote and Lembo. They confirmed Galileo's observations, apart from some

[23] About the relation between Galileo and the Jesuits there is an abundant literature: Pierre de Vrégille, Galilée et les jésuites, *Études* 110 (1907), 460–478, 584–600; Bellino Carrara, I gesuiti e Galileo. *Rivista de Apologia Cristiana* (July–August) (1914),1–62; García Villoslada *Storia*, 194–213 (Galileo e i professori del Collegio Romano); Lattis, *Between Copernicus and Galileo*, 180–204; Ignacio Núñez de Castro, De la amistad y desencuentro de Galileo con los jesuitas. *Archivo Teológico Granadino*, 68 (2005), 69–109.

reserve with respect to the mountains on the Moon and the stars of the Milky Way, but without arriving at any cosmological conclusions. The astronomers of Clavius' group soon made use of the telescope for their astronomical observations and could verify Galileo's discoveries. Lembo and Grienberger built the first telescope at the Roman College and they kept improving these instruments. Galileo was very satisfied with the verification of his observations by the Jesuits. These good relations were soon marred by controversies which Galileo had with two Jesuit astronomers, and by the part he thought some Jesuits had in his trial and condemnation.

Christoph Scheiner (1575–1650), professor from 1605 at the University of Ingolstadt, enjoyed great prestige and influence in the Imperial Court because of his knowledge of mathematics.[24] He had been especially interested in optical problems. In March 1611, he began to observe Sun spots for the first time and decided to make a systematic study of this phenomenon. He used for these observations a telescope with an equatorial mounting designed by Grienberger, which he called a "heliotropic telescope", such that he could project the image of the Sun on a surface. In this way he could follow the evolution of Sun spots, their motion over the solar disc and other characteristics. The following year, 1612, he published his results in three letters (*Tres epistulae de maculis solaribus,* Three letters about Sun spots) addressed to the Augsburg nobleman, Mark Welter. Galileo, when he saw the publication, felt offended and upset because he claimed that he had observed the Sun spots earlier in 1610 and although Scheiner knew about this, he had not mentioned it. Later, Scheiner maintained not the priority, but the independence of his observations. In fact, Galileo published his observations in 1613, and in his work *Il saggiatore* (The Assayer), published in 1623, insinuated without mentioning his name that Scheiner knew about his observations of Sun spots, practically charging him with plagiarism. In his letters Galileo was more explicit and accused Scheiner of taking advantage of him. In fact the Sun spots had been observed at the same time also by David Fabricius and his son Johannes in Wittenberg, who published this in 1611 in a little known booklet and by Thomas Harriot in London, who had not published it. Scheiner published his complete and definitive work about the Sun in 1630 with the title, *Rosa Ursina sive de Sole* (Rosa Orsina or about the Sun; the title refers to the Orsini family who had subsidized the publication) (Fig. 1.3). Scheiner still held to some aspects of Aristotelian physics, which prevented him from putting the spots on the Sun itself. He considered that they were objects or clouds rotating around the Sun, since the Sun itself, as a heavenly body, according to Aristotle, could not have any defects. In addition, he thought that there was not yet sufficient evidence to

[24] Antonio Due Rojo, El fundador de la heliofísica. *Urania* 223–224 (1950), 121–137; William R. Shea, Galileo, Scheiner and the interpretation of sunspots. *Isis* 61(1970), 498–519; Franz Daxecker, *Das Hauptwerk des Astronomen P. Christoph Scheiner SJ "Rosa Ursina sive Sol" eine Zusammenfassung*. Berichte nat-med Verein Insbruck. Supplement 13 (1996). Franz Daxecker, *The physicist and astronomer Christopher Scheiner. Biography, letters, works.* Innsbruck: Innsbruck University (2004); Juan Casanovas, The discovery of sunspots. In: W. Schröder (ed.) *Historical events and people in aeronomy, geomagnetism and solar-terrestrial physics. Beitr. Gesch. Geophysik und Kosmischen Physik* 7 (2005), 243–260.

Fig. 1.3 Christoph Scheiner (1575–1650); observation of sunspots, (*Rosa Ursina*, 1630) (S. Maffeo, Vatican Observatory)

decide if the spots were objects rotating around the Sun or were on the surface of the Sun itself. Out of respect for his observations, Scheiner abandoned some Aristotelian doctrines and maintained that the heavens are fluid. He accepted the cosmological system of Tycho Brahe. He defended himself against Galileo's suggestions and affirmed his good faith and the independence of his observations. The importance of Scheiner's work as a true pioneer of heliophysics is independent of his controversy with Galileo.

Another controversy – with Orazio Grassi about comets – also marred the relations of the Jesuits with Galileo. In 1618 three comets were observed in Rome, one of them very bright which could be seen for several months. Grassi observed them very carefully and organized an academic contest about them. He published the conclusions in an anonymous work (*De tribus cometis anni 1618 disputatio astronomica*, 1619, Astronomical disputation about the three comets of the year 1618). In this work Grassi followed the ideas already proposed by Tycho Brahe and Kepler and showed that comets are heavenly bodies in the sky, described their paths and explained how the tails are always pointed away from the Sun. Thus he disputed the Aristotelian doctrine that considered comets to be meteors, i.e. atmospheric phenomena. Galileo, who because of a sickness had not been able to observe carefully the comets, felt that he had to intervene since he could not accept that something in the skies had been unnoticed by him. He published an answer to Grassi signed by Mario Guiducci, one of his disciples. In this publication Galileo complained that Grassi did not make any reference to him and made reference to Tycho Brahe, for whom he had little esteem because Brahe had not accepted completely the Copernican system. Galileo defended the traditional doctrine of comets as atmospheric phenomena and denied the validity of Grassi's calculations which put them at a greater distance than the Moon. Grassi answered with a publication (*Libra astronomica ac philosophica*, 1619, Astronomic and philosophical scales) signed with the pen name Lotario Sarsi, a supposed student, where he refuted Galileo's opinions. The second answer from Galileo, signed with his own name, was the well-known work *Il Saggiatore* (1623) where he continued his attacks against Grassi and the professors of the Roman College. In this work, Galileo went beyond the subject of the controversy and wrote a defense of the nature of the new science, one based on observations not on authority. This polemical work written in Italian was very popular. It was dedicated to the Cardinal Maffeo Barberini, who had just been elected Pope (Urban VIII) and who was pleased with the book. Grassi could not remain silent in view of Galileo's abuse against himself, against Scheiner and in general against the professors of the Roman College, and he answered with a new book, also signed with the pen name of Sarsi, in which he defended his observations and correct interpretation of the comets as heavenly bodies and rejected Galileo's attacks. There were no further answers by Galileo, but in his letters he refers to Grassi in harsh terms. He was convinced that he alone was the only one who could discover anything new in the heavens and that this was a truth which neither wickedness nor envy could deny.

These two controversies caused strong hostility in Galileo toward the Jesuits of the Roman College. Adding to this situation came the role played by the Jesuits in the condemnation of heliocentrism in 1616 and of Galileo himself in 1633.[25] Round about 1613, Galileo's public defense of the Copernican system began to raise suspicions of ecclesiastical authorities, who considered this system contrary to the literal interpretation of Holy Scriptures, where some texts talk about the motion of the Sun and the stability of the Earth. In these years, Cardinal Bellarmino, who was a member of both the Congregations, of the Index and that of the Holy Office, had recommended prudence in a letter to Paolo Foscarini, a Carmelite friar, and a friend of Galileo, who had published a book defending the Copernican system and its compatibility with the texts of the Scriptures. Bellarmino stated that the Copernican system could be proposed as a hypothesis, but not as representing the real situation, until there was a conclusive demonstration of the motion of the Earth, a demonstra-

[25] About the trail of Galileo and the role played by the Jesuits: Annibale Fantoli, *Galileo for Copernicanism and for the Church*, Vatican: Vatican Observatory Publications (1996); Ernan McMullin (ed.), *The Church and Galileo*. Notre Dame (Ind.): University of Notre Dame Press (2005). Carmelo Oñate, Galileo científico y procesado, su relación científica y procesal con los jesuitas. *Letras de Deusto*, 27 (1997), 105–129.

tion that he thought did not exist. In consequence the texts of the Scriptures should be literally interpreted, as had been done traditionally. In 1616 a commission of the Holy Office declared the system of Copernicus to be philosophically absurd and contrary to the letter of the Scriptures. In consequence Copernicus' book was put on the Index of forbidden books until it was corrected, together with Foscarini's book. The Jesuit Benedetto Giustiniani (1550–1622), who had been a Rector of the Roman College, took part in this commission. In 1616, the Holy Office entrusted Bellarmino to inform Galileo that he should neither teach nor defend publicly Copernicus' doctrine, except as a mere hypothesis. On this occasion Galileo asked and obtained from Belarmino a document which clearly stated that he had not been accused or condemned, as rumor in Rome had it, by the Inquisition. Galileo presented this document later in his trial, but to no avail. In 1618 the Congregation of the Index entrusted Grienberger and Grassi to examine the corrections made by Francesco Ingoli to Copernicus' book. Bellarmino maintained a cordial and favorable attitude towards Galileo and, since he had died 12 years before, did not take part in the trial of 1633. The only Jesuit who took part in Galileo's trial was Melchior Inchofer (1585–1648), a theologian consultant of the Roman Curia, who was one of the three theologians who examined Galileo's book, *Dialogo delli due massimi sistemi del mondo* (Dialogue on the two great world systems, 1632). The publication of this book had been the cause of the trial. Inchofer was a staunch supporter of geocentrism and he was very critical of the defense Galileo made in his book of the Copernican system. However, one cannot say that his opinion was decisive in Galileo's condemnation. The same year 1633, Inchofer, commissioned by the Pope, published *Tractatus syllepticus* (A Treatise Using Syllepsis), a defense of geocentrism and of the condemnation of Galileo. Galileo believed that the Jesuits had been behind his condemnation and wrote in a letter to his friend Elia Diodati in 1634 that "war did not come to me for my opinions, but because I fell from grace with the Jesuits".

The two controversies mentioned, those with Scheiner and Grassi, and the part played by the Jesuits in Galileo's trial have led to exaggerated views of the situation and to the supposition of a continuous hostility between the Jesuits and Galileo. Blaise Pascal affirmed that it was the Jesuits who instigated the decree from Rome which condemned Galileo's opinion about the Earth's movement. As has been seen, however, Galileo was well received at the Roman College and his relations with Clavius were always friendly. There is an abundant bibliography on Galileo's trial and condemnation, in which many factors were involved, and it not fair to put the blame on the Jesuits. Jesuit professors of philosophy were proponents of the Aristotelian doctrine and Jesuit theologians defended the literal interpretation of the Bible, two issues with which Galileo's new cosmology collided. A strong defense of the traditional doctrine is to be found in Scheiner's posthumous work, *Prodomus de sole mobili et stabili terra contra Galilaeum* (Introduction about the moving Sun and stable Earth against Galileo, 1651), where he clearly adopts a position against that of Galileo. However, some professors of mathematics were more sympathetic to Galileo's ideas. For example, Biancani accepted Galileo's theory concerning the flotation of bodies in liquids. It is incorrect to assume a unanimous Jesuit position against Galileo. Grassi, himself an interested party, wrote in 1633 to his friend

Girolamo Bardi, professor at the University of Pisa: "concerning Galileo's afflictions, I tell you sincerely that I am very upset about them because I had more affection for him than what he has ever shown for me".[26] Grienberger is said to have commented after Galileo's condemnation: "If Galileo had maintained the affection of the fathers of this College, he could have lived peacefully and suffered none of his misfortunes and could have written about any subject, even about the motion of the Earth".[27] Grienberger was open to the new ideas, although out of prudence he never accepted them publicly. One has to bear in mind, with respect to Galileo's trial, the obligation (already mentioned) that Jesuits were under to defend Aristotelian doctrine in philosophy and also the problem posed by the interpretation of Scripture. After the inclusion of Copernicus' book in the *Index* and the condemnation of Galileo, the acceptance of heliocentrism became a very serious problem for Jesuit astronomers. Until the middle of the eighteenth century, they were obliged to defend the stability of the Earth and its central position in the universe.

[26] Villoslada, *Storia*, 212.

[27] Vrégille, "Galilée et le jésuites", 474.

Chapter 2
Mathematics, Astronomy and Physics in Colleges and Observatories

2.1 From the Geometry of Euclid to the Integral Calculus

As the network of Jesuit colleges was spreading throughout Europe, chairs of mathematics were established, as was required by the norms of the *Ratio studiorum*. However, mathematics was already taught in many schools before the institutionalization of such chairs. The numbers of chairs of mathematics increased at the beginning of the seventeenth century, as the norms of the *Ratio* were put into practice, and in the eighteenth century, with the generalization of these studies. According to Karl Adolf Fischer, in Europe during these two centuries there were 91 chairs of mathematics in Jesuit colleges: 28 in France, 23 in Germany, 18 in Italy, 12 in Austria, 4 in Spain, 3 in Portugal, and 3 in Bohemia.[1] Since there were about 625 colleges, this is a relatively small proportion. Chairs of mathematics were only required in those colleges with complete programs of philosophy, the so-called "major colleges". These were only a small fraction, as most colleges taught only grammar or humanities with a little philosophy. For example in Spain in the eighteenth century there were 130 colleges, but only three had permanent chairs of mathematics. As already mentioned, other sciences were included under mathematics, such as astronomy, geography, cosmography, optics, mechanics, hydrography and civil and military architecture. Following Clavius' main interest in geometry, this discipline occupied first place in the teaching of mathematics in Jesuit colleges. For many years, teaching was based on the *Elements* of Euclid, but in most cases, only the first six books, followed by the eleventh and twelfth, were taught, with Clavius's commentary as textbook. In later times, the emphasis changed from geometry to algebra, especially in France, and finally to mathematical analysis with the introduction of differential and integral calculus in the eighteenth century. Outstanding in the teaching of mathematics were in France the colleges of Avignon, Marseille, Toulouse, Tournon,

---

[1] Karl Adolf Franz Fischer, Jesuiten-Mathematiker in der Deutschen Assistenz bis 1773. *Archivum Historicum S.I.*, 47 (1978), 159–224, and Jesuiten-Mathematiker in der Französischen und Italienischen Assistenz bis 1762 bzw. 1773. *Archivum Historicum S.I.*, 52 (1983) 52–92.

A. Udías, *Jesuit Contribution to Science: A History*,
DOI 10.1007/978-3-319-08365-0_2

Pont-à-Mousson, La Flèche and Paris (first Clermont and later Louis-le-Grand); in Austria and Bohemia those of Vienna, Graz, Prague and Olmutz; in Belgium, Antwerp and Louvain; in Germany, Ingolstadt, Freiburg, Cologne and Aachen; in Italy, Rome, Naples, Parma, Genoa, Milan and Palermo; in Portugal, Lisbon, Evora and Coimbra and in Spain, Madrid, Barcelona and Cadiz. In France Jesuits also held the royal chairs of mathematics and hydrography in Marseille, Nantes, Toulon and Brest. Among the students in Jesuit colleges we find a number of those who later excelled as scientists, such as Evangelista Torricelli, Descartes, Mersenne, Alessandro Volta, Giovanni Domenico Cassini and Claude Helvetius, to mention only a few.

Many professors taught mathematics only for a few years, and only a relatively small number dedicated themselves fully to it. Some of the most outstanding Jesuit professors of mathematics[2] merit a brief mention. Most of the first generation of Jesuit professors of mathematics were direct students of Clavius. In France teaching mathematics began in the college of Tournon in 1577, with professors who had been students of Clavius in Rome. Pierre Bourdin (1595–1653), professor in the colleges of Paris (Clermont) and La Flèche, belonged to this first generation of Jesuit mathematicians and his teaching centered above all on geometry (*Prima geometriae elementa*, First elements of geometry, 1640). French Jesuits were soon confronted with the ideas of Descartes. Descartes had been educated by the Jesuits for 8 years in the college of La Flèche and acknowledged the teaching of mathematics he had received. After giving up Aristotelian philosophy, he presented a new philosophy in line with the new modern science based on the method of the methodical doubt. Bourdin disputed with him about the validity of his philosophical method, but otherwise kept a good relation with him. Claude François Milliet de Chales (1621–1678) was professor of mathematics in the colleges of Lyon and Marseille, where he also taught hydrography and published several books on geometry, among them, *Euclidis elementorum libri octo* (Eight books of Euclid's Elements, 1660) and *Cursus seu mundus mathematicus* (Course or world of mathematics, 1674). In the second book he included also astronomy, cartography and navigation. The Frenchman Claude Richard (1589–1664), was professor at the *Colegio Imperial* of Madrid during 34 years and wrote a commentary to the complete works of Euclid, *Euclidis elementorum geometricorum libri tredecim* (The 13 books of Euclid's Elements, 1645), where he also included commentaries on the works of other Greek authors like Proclus, Hypsicles of Alexandria and Isidore of Miletus. He also wrote a commentary on the book of Apollonius of Perga about conic sections (ellipse, parabola and hyperbola), *Apolloni Pergei conicorum libri VI* (The six books on the conics of Apollonius of Perga, 1655). The Czech Jakub Kresa (1648–1715), also professor at

[2] A good summary of Jesuits mathematicians is found in Joseph MacDonnell, *Jesuit geometers*. Saint Louis: Institute of Jesuit Sources (1989). Their contribution to the birth of modern science in George D. Bishop, *Jesuit pioneers of modern science and mathematics*. Anand (Gujarat) : Gujarat Sahitya Prakash (2005). For professors in French colleges: François de Dainville, L'enseignement des mathématiques dans les collèges jésuites de France du XVI au XVIII siècle. *Revue d'Histoire des Science* 7, (1954) 6–21, 109–123; Romano (1999), *La contre-réforme*, 287–352.

the college of Madrid, wrote a Spanish translation of Euclid's *Elements* (*Elementos geométricos de Euclides*, The geometric elements of Euclid, 1689) where he added a number of his own demonstrations.[3] These are some examples of work of Jesuits on Euclid's geometry. Better known for his work in architecture was the Spanish mathematician Juan Bautista Villalpando (1552–1608) who had studied under the famous architect Juan de Herrera and was professor of mathematics in the college of Cordoba before moving to Rome in 1592. His main work is a detailed reconstruction of the temple of Jerusalem, *In Ezechielis explanationes et apparatus urbis et templi Hierosolymitani* (The description and structure of Ezequiel of the city and temple of Jerusalem, 3 vols., 1594–1605) published in Rome with Jerónimo de Prado (1547–1595). This work was used by Newton in his study of the temple of Jerusalem.

The study of some special curves, such as spirals and cycloids, and the determination of the center of gravity of plane and solid figures were two topics of geometry that attracted the attention of Jesuit professors. Antoine de la Loubère (1600–1664), professor at Toulouse, contributed to the study of the geometry of cycloids with his work, *Veterum geometriae promota in septem de cycloide libri* (Introduction to the geometry of early mathematicians about the cycloid in seven books, 1660). He disputed with Blaise Pascal about the solution of problems regarding the geometry of cycloids. Another of his works, *Quadratura circuli et hyperbolae segmentorum* (Quadrature of segments of the circle and hyperbola, 1660), is considered to be a precursor of the integral method to determine areas limited by curves. Pierre Nicolas (1642–1714), professor at the colleges of Tournon and Toulouse, published *De lineis logarithmeticis et spiralibus hyperbolicis exercitationes geometricae* (Geometrical exercises on logarithmic curves and hyperbolic spirals, 1699) and *De conchoidibus et cissoidibus exercitationes geometricae* (Geometrical exercises on conchoids and cissoids curves, 1697), both dedicated to the study of the properties of these curves. Paul Guldin (1577–1643), professor at Graz and Vienna, treated the problem of the determination of the barycenter or center of gravity of different types of geometrical figures. He proposed the theorem, known today as Guldin's theorem, about the properties of the center of gravity of bodies in rotation (*De centro gravitatis*, vol I–IV, 1635–1641). He had a close relationship with Kepler and helped him in his monetary problems and in his difficulties with the Evangelical Church. Jean Karen della Faille (1597–1652), a Belgian Jesuit disciple of Grégoire de Saint-Vincent, was professor at the *Colegio Imperial* in Madrid and the first Jesuit to be appointed Royal Cosmographer of the Indies Counsel (*Cosmógrafo Real del Consejo de Indias*) by King Phillip IV. He published *Theoremata de centro gravitatis* (Theorems about the center of gravity, 1632) where he dealt with the determination of the center or gravity of circular and elliptic sectors. The center of gravity was also the subject of the most original work of José Zaragoza (1627–1685), also professor at the college of Madrid, *Geometria magna in minimis* (Geometry large in the smallest, 1674). In this work he defined the barycenter in purely geometric terms and discussed its general properties and its determination in many types of figures.

[3] Agustín Udías, Los libros y manuscritos de los profesores de matemáticas del Colegio Imperial de Madrid, 1627–1767. *Archivum Historicum S. I.*, 74 (2005), 369–448 (Richard, 383–384; Kresa, 391–392).

Zaragoza is considered to be the most outstanding mathematician of seventeenth century Spain.[4]

Grégoire de Saint-Vincent (1584–1667), a disciple of Clavius in Rome and founder with François de Aguilón of the school of higher mathematics at the college of Antwerp (moved to Louvain in 1621),[5] deserves a special mention. Saint-Vincent erroneously believed that he had discovered the geometric solution of the problem of the quadrature of the circle, a problem without solution, and this has tarnished his memory. His main work *Opus geometricum quadraturae circulis et sectionum coni* (Geometric work on the quadrature of the circle and of the section of the cone, 1647), divided into 12 books, contains an extraordinary wealth of geometric knowledge about many subjects, especially about the properties of the ellipse, parabola and hyperbola and their quadrature. He presented his own theory on the method of *exhaustions* or indivisibles for the determination of areas, independently of the similar contemporary proposals by Bonaventura Cavalieri. He took the word *exhaustion* in its literal sense allowing the subdivision to continue to infinity. His infinite subdivision brought him to the notion of the limit of an infinite geometrical progression. In this respect Saint-Vincent can be considered one of the precursors of the integral calculus. His claim to have found the solution to the quadrature of the circle brought him many criticisms, including some from his Jesuit brothers. Grienberger was given the task of revising Saint-Vincent's book, and only gave his approval after a year of many doubts and deliberations. The book was only published after Grienberger's death. The Superiors rightly thought that the Society's reputation could be discredited by its publication. Christian Huygens criticized Saint-Vincent's claim about the quadrature of the circle; nonetheless both kept up a correspondence and through Huygens, Saint-Vincent's work had an influence on Gottfried W. Leibniz.

Algebra had only a secondary importance for Clavius in relation to geometry. After the publication by Descartes of his analytic geometry (*La géométrie*, 1637) – where he united geometry and algebra, thought until then to be two separate mathematical disciplines without any interrelation – interest for mathematical analysis grew among Jesuits, especially, in France. Jacques de Billy (1602–1679), professor at the colleges of Pont-à-Mousson, Dijon and Rheims, moved away from the Jesuit traditional emphasis on geometry and centered his interest on algebra. This interest is present in his first work *Abrégé de préceptes d'algébre* (A summary of the rules of algebra, 1637), which was translated into Latin and English. After this book, two more textbooks on algebra followed: *Nova geometria clavis algebrae* (New geometry key of algebra, 1643) and *Doctrinae analyticae inventum novum* (New discovery of the analytical doctrine, 1670). Billy maintained friendship and correspondence with the famous mathematician Pierre de Fermat, and included some of Fermat's theorems in his *Doctrinae analyticae.* Billy's work had a lasting influence on the development of algebraic notation. In the *Collège Louis-le-Grand* of Paris Louis

[4] Udías, "Los libros", 384–391.

[5] Charles Naux, Gregoire de Sant-Vicent: A pioneer of the calculus. *The Mathematical Gazette* 78 (1994) 315–319.

Bertrand Castel (1688–1757) introduced into his teaching the Cartesian synthesis of geometry and algebra or analytic geometry, which Descartes had called *mathesis universalis* (universal mathematics), and included it in his work, *Mathématique universelle abrégée* (Abridged universal mathematics, 1724). In this way mathematical analysis took precedence over geometry in the Jesuit colleges of France.

The development of algebra opened the way for the proposal of the differential and integral calculus by Newton and Leibniz. Differential and integral calculus was first developed by Newton between 1666 and 1676, although it was first published by Leibniz in 1684 and 1686. Jesuit professors of mathematics incorporated these developments into their teaching, but publications about them came later. Among the first were those of Vincenzo Riccati (1707–1775), professor at Bologna, the son and brother of mathematicians. He published in 1752, *De usus motus tractoris in constructione aequationum differentialium* (About the use of the motion of bodies in the setting of differential equations), which was followed by a more extensive book about the development of differential and integral calculus, *Institutiones analyticae* (Analytical institutions, 3 vols. 1765–1775). Josef Stepling (1716–1778), professor at the college (*Collegium Clementinum*) and university of Prague, studied carefully the works of Leibniz and of his disciple Christian Wolff, with whom he maintained a long correspondence, as well as with the great German mathematician Leonhard Euler.[6] His work, *Differentiarum minimarum quantitatum variantium calculus directus vulgo differentialis* (Direct or commonly called differential calculus of the differences between minimal quantities of variables, 1764), was very well received. Stepling was Director of Higher Education in Bohemia and put in charge by the Empress Maria Teresa of the reform of university teaching with emphasis on mathematical sciences. The Empress appointed him as *Regius Director Facultatis Philosophicae* (Royal Director of the Philosophical Faculty) and ordered a monument to be placed in his honor in the university library. Stepling's disciple and assistant Stanislas Vydra (1741–1804), also professor of mathematics at the University of Prague, was highly esteemed for his teaching. He published in 1773, *Primae calculi differentialis et integralis notiones* (First notions of differential and integral calculus), an introductory textbook of calculus. Karl Scherffer (1716–1783) taught mathematics in the colleges of Graz and Vienna using the Euler's books on mathematical analysis.

Giovanni Girolamo Saccheri (1667–1733) is one of the most original Jesuit mathematicians.[7] He was first professor of philosophy and theology until 1699, when he became a professor of mathematics at the University of Pavia. He had been a student with Tommaso Ceva (1648–1737), Jesuit mathematician and professor at the college of Brera in Milan, author of *Opuscula mathematica* (Short works of mathematics, 1691). Through him Saccheri got in contact with his brother, Giovanni

[6] Franz Delzel, Joseph Stepling. *Abbildungen Böhmischer und Mährischer Gelerten und Küntler* (Prague) 4, (1782), 164–172.

[7] Henry A. Boyle, Girolamo Saccheri, S. J. and Euclid's parallel postulate. *Bulletin American Association of Jesuit Scientists* 17 (1939) 93–99; Alberto Dou, Logical and historical remarks on Saccheri's geometry, *Notre Dame Journal of Formal Logic*, 11 (1970) 385–415.

Ceva, a famous mathematician, author of the theorem known by his name. Saccheri is considered to be one of the pioneers of mathematical logic with his work *Logica demostrativa* (Demostrative logic, 1697). His most innovative work is *Euclides ab omni naevo vindicatus* (Euclid vindicated from all stain, 1733) where he makes an effort to base on axioms all the mathematical sciences and thus envisions the possibility of non-Euclidian geometries. Faced with Euclid's fifth postulate about parallel lines, he considered the possibility of its negation, presenting two alternative hypotheses which suggest the two non-Euclidian plane geometries, elliptical and hyperbolic. Without really knowing it, Saccheri had opened the door to new geometries.

Some professors devoted themselves mainly to pedagogical work in mathematics with the publication of popular textbooks. Jesuit superiors concerned about the quality of the schools considered this type of work more important than research into new development of mathematics. André Tacquet (1612–1660), professor at Antwerp and Louvain and follower of the work of Saint-Vincent and de Aguilón, stands out with the publication of two very popular textbooks used in many Jesuit colleges: *Elementa euclidea geometriae planae et solidae* (Euclidian elements of plane and solid geometry, 1654) and *Arithmeticae teoría et praxis* (Theory and practice of arithmetic, 1656). The first underwent more than 30 editions during the seventeenth and eighteenth centuries and was translated into English with some additions and three editions by William Whiston, successor of Newton at the Lucassian chair of mathematics at Cambridge University. Henry Oldenburg, Secretary of the Royal Society, considered it to be one the best textbooks of mathematics ever written. Tacquet also wrote a research book, *Cylindricorum et anularium libri IV* (Four books about cylinders and annularies, 1651), a precursor of the infinitesimal calculus which had a certain influence on the works of Pascal. Tacquets's complete works, *Opera mathematica* (1669–1707) in eight volumes, were published posthumously.

Jesuit professors in Spain, faced with the lack of good textbooks of mathematics, tried to solve this situation (Fig. 2.1).[8] Zaragoza, at the college of Madrid, decided in 1660 to write a complete course of mathematics divided into eight volumes dedicated to arithmetic, geometry, spherical geometry, trigonometry, astronomy, optics, mixed mathematics and mathematical physics. He only completed the first four volumes. A century later, Tomás Cerdá (1715–1791), professor in Barcelona and Madrid, was also concerned with the lack of good textbooks of mathematics in Spanish and began a similar ambitious project of a complete course of mathematics with the title, *Lecciones de mathematicas para uso de la clase* (Lectures in mathematics for the use of schools). Cerdá had received his training in mathematics in Marseille with the Jesuit, Esprit Pezenas, and through him he had come into contact with the works of English mathematicians of the school of Newton. Cerdá divided his work into five volumes treating: arithmetic and algebra, algebraic equations, geometry and trigonometry, analytic geometry and differential and integral calculus. Between 1758 and 1760, he published the first three volumes, but the banning

[8] Udías, "Los libros", 384–388 (Zaragoza); 399–401 (Cerdá).

HVGONIS SEMPILII
CRAIGBAITÆI SCOTI
E SOCIETATE IESV
DE
MATHEMATICIS
DISCIPLINIS
LIBRI DVODECIM
AD
PHILIPPVM IV.
HISPANIARVM ET INDIARVM
REGEM CATHOLICVM.
ANTVERPIÆ,
EX OFFICINA PLANTINIANA
BALTHASARIS MORETI.
M. DC. XXXV.

Fig. 2.1 *Mathematicis Disciplinis* by Hugh Semple, professor at Colegio Imperial, Madrid (Biblioteca Comunidad ICAI, Madrid)

and expulsion of Jesuits from Spain in 1767 prevented the publication of the last two volumes, which were left in manuscript form. In Portugal, Jesuit teaching of mathematics was concentrated in the colleges of Lisbon, Coimbra and Evora. In 1685 Father General Charles de Noyelle (1615–1686) had stressed the importance of the study of mathematics in Portugal, especially because of the need for this among those who were sent to the Chinese mission. Inácio Monteiro (1724–1812) stands out among the professors of Coimbra with his complete course of mathematics, *Compendio dos elementos de mathematicas* (Short course of the elements of mathematics, 1754–1756).

2.2 The First Astronomical Observatories

The development of astronomy required the establishment of fixed places of observation and led to the creation of astronomical observatories. Prior to the use of telescopes, Tycho Brahe in 1580 established the first European modern observatory on the island of Hven in Denmark. After Galileo, in 1609, began to use the telescope to observe the heavenly bodies, this instrument became the necessary means for astronomical observations. As telescopes became larger, up to several meters in length, it was necessary to have them installed in a permanent way in observatories. As useful dates of reference, the first most important European observatories were those of Paris founded in 1667 and Greenwich in 1676. Jesuits soon began to use telescopes for astronomical observations: Clavius and Grienberger were using them in 1612 at the Roman College, and soon the use of telescopes became a common practice in the teaching of mathematics in Jesuit colleges, where astronomy was included as an important feature. However, it took some time before proper observatories as such were established in some colleges, where instruments were installed in permanent form and a person appointed as director.[9] The instruments installed in the observatories were mainly telescopes, sextants, quadrants, clocks, etc. At first, the instruments were made in the colleges themselves, but with time, as they became more complex, they were acquired from specialized shops. In many observatories, in addition to astronomical observations, meteorological observations were made, such as temperature, atmospheric pressure, air humidity and rainfall. This required the use of proper instruments such as thermometers, barometers and pluviometers. Thus, the establishment of observatories was a natural consequence of the importance given to the teaching of astronomy and sciences in general in the Jesuit colleges. At the middle of the eighteenth century, there were 31 observatories in the Jesuit colleges of Europe. This represented one third of all the observatories at that time in Europe. Seven were in France, thirteen in Central Europe, eight in Italy, one in Spain and two in Portugal (Fig. 2.2).

The first Jesuit observatories seem to have been those established about 1630 in Ingolstadt (Germany) and Avignon (France). The observatory of Ingolstadt was installed first in a tower of the Kreuzkirche church. In 1650 a wooden tower was built called *Turris mathematica* (mathematical tower), where the telescopes and other instruments were installed and observations made. The tower was destroyed by a fire in 1684. Nothing is further known about the observatory until between 1720 and 1730, when two professors of mathematics of the university, Nicasius Gramatici (1684–1736) and his successor Joseph Schreier (1681–1754), renewed astronomical observations. The observatory of Avignon was founded by Athanasius Kircher (to be mentioned in Chap. 3) during his stay in the college between 1631 and 1633. Observations were made from the La Motte tower. One of the directors of the observatory was Jean Bonfa (1638–1724), professor of mathematics, who made accurate observations of eclipses of the Sun and the Moon and of several comets. Bonfa corresponded with Giovanni Domenico Cassini, the first director of the Paris observatory. The observatory was renovated at the middle of the eighteenth century by Jean Claude Morand (1707–1780) and observations continued there till the expulsion of the Jesuits from France in 1764.

Most Jesuit observatories were founded during the eighteenth century. At the *Collège de la Trinité* at Lyon, one of the colleges in France with the strongest

[9] Agustín Udías, *Searching the heavens and the earth. The history of Jesuit observatories.* Dordrecht: Kluwer (2003), 21–34.

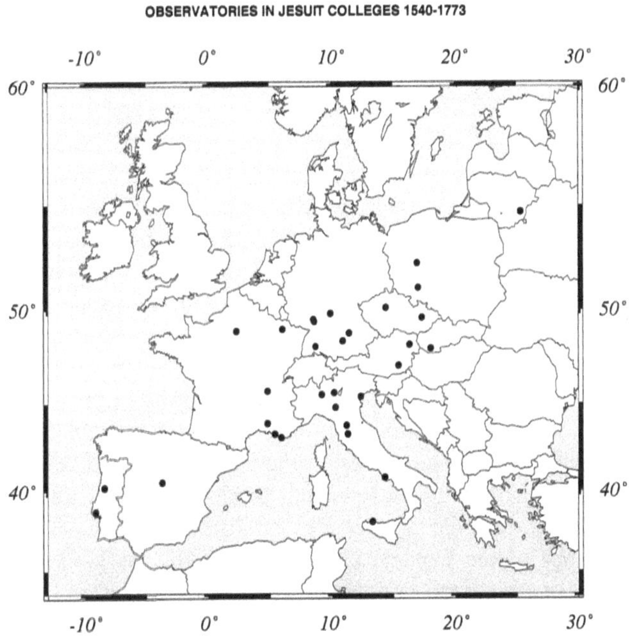

Fig. 2.2 Astronomical observatories in Jesuit colleges, 1540–1773 (Udías 2003)

tradition in teaching mathematics from 1605, an observatory was founded in 1701.[10] Jean de Saint-Bonnet (1652–1702) oversaw the construction, built upon one tower of the church of the college, but he died in an accident during the works. The first director was Pierre Thaillandier (1676–1713), who made observations until 1707, when he was sent to India, where he made astronomical observations in Pondicherry (Sect. 4.7). In 1740, Laurent Béraud (1702–1777) was appointed director. He was an excellent teacher and had among his students Jean Étienne Montucla, author of a famous history of mathematics and the renowned French astronomer Joseph Jérôme Lalande, director of the Observatory of Paris. Lalande attributed to Béraud the awakening of his interest for astronomy. Béraud made numerous observations of comets, solar eclipses and sunspots and especially of two transits of Mercury on the solar disk, which he presented to the academies of science of Lyon and Paris of which he was a correspondent member. Antoine de Laval (1664–1728) founded in 1702 an observatory in Marseille which was part of the Royal School of Hydrography of the French Navy, where he was professor.[11] Laval in 1718 founded another

[10] Pierre de Vreguile, *L'observatoire du Collège de la Trinité à Lyon*. Paris: Victor Retaux (1905).

[11] Suzane Débarbat and Simone Dumont, Antoine-François Laval (hydrographe du Roi, jésuite et astronome. *Histoire des Sciences et des Techniques. 115 Congress Nat. Soc. sav. Avignon* (1990), 17–26.

observatory in Toulon, also in a school of the French Navy. In both observatories he carried out a large program of observations, such as those of the Jupiter's satellites, sunspots and the occultation by the Moon of stars and planets. In 1728 Esprit Pezenas (1692–1776) was appointed professor of hydrography and director of the observatory at Marseille.[12] He reorganized the observatory, installed new instruments and established a program of observations. In 1749 the school was closed, but not the observatory which acquired new importance with the title of *Observatoire Royal de la Marine* (Royal Observatory of the Navy). Pezenas continued as director and pursued his observations, especially a detailed study of sunspots, which he published, as correspondent member, in the proceedings of the Academy of Science of Paris. He contributed to the introduction in France of the differential and integral calculus developed by English authors with the translation of the mathematical works of Collin Maclaurin. He published a treatise on the solution of spherical triangles (*Manière de réduire en tables la solution de tous le triangles sphérique*, A way to reduce in tables the solution of all spherical triangles, 1772). Several Jesuits like Louis Lagrange, Martin O. Poczobut and Tomás Cerdá were trained by Pezenas in his observatory at Marseille.

The observatory of the College of Vienna, founded in 1743 by Josef Franz (1704–1776) was one of the most important Jesuit observatories. In 1755 the empress Maria Theresa gave it the title of "Imperial Observatory" along with her support. In 1756 Maximilian Hell (1720–1792), born in Selmecz (Hungary), was appointed director.[13] Besides his work in astronomy, he was also professor of mathematics and interested in the problems of electricity and magnetism. Joseph Liesganig (1719–1799), a member of the observatory, was in charge of geodesic measurements in Austria.[14] In 1757, following the example of the Observatory of Paris, Hell began to publish yearly astronomical ephemerides in Vienna. He published 37 volumes up until 1792, with the help of two Jesuit astronomers, Franz Trisnecker (1745–1819), who succeeded him as director and Anton Pilgram (1730–1793). The *Collegium Clementinum* of Prague was a magnificent school founded in 1556 with the support of the Emperor Ferdinand I. In 1721 an observatory was established on a tower of the college. In 1751 Josef Stepling, appointed director, installed new instruments and established a very extensive program of astronomical and meteorological observations. Stepling began to publish very detailed and accurate observations of atmospheric pressure, temperature and air humidity. He continued as director of

[12] Guy Boistel, Inventaire chronologique des oeuvres imprimées et manuscrites du père Esprit Pezenas (1692–1776), jesuite, astronome et hydrographe marseillaise. *Revue d'Histoire des Sciences,* 56 (2003), 221–245.

[13] Oswald Gehlert, Jesuiten als Naturwissenschafter, *Kalksburger Korrespondence* 218 (1990) 28–36; Per Pippin Aspaas, *Maximilian Hell (1720–1792) and the eighteenth century transits of Venus: A study of Jesuit science in Nordic and Central European context.* Doctoral Thesis University of Tromsø, Tromsø (2012). Nora Pärr, *Maximilian Hell und sein wissenschaftliches Umfeld in Wien des 18. Jahrhunderts.* Religiongeschichte der frühen Neuzeit vol. 14. Vienna: Traugott Bautz (2013).

[14] Oswald Gehlert, "Jesuiten als Naturwissenschaftler", 32–34.

the observatory until his death in 1778, after the suppression of the Society of Jesus in 1773.

In 1578 a college had been founded in Vilnius (Lithuania) which from 1639 acquired great importance with the support of the kings of Poland. The observatory was established in 1753 in two towers added to the college building with the financial support of the Countess Elzbieta Oginska-Puzynina who had a great interest in astronomy. Tomasz Zebrwoski (1714–1758), professor of mathematics, who had been a student of Stepling in Prague was the first director. He introduced a new program of mathematics in the college and was an accomplished architect who built several churches in Vilnius. In 1758 Martin Odlanicki Poczobut (1728–1810), who had studied astronomy in Italy and in France with Pezenas, succeeded Zebrwoski as director. His astronomical observations, published in 34 volumes, were highly esteemed by the astronomers of his time. In 1769 he was elected member of the Royal Society of London and in 1776 correspondent member of the *Académie de Sciences* of Paris. After the suppression of the Society he continued as director of the observatory and professor of mathematics at the state university, where he was rector between 1780 and 1807. In 1808 he joined the Jesuits who had survived during the suppression under the protection of the Czarina Catherine the Great in the territory of Belarus. In Germany there were several small observatories. In 1757 Franz Huberti (1715–1789) established an observatory at Würzburg, on the tower of the Neubaukirche, where he made astronomical and meteorological observations. He continued his work after the suppression until his death. He was succeeded at the observatory by his assistant Franz Xavier Trentel (1730–1804) who had studied with Hell in Vienna and who published several textbooks of mathematics. Two observatories were founded in Schwetzingen in 1764 and in Mannheim in 1772 where Christian Mayer (1719–1783), made observations. He is credited as having been one of the first to study binary stars. In Ingolstadt Johann Baptista Cysat (1587–1657) carried out important observations of a comet in 1618 and determined its position beyond the Moon, its motion and magnitude (*Mathematica astronómica de loco, motu, magnitudine et causis cometae*, Mathematical and astronomical study about the position, motion and magnitude of the comet, 1619). Professor of mathematics at the University of Insgolstadt, Cysat had been disciple and assistant of Scheiner, and had made many of the observations of sunspots, that Scheiner used in his book about the Sun.

We have seen the importance of the Roman College in the introduction of teaching mathematics and astronomy in Jesuit colleges. However, no observatory was formally established there. Borgondio and Boscovich, both professors at the college, made plans for building one on the roof of the Church of Saint Ignatius, which was part of the college. The design by Boscovich in 1744 was very ambitious occupying a great part of the roof. Although Pope Benedict XIV approved the project it was never carried out. As we will see, it was after the suppression of the Society that an observatory was built at the college, which was later run by the Jesuits. One of the first Jesuit observatories in Italy was that of the college of Brera at Milan. Boscovich designed it and oversaw its construction, but its first director in 1760 was Louis Lagrange (1711–1783), who has been trained by Pezenas at Marseille.

Besides his observational work, Lagrange dedicated himself to training young astronomers, Jesuits and laymen. Some of them continued working at the observatory after the suppression of the Society. Boscovich carried out some of his observations there (as we will see in Chap. 3). This observatory was considered to be the best of its kind in Italy. Lalande in 1776 praised this observatory as the one with the best design for the practice of astronomy. In 1757 Giacobo Belgrado (1704–1789) established an observatory at the college of Parma. He was more interested in meteorology than astronomy and published some of his findings about the thickness of the atmosphere. An interesting case was the observatory established by Leonardo Ximenes (1716–1786) in the *Collegio di S. Giovannino* at Florence, which remained linked to his person after the suppression of the Society and continues today with the name of *Osservatorio Ximeniano*.

In Spain an observatory was established in 1752 at the *Colegio Imperial* of Madrid, where the professors of mathematics had the official title of *Cosmógrafos Reales* (Royal cosmographers). The observatory had several telescopes acquired in England. This was the first astronomical observatory in Spain, a year before the Naval Observatory of Cadiz generally considered to be the first. Johann Wedlingen (1715–1790), a Czech professor of mathematics, made observations between 1757 and 1767 and published some in the *Philosophical Transactions* of the Royal Society of London.

In Portugal, in 1722 the Neapolitan Giovanni Battista Carbone (1694–1750), established an observatory with the support of the King John V at the *Real Colégio de Santo Antão-o-Novo* of Lisbon.[15] Carbone was named Royal Mathematician and carried out observations till 1749. He was succeeded by Eusebio da Veiga (1717–1798) who reorganize the teaching of astronomy at the college and began publishing the first astronomical ephemerides in Portugal. The observatory was destroyed during the great Lisbon earthquake of 1755 and 4 years later the Jesuits were expelled from Portugal. Da Veiga went to Rome and took up the position of director of the observatory, *Specola Caetani*, which had been created in 1778 by Francesco Caetani, Duke of Sermoneta. In 1784 he began the publication of astronomical ephemerides, the first published in Rome.

Outside Europe, Jesuits recorded astronomical observations in China and India, as we will see in Chap. 4. In America, although there were a considerable number of colleges and universities, only a small observatory was established in the mission of Saint Cosme and Damian, one of the Indian settlements (Jesuit reductions) of Paraguay. The observatory established there in 1706 was the personal enterprise of Buenaventura Suárez (1679–1750), born in Santa Fe (Argentina), and only lasted during his lifetime.[16] He handmade his first telescopes grinding the lenses by himself and only after 1745 acquired from Lisbon two English telescopes and clocks. Using his observations Suarez published in 1743, *Lunario de un siglo*

[15] Henrique Leitão, *A ciência na "Aula da esfera" no colégio de Santo Antão, 1590–1759*. Lisbon: Ministério da Cultura (2007).

[16] Guillermo Furlong Cardiff, *Glorias Santafesinas*. Buenos Aires: Surgo (1929), 100–132.

(Lunar ephemerides for a century), which was well considered by astronomers in Europe and had several editions. It gave the ephemerides of the moon and sun, dates of eclipses and other information for the 100 year period 1740–1840. Although he was in remote Paraguay, Suárez kept in contact with astronomers in Europe, especially Gramatici in Ingolstadt, and with Kögler in China. In other parts of America Jesuits made astronomical observations, but without establishing permanent observatories. This was the case of the Czech, Valentin Stansel (1621–1705), in Brazil who published his astronomical observations in *Coeli Brasiliensis Oeconomia* (Arrangement of Brazilian heaven, 1666) with a preface by Kircher whom he had known in Rome. He also wrote a book defending the Tychonic system (*Uranophilos caelestis peregrinus,* Uranophilos, the celestial pilgrim), following the style of *Iter exstaticum* of Kircher. The Italian Nicolò Mascardi (1624–1674), who had arrived in Santiago in 1652, carried out astronomical observations in the Chiloe Island in southern Chile. He sent his observations to Kircher, who had been his professor in Rome and with whom he maintained a long correspondence. He made detailed observations of the comet of 1664 and wrote a report, *Observatio Cometarum* (Observations of comets) which was included in the book by the Peruvian astronomer Francisco Ruiz Lozano, *Tratado de los Cometas* (A treaty on comets, 1665), the first astronomical publication in America.

As we have seen, most Jesuit observatories began to work in a stable form, with proper buildings and instruments and directors fully dedicated to them, toward the beginning of the eighteenth century. They were soon affected by the expulsion of Jesuits from Portugal in 1759, from France in 1764, from Spain in 1767, and finally by the suppression of the Society in 1773. They had at most some 60 years of existence, and we cannot know what would have been their development if they had been active for a longer time. The future of the observatories after the Jesuits were no longer present varied greatly. Many simply disappeared due to the carelessness of those supposed to have taken care of them. Some continued in operation as state observatories like those of Vienna and Prague. In some of them, as we have seen, the same Jesuit directors remained in charge after the order was suppressed. Thus, the great efforts made in founding and managing the first Jesuit observatories finished tragically.

2.3 Acceptance of the Heliocentric System

The publication of Copernicus's work in 1543 brought forward the end of geocentric astronomy, which had been in force in the West since Greek antiquity to the sixteenth century. Ptolemy's work, known in the West by its Arabic title *Almagest* and translated into Latin in the twelfth century, was the fundamental and generally accepted proponent of geocentric astronomy. At first, the acceptance of Copernicus' heliocentric system met with serious difficulties and nearly 150 years passed before it was universally accepted. Many contemporary astronomers accepted it as a useful astronomical tool, which provided better tables of the positions of the planets, but without agreeing to its cosmology. For example, the Italian astronomer Magini published in 1589 astronomical ephemerides using Copernicus's work, but without agreeing to the heliocentric system. The difficulties were not only of a scientific character – like finding a clear proof of the movement of the Earth around the Sun given that no stellar parallax was observed – but also of a philosophical and theological nature. The generally accepted Aristotelian physics required the central position of the Earth and there are texts in the Bible which talk about the motion of the Sun and the stability of the Earth. The first Jesuit astronomers were caught up in this debate.[17] As we have seen, Clavius remained faithful to the traditional view, with a clear defense of the geocentric astronomy of Ptolemy. The biblical argument also influenced him since in the front page of his complete works there are two engravings representing the Bible stories of Joshua and king Ezechias which mention movement of the Sun. However, a number of recent astronomical observations had already pointed to the inadequacy of both Ptolemaic astronomy and Aristotelian physics. Between 1570 and 1597, from his careful observations at his observatory of Uraniborg of the position of the planets, of a supernova star in 1572, and of a comet in 1577, Tycho Brahe had become convinced that the heavens were not incorruptible, as affirmed by Aristotle, that the supposed celestial crystalline spheres, which move the heavenly bodies, could not exist and that the planets must all rotate around the Sun and not around the Earth. In spite of this, Brahe never accepted the system of Copernicus and between 1583 and 1587, proposed his own geo-heliocentric system, in which the Earth remains at the center of the universe and the Moon and the Sun rotate around it, but the planets rotate around the sun. This system is equivalent to that of Copernicus, as indicating the position of the planets as seen from the Earth. As a good observer Brahe did not accept the Copernican system because he could not observe the parallax of the stars due to the revolution of the Earth around the Sun.

Some years later, between 1619 and 1629, the German astronomer, Johannes Kepler, adopted completely the Copernican system and using Brahe's observations presented the three laws of planetary motion. He abandoned the circular motion of the planets that had been accepted for all heavenly bodies since the time of Pythagoras, and proposed elliptic orbits. We have seen how Galileo's use of the telescope for astronomical observations led to many discoveries which backed the heliocentric theory. Clavius admitted in the last edition of his *Comentarius* in 1611, a year before his death, that, given these new observations, astronomers might revise the order of the celestial orbits. Jesuit astronomers, successors of Clavius, were soon faced with the fact that they could not continue holding to traditional astronomy.[18] But the new Copernican theory was not compatible with Aristotelian physics that

[17] Johann Schreiber, Die Jesuiten des 17 und 18 Jahrhunderts und ihre Verhältnis zur Astronomie. *Natur und Offenbarung* 49 (1903), 129–143, 208–221; Manuel M. Sánchez Navarro-Neumann, Jesuitas astrónomos. *Broteria* 24 (1937), 423–436.

[18] For the change in the image of the universe and the Jesuits see: W. G. L. Randles, *The unmaking of the medieval Christian Cosmos, 1500–1760. From solid heavens to boundless aether*. Aldershot (U.K.): Ashgate (1999).

had to be taught at Jesuit colleges. This created tensions between the professors of mathematics and those of natural philosophy. The situation was further complicated in 1616 by the inclusion by the ecclesiastic authorities of the works of Copernicus and of all books which defended the motion of the Earth in the Index of Forbidden Books and by the condemnation of Galileo in 1633. This condemnation made clear that this doctrine was contrary to the literal interpretation of Sacred Scriptures. In consequence, the Copernican system could only be accepted as a hypothesis, as had been already proposed by Bellarmino.

The principles of the traditional Ptolemaic-Aristotelian cosmological model first needed to be revised. The incorruptibility of the heavens, which meant that no changes were there possible had to be abandoned together with the existence of the solid crystalline spheres where the heavenly bodies were located and which were responsible for their motion. The incorruptibility of the heavens followed from their being made, according to Aristotle, of a special substance, called the "ether" or "fifth element" (terrestrial bodies were made of the four elements: earth, water, air and fire). The first Jesuit who abandoned the traditional model of the celestial spheres was not one of the astronomers or mathematicians, but paradoxically, Bellarmino, when he taught natural philosophy in 1570–1572 at Louvain. Basing himself on the writings of the Fathers of the Church, Basil and John Damascene, Bellarmino denied the existence of a fifth element and of the existence of the solid spheres and proposed a theory of the fluid nature of the planetary heaven, in which the planets move "like the birds in the air and the fishes in the sea". Only the last sphere of the fixed stars was solid and it marked the limit of the material universe. Bellarmino divided the universe into three heavens which he called *Aereum*, *Sidereum* or *Aethereum* and *Empyreum*: the first was of a fluid nature and contained the planets; the second was solid and was where the fixed stars were located; and the third was where the blessed saints inhabit. In 1614 the authorities of the Jesuit order, and especially Father General Claudio Acquaviva, judged this opinion to be contrary to the Aristotelian doctrine. However, little by little, it gained followers among Jesuit astronomers. One of them was Cristoforo Borri (1583–1632), who taught from 1610 the incorruptibility and fluidity of the heavens, first in Italy and then in Portugal. This opinion, considered to be a novelty and outside traditional doctrine, brought Borri problems with his superiors and caused him to be sent to India and China. Faced with the difficulty of still holding to the Ptolemaic system still defended by Clavius, Borri proposed in his work *De tribus coelis* (About the three heavens, 1630) the acceptance of the system proposed by Tycho Brahe. He had already taught this system in Portugal and was one of the first Jesuits to do so.[19] Pietro della Valle, an Italian traveler, translated a summary of Borri's work into Persian and sent it to the Persian astronomer Zain- al-Din. This was the first European astronomical book used in Persia. Borri mentioned that this system was not a novelty,

[19] Michel-Pierre Lerner, L'entrée de Tycho Brahe chez les jésuites ou le chant du cygne de Clavius. In Luce Giard (ed.), *Les jésuites à la Renaissance*. Paris: Presses Universitaires de France (1995), 145–185.

since the philosopher Martianus Capella in the fifth century had already proposed that Mercury and Venus rotate around the Sun. Scheiner also in 1614 had defended the system of Brahe and suggested, we do not know on what basis, that Clavius himself had been inclined to this doctrine at the end of his life. The acceptance of Tycho Brahe's cosmology implied also the location of comets above the Moon, the non-existence of the heavenly spheres, the fluidity of the heavens and the autonomous movement of the planets. All of these views were contrary to Aristotelian doctrine. For this reason, Grienberger considered that in this area one should act with prudence. Georg Schönberger (1597–1645), a disciple of Scheiner in Ingolstadt, had also defended in his book, *Sol illustratus et propugnatus* (The Sun, illustrated and defended, 1627), basing himself on purely astronomical grounds, the fluidity of the planetary heaven and that the orbits of Venus, Mercury and Mars rotate around the Sun.

Biancani, a disciple of Clavius and professor at Parma, proposed most strongly the Tychonian system in his work, *Sphaera mundi seu cosmographia* (The sphere of the world or cosmography, 1620), a very popular book which had several editions in 1630, 1635 and 1653.[20] Biancani had defended Clavius's position that mathematics is a true science and now justified his disagreement with Clavius, who was faced with the impossibility of explaining the new astronomical observations with the traditional geocentric system. Moreover, he suggested, like Scheiner, that Clavius at the end of his life had accepted this position in a sort of "swan song". Grienberger, who had been the censor of the book before its publication, denied this claim. According to Grienberger, who knew Clavius better, he had never departed from the traditional system which he had defended in all editions of his *Comentarius.* Biancani considered his book to be a necessary substitute to Clavius' work. The acceptance of Brahe's system by Jesuit astronomers was not free from debates and controversies with the professors of philosophy, who remained tied to the principles of Aristotelian physics.

Giovanni Battista Riccioli (1598–1671), born at Ferrara, a disciple of Biancani and professor at Parma and Bologna, was the Jesuit who presented the definitive defense of the geo-heliocentric system.[21] Riccioli had hesitated whether to dedicate himself to theology or to astronomy, but finally he was convinced by his love for the latter. In his work, *Almagestum novum* (The new Almagest, 1651) and later in the shorter *Astronomia reformata,* (Reformed astronomy, 1665), Riccioli developed in great detail and with great erudition, based on his own experiments and observations, the different astronomical systems. He considered the Copernican system to be "the most beautiful, most simple and best imagined", but he could not accept it for many reasons. He opted for the system of Tycho Brahe, but with the modification that he considered Jupiter and Saturn to rotate at a large distance around the

[20] Baldini, *Lege Impone,* 401–450.

[21] M. T. Borgato (ed.) *Giambattista Riccioli e il merito scientifico dei gesuiti nell'età barroca.* Florence: L.S. Olschki (2002), 36–40; Carmelo Oñate, "El gran astrónomo jesuita G. B. Riccioli (1598–1671)" (manuscript, 2000).

Earth not around the Sun. In book IX, entitled "The system of the world", Riccioli dedicates 300 pages to explain the different astronomical systems, the greatest part (200 pages) being a refutation of the Copernican system with 120 arguments. In his second work, Riccioli concluded that the heavenly phenomena can be explained equally well with the Earth stable or moving. He added that only the authority of Sacred Scriptures move us to affirm the motion of the Sun and the stability of the Earth, though the reasons for one or the other are equally convincing. However, he praised the works of Kepler and accepted the elliptical orbits for Mercury, Venus and Mars around the Sun and the Sun around the Earth. He also accepted other modern discoveries, like the satellites of Jupiter observed by Galileo and the rings around Saturn proposed by Christian Huygens. He defended the fluid nature of planetary space, but maintained a solid sphere for the fixed star as the boundary for the material finite universe. Behind this sphere the "Empyrean heaven" was located, the habitat of the blessed.

The Copernican system opened the door to the consideration of an infinite universe. One of the first to propose this idea was the Englishman Thomas Digges in 1576, who placed the stars as distributed through a boundless infinite space, instead of on the surface of a sphere, the limit of the universe. Acceptance of an infinite universe raised serious difficulties among Jesuit astronomers because of the philosophical and theological consequences. Riccioli also presented in his *Almagestum novum* a very detailed map of the topography of the Moon based on the careful observations made with the telescope by his young assistant Francesco Maria Grimaldi (1613–1663). In this map names were given to topographic features, using the names of great astronomers, among them 18 names of Jesuits. Many of these names have been maintained in the present maps of the Moon in which there are 34 names of Jesuits. Earlier, in 1647, Johann Hevelius had published a map of the Moon (*Selenographia*) also giving names to the topography, but the names given by Riccioli and Grimaldi have been preferred in modern maps. There is a debate whether Riccioli was convinced of his rejection of Copernicus' system or if he did this only because of the need to obey the ecclesiastic censure and the imposition by his Order of the Aristotelian doctrine. However, it is difficult to accept that he was not entirely sincere about what he wrote. Riccioli's work was very popular among Jesuits and it replaced Clavius' work as the textbook of astronomy. Zaragoza, published in Spain his *Tratado de la esfera, en común, celeste y terráquea* (Treatise about the sphere, whether common, celestial or terrestrial, 1675), the first textbook in Spanish on astronomy and geophysics.[22] He relied on Riccioli and defended both the fluid nature of planetary space and that Mercury and Venus rotate around the Sun.

With time, as more evidences were presented supporting the heliocentric system, its not acceptance by Jesuit astronomers became more difficult. However, the ecclesiastical condemnation carried a heavy burden and prohibited its public teaching,

[22] Agustín Udías Vallina, El P. José Zaragoza y su obra. In: José Zaragoza, *Esphera en comun, celeste y terraquea* (1675) (Facsimile edition). Madrid: Real Sociedad Geográfica y Escuela T. S. Ingenieros de Topografía, Geodesia y Cartografía (2006), 11–46.

except as a hypothesis. By the end of the seventeenth century, the system of Copernicus was widely accepted although still not by all astronomers. This was the case, for example, in 1659 with Cassini, the first director of the Observatory of Paris. With the publication of Newton's, *Principia Mathematica* (1687) and the law of universal gravitation, it was clearly established that the Sun could not rotate around the Earth and that all the planets must rotate around the Sun. The first experimental demonstration of the orbital motion of the Earth around the Sun came later in 1729 with the English astronomer James Bradley's observation of the aberration of the light coming from a star. By that time, all astronomers had already accepted the Copernican system. In the first part of the eighteenth century, the heliocentric system and Newtonian astronomy were defended in some Jesuit colleges. Boscovich, as we will see in the next chapter, fully accepted Newtonian astronomy and it seems that he may have influenced the decision of Pope Benedict XIV in 1757 to remove from the *Index* the prohibition against all books which defend the motion of the Earth. Other steps in the same direction were also taken in the Church with the publication with ecclesiastical approval of the complete works of Galileo in 1741. Because of an oversight, the book of Copernicus was not definitively removed from the *Index* until 1823. The expulsion of the Jesuits from Portugal, France and Spain between 1759 and 1767 and the suppression of the order in 1773 prevented the development of the work of Jesuit astronomers within the context of heliocentric and Newtonian cosmology.

2.4 The Transits of Venus

An example of the Jesuits' observational work in astronomy can be found in the observations of the transits of Venus across the solar disk of 6 June 1761 and 3 June 1769.[23] The transits of Venus are interesting astronomical phenomena. They occur in a pattern that generally repeats itself every 243 years, with pairs of transits 8 years apart separated by long gaps of 121.5 years and 105.5 years. In the transits, Venus appears as an opaque little black circle which crosses the surface of the Sun. Observations of the transits from different points of the Earth provide valuable astronomical data which, among other things, allows an accurate determination of the distance from the Earth to the Sun. Edmond Halley in 1677, had predicted the dates of the transits of 1761 and 1769 and the astronomers prepared themselves to make observations from different observatories in Europe and also by organizing expeditions to Siberia, Madagascar, South Africa and Baja California (Fig. 2.3).

Jesuit astronomers could not overlook this opportunity and contributed with observations from many of their observatories. Béraud made observations of the transit of 1761 at the observatory of the *Collège de la Trinité* of Lyon. Jesuits were

[23] Harry Woolf, *The transits of Venus. A study of eighteenth century science*. Princeton: Princeton University Press (1959).

Fig. 2.3 Maximilian Hell (1720–1792) during his observations of the transit of Venus in Lapland (Udías 2003)

banned from France in 1764, but Béraud returned privately to Lyon in order to observe the transit of 1769 and installing his telescope on the hill of Fourvière thus completed his observations. He wrote a report on his observations but was not able to publish them. Jean Nicolas Cairon de Merville (1714–1768) also observed the transit of 1761 in the *Collège Louis-le-Grand* of Paris. Mayer observed it in Schwetzingen and the Czarina Katherine II of Russia invited him to observe the transit of 1769 in Saint Petersburg, where he published the results of his observations. Other Jesuits who observed the transit of 1761 were Huberti in Würzburg, Ximenes in Florence and Liesganig in Vienna. Stefan Luskina (1725–1793) who had studied in Vienna observed the same transit in Warsaw and wrote a long report (*Observation astronomique du passage de Venus par le disque du Soleil*, Astronomical observation of the transit of Venus across the solar disc, 1761). Luskina later became a journalist and was director of the Gazette of Warsaw. The Austrian Christian Rieger (1714–1780) observed the transit of 1761 in the observatory of the *Colegio Imperial* of Madrid. In 1761 Boscovich travelled to Istanbul to make observations of the transit, but problems during the journey delayed his arrival and he could not make them. The *Royal Society* invited him to observe the transit of 1769 in California, but since this was Spanish territory he was not allowed to travel there. Jesuits were banned in 1767 from all Spanish territories. Thus, in spite of his great interest for these observations, Boscovich could not participate in this international astronomical program.

The most memorable Jesuit observations were those of Hell, director, as we saw, of the Vienna Observatory.[24] His recognized competence as an astronomer moved Christian VII, King of Denmark and Norway to invite him to observe the transit of 1769 in the north of Lapland. This was really a surprising invitation, since this Protestant King had forbidden the Jesuits in his territory. Hell travelled accompanied by Johann Sajnovics (1733–1785), a Hungarian Jesuit, director of the observatory founded by Hell at Tyrnau (Hungary). After a long journey from Vienna to Copenhagen and then by sea along the coast of Norway, Hell and Sajnovics arrived in October 1768 at the island of Vardø in Lapland, at northern latitude of 70°, where they stayed for over a year. Besides observing the transit of Venus, they made many other astronomical observations and observations of the northern lights, magnetic declination and tides. They took also note of the language, culture and customs of the Laplanders. Sajnovics was the first to recognize the similarities between the Lappish and Hungarian languages. From his observations Hell determined the solar parallax (8.70″) and the distance from the Earth to the Sun (152.216 million kilometers, today the accepted value is 149.597 million kilometers). Differences between his observations and those of other observers led some astronomers to doubt them. Later, in 1835, the astronomer, Carl Ludwig Littrow, accused Hell of having falsified the observations. This brought Hell's work under suspicion, until in 1883 the American astronomer Simon Newcomb, examining Hell's original notes, demonstrated they were correctly made and were in fact more exact than those of other contemporary astronomers.[25] Hell presented his observations to the Danish Royal Academy and published them in *Observatio transitus Veneris ante discum Solis die 3 junii anno 1769, Wardoehusii*, (Observation of the transit of Venus in front of the disk of the Sun the 3rd of June of 1769 in Vardø, 1770). The complete report of his studies and observations carried out at Vardø and during the journey, which had the title *Expeditio literaria ad Polum arcticum*, (Literary expedition to the Arctic Pole) and was divided into three volumes, *Historicus*, *Physicus* et *Mathematicus et Astronomicus* (Historical, Physical and Mathematical and Astronomical), but was never published and survives only in manuscript form.

Although the information is not very conclusive, it seems that Jesuit astronomers also made observations outside Europe. Augustin Hallerstein (1703–1774), director of the Imperial Observatory of Beijing, probably observed the two transits of Venus. Other Jesuits of the French mission of Beijing made observations at their small observatory. There is no direct evidence, but it is also probable that some French Jesuits in the missions of Chandernagore and Pondicherry in India made similar observations. All these observations by Jesuit astronomers show that they participated actively in this early international program of astronomical observations.

[24] Per Pippin Aspaas, *Maximilian Hell* (2012), 184–218.

[25] George Sarton, Vindication of Father Hell. *Isis*, 35 (1944), 97–105.

2.5 Mathematical and Experimental Physics

As has been already mentioned, the teaching of mathematics included at that time the application of mathematics to the study of physical phenomena, which was known as applied or mixed mathematics and which today we simply call physics. Specific subjects of applied mathematics were astronomy, mechanics, optics, hydrology, cosmography and the study of other phenomena such as magnetism and electricity. In fact, one of the most important elements at the beginning of modern science was the introduction of mathematical formalism or the "mathematization" of natural phenomena, together with the recourse to experiments and observations. This involved the abandonment both of qualitative analysis and of the text commentaries which were characteristics of Aristotelian *physica* (we will refer to it by the Latin word *physica* to distinguish it from what is today known as physics). These two elements of modern science, namely, mathematical formalism and empirical foundation, created problems for the teaching by Jesuits, who were obliged to be faithful to the Aristotelian doctrine. In Jesuit schools professors of natural philosophy taught Aristotelian *physica*, while professors of mathematics were those who introduced the teaching of the new mathematical and experimental physics. While the pioneers of modern science, like Galileo, Kepler, Descartes, Huygens and Newton, totally abandoned Aristotelian *physica*, the Jesuits, who were obliged to teach it, chose a middle road. They took an eclectic position which mixed elements of the two. Professors of natural philosophy continued the teaching of Aristotelian *physica*, but little by little they introduced elements of modern physics. Professors of mathematics, on the contrary, were free to teach the new approaches of mathematical and experimental physics.[26]

However, the Jesuits were not literal in their following of Aristotle and Saint Thomas Aquinas in philosophy. Moreover, from, the middle of the fourteenth century Aristotelianism was not a monolithic body of doctrine and was split up into several schools of thought. Francisco Suárez (1548–1617), professor of philosophy and theology in Rome, Alcalá, Salamanca and Coimbra, opened the way with his very influential work *Disputationes metaphysicae* (Metaphysical disputations, 1579), for a liberal interpretation of the traditional positions. He dared to move away from Saint Thomas' opinions with great freedom, even in fundamental questions. This interpretative freedom influenced especially the texts of natural philosophy of the *Cursus Collegii Conimbricensis Societatis Iesu* (Course of the College of the Society of Jesus of Coimbra), which were used as a textbook in most Jesuit colleges. Pedro da Fonseca (1528–1599), known as the "Portuguese Aristotle", was the initiator

---

[26] John L. Heilbron. *The elements of early modern physics*. Berkeley: University of California Press (1982), (Chapter II. 1. Jesuits. Their magisterium, their eclecticism, their constraint, 93–106); Peter Dear, Jesuit mathematical science and the reconstruction of experience in the early seventeenth century. *Studies in History and Philosophy of Science*, 18 (1987), 133–175; William A. Wallace, The problem of apodictic proof in early seventeenth century mechanics. Galileo, Guevara and the Jesuits. *Science in Context*, 3 (1989), 67–87.

and driving force of this work, which covered the whole course of philosophy in eight volumes published between 1592 and 1606. Another influential person was the Spaniard Rodrigo de Arriaga (1592–1667), who was very famous as a professor of philosophy and theology at Prague, so much so that the saying was, "Go to Prague and listen to Arriaga". His work, *Cursus philosophicus* (Course of philosophy, 1632), edited several times, kept within the Aristotelian tradition, but with great freedom of thought.[27] Regarding the structure of the universe, he accepted the fluid nature of planetary space, though he rejected the arguments from astronomical observations. Similar opinions were defended in the commentaries to Aristotle's *De Coelo* by Pedro Hurtado de Mendoza (1578–1651), professor at the University of Salamanca and by Francisco de Oviedo (1602–1651).

Nicolò Cabeo (1586–1650), a disciple of Biancani and professor at Parma and Genoa, wrote a commentary on Aristotle's *Meteorologica* in 1646 which was reedited in 1686 with the title *Philosophia experimentalis* (Experimental philosophy). It was actually an original presentation of the physics of terrestrial phenomena based on observations and experiments. In some works of this type by Jesuit authors, elements of experimental physics are integrated along with Aristotelian doctrine. For example, Melchior Cornaeus (1598–1665), professor at Würzburg, wrote his *Curriculum philosophiae peripateticae* (Course of peripatetic philosophy, 1657), where he introduced the mathematical treatment of some problems, such as the motion of heavy bodies, and experimental elements, such as the vacuum pump[28] among others. This mixture of modern and traditional elements was not well received by his superiors, who resented that he was moving away from accepted doctrine. Riccioli in his *Almagestum Novum* presented his experiments on the fall of bodies (made from the top of the Asinelli tower almost 100 m high in Bologna), where he confirmed the law proposed by Galileo.[29] Pierre Gautruche (1602–1681), professor at the college of La Flèche, also took this mixed approach in his popular work, *Philosophiae ac mathematicae totius institutio* (Explanation of the whole of philosophy and mathematics, 5 vols., 1656). Gautruche was faithful to a strict Aristotelianism, but he was open and interested in the advances of experimental physics. He was also very interested in mathematics and the part of his work about mathematics was edited in London in 1683. In astronomy, he accepted the geo-heliocentric system of Brahe, as was common among Jesuits of his time, but he was contrary to the geocentrism of Aristotle. Antón Mayr (1673–1749), professor at Ingolstadt, presented this integration of elements of experimental physics together with the general lines of Aristotelian *physica* in *Philosophia peripatetica antiquorum principiis et recentiorum experimentis conformata* (Peripatetic philosophy derived from the principles of the past and modern experiments, 1739). This is the case also

[27] Carmelo Oñate, Rodrigo de Arriaga, filósofo del barroco ¿también científico? *Letras de Deusto*, 22 (1992), 77–94.

[28] Marcus Hellyer, *Catholic Physics. Jesuit natural philosophy in early modern Germany*. Notre Dame (Ind.): University of Notre Dame Press (2005).

[29] Christopher M. Graney, Anatomy of a fall: Giovanni Battista Riccioli and the story of g. *Physics Today*, 65 (2012), 36–40.

of the German Maximus Mangold (1722–1792), author of *Philosophia recentior praelectionibus publicis accomodata* (Recent philosophy for public lectures, 1763). In this work he makes an explicit reference to the use of experiments to explain natural phenomena, moving away from the traditional explanations of Aristotelian doctrine.

Jesuit mathematicians established from the beginning what separated them from the *physici* (physicists) teachers of natural philosophy who followed the Aristotelian *physica*. The mathematicians made clear that they only considered the quantitative aspects of the phenomena, leaving the consideration of their essential nature and their causes to the *physici*. This distinction allowed them to propose experiments and measurements which interfere with natural processes, to use instruments, and to ignore the strict separation between the natural and the artificial, which was part of the Aristotelian doctrine. This was also a pragmatic solution which allowed them to dedicate themselves to the new mathematical and experimental physics without openly denying Aristotelian *physica*, which was required by the Jesuit authorities and was linked with some theological interpretations, such as those of the Eucharist in terms of the doctrine of transubstantiation. This approach permitted them also to stay away from thorny problems about physical reality or essential causes which they left to the *physici*.[30]

In order to justify their position,[31] Jesuit mathematicians also proposed consideration of what we can call philosophy of science. Scheiner in his work *Oculus* (The eye, 1619) defended the true character of the science of optics, arguing that it studies visible objects and the organ of vision from the mathematical point of view, while *physici* do so from that of causality. He admitted, however, that the two approaches are related since both consider objects which are apprehended through the senses. An important aspect of the ideas proposed by Jesuit mathematicians concerned the way in which formulations of mathematical physics are justified by observations and experiments, linking them in this way with the experimental physics. Biancani, who had defended mathematics as true science, made an interesting distinction in order to back up this position in his work on astronomy *Sphaera mun*di (1620). He distinguished between suppositions, phenomena or experiences, and observations or experiments. Suppositions are principles which do not need demonstration. Phenomena or experiences are known by all and are considered to be evident. Observations on the contrary are obtained through experiments and by the use of instruments and are not known directly by all. For example, in astronomy, observations depend on the specialized work of astronomers and not all have access to them. *Physici* admit only the evidence obtained through experience, but not that from experiments or observations, which they consider to be something artificial and not natural. Mathematical sciences can use demonstrations based on "artificial" experiments because they consider only the quantitative aspects of reality.

[30] Heilbron, *The elements*, 93–100.

[31] Dear, "Jesuit mathematical", 148–152.

Similar arguments can be found in the works on optics by Scheiner (*Oculus*, 1604) and by Aguilón (*Opticorum libri sex*, 1613).

This sharp distinction drawn by Jesuit mathematicians between mathematical and physical sciences soon began to be abandoned.[32] Galileo had endeavored to do away precisely with this distinction, since he considered himself a philosopher and gave demonstrative character to his mathematical demonstrations. Cabeo, who considered himself a philosopher, affirmed that in all his studies of the magnetic properties there is nothing which is not confirmed by repeated experiments. He maintained the distinction between experiences and experiments and admitted that magnetic properties are not accessible through direct experience, but through specific experiments designed for each case. However, he thought that through repetition of these experiments one can obtain the same evidence as in direct experience. Scheiner, in his work about the Sun (*Rosa Ursina*, 1630) had already affirmed that the results from experiments in mathematical sciences are also relevant in the *physica,* as are also the conclusions from mathematical demonstration.

The introduction of these ideas, considered to be novelties, by Jesuits in their teaching caused some alarm among superiors. I mentioned earlier the rules about teaching Aristotelian doctrine in philosophy, and about the required soundness and uniformity of the doctrine that was to be presented. Father General Claudio Acquaviva had already insisted in a letter in 1611 (*De soliditate et uniformitate doctrinae*, About the soundness and uniformity of doctrine), which he repeated in another letter in 1613, about the need to keep uniformity in the teaching of the Society, which included following Aristotle and Saint Thomas[33] in philosophy. In order to ensure both things, soundness and uniformity of doctrine, these recommendations were repeated by other Fathers General who published lists of propositions which should not be taught by Jesuit professors.[34] Father General Francesco Piccolomini (1582–1651), published in 1651 a first list of forbidden doctrines (*Ordinatio pro studiis superioribus*, Ordinance for higher studies). The list contained 67 propositions of philosophy, which included subjects such as the atomic composition of matter, the diurnal motion of the Earth, the nature of light, weight of the air and condensation and rarefaction of gasses. The document did not question their scientific merit, but forbade the teaching of them to guarantee uniformity of doctrine. It left a door open since it admitted that propositions generally accepted by a majority of authors could be taught. Father General Tirso González (1624–1705) drew up a new list of forbidden propositions, following a ruling of General Congregation 14 (1696), Decree 5, which increased the list to 89. This list was not published because of the opposition from several Jesuit Provinces, especially those of France, which demanded greater freedom of opinion regarding propositions

---

[32] Dear, "Jesuit mathematical", 166–173.

[33] *Epistolae selectae Praepositorum Generalium ad Superiores Societatis.* Roma: Polyglottis Vaticanis (1911), 207–214.

[34] José Luis Orella, Un elenco jesuítico de proposiciones filosóficas. *Pensamiento* 23 (1967), 273–306.

of physics. In 1706 the new Father General Michelangelo Tamburini (1648–1730) published a new list of 30 forbidden theses. The target of this list was the philosophy of Descartes which was quite common in French Jesuit schools. The list included the Cartesian theory of matter, which was not allowed to be presented even as a hypothesis. The same concern was present in the decrees of other General Congregations.[35] For example General Congregation 16 (1731), Decree 36, tried to control excessive freedom in philosophical subjects and also to avoid limiting them to purely speculative problems. Thus the Congregation proposed that Aristotelian doctrine should be complemented with experimental physics and mathematics. This showed only a moderate progress, since the same Congregation ordered that those professors who taught "novelties" (*novitates*) or moved away from the Aristotelian doctrine should be removed from teaching. This conservative position was also represented in General Congregation 17 (1751), which confirmed what was established in the previous one (Decree 13) and added that in philosophy priority should be given to the methodology and questions related with scholastic theology. The Congregation insisted that the scholastic syllogistic method should be used even in questions of experimental physics. It added, however, that mathematical demonstrations might be used when necessary in questions of physics. These regulations show the concerns of superiors faced with the advance of modern science. However, in spite of all these attempts to stop it, the decline of Aristotelian *physica* and the acceptance of modern physics were inevitable.

2.6 Optics and Magnetism

One of the branches of physics in which Jesuits made important contributions was optics. Scheiner in his work, *Oculus* (The eye, 1619), analyzed the complete problem of vision using geometrical arguments to describe the paths of light rays. He insisted that optics, traditionally considered a part of mathematics, is a true and proper science distinct from the *physica*, but with many things in common. François de Aguil6n (1567–1617), born in Brussels of a Spanish family, was professor at the college of Antwerp and founder with Saint-Vincent of a school of higher mathematics.[36] His work on optics, *Opticorum Libri Sex* (The six books of optics, 1613), was a very complete textbook, which influenced the work on optics by Willebord Snell, but was soon surpassed by the works of Kepler and Descartes. The book includes engravings on the title pages of each book by the famous painter Paul Rubens. The author explains in an original way the theory of colors and binocular vision, and proposes a law for the variation of the intensity of light with distance. He also explains the

[35] *Institutum Societatis Iesu*, vol. II, Florencia: SS. Conceptione (1893), 413 (C.G. XIV, Dec, 5), 431–432 (C.G. XVI, Dec. 36), 436–437 (C.G. XVII, Dec. 13).

[36] A. Ziggelaar, *François de Aguil6n S.J. (1567–1617) Scientist and Architect*. Rome: Institutum Historicum S. I. (1983).

different types of projection, and gave its name to stereographic projection. Aguilón was also an accomplished architect and drew the plans for several Jesuit churches in Belgium. The most original Jesuit contribution to optics was that of Francesco Maria Grimaldi, professor at Bologna and disciple and assistant of Riccioli. His work *Physico-mathesis de lumine, coloribus et iride* (Physics and mathematics about light, colors, and the rainbow, 1665) was published after his early death by his mentor and friend Riccioli.[37] The book deals with the problems concerning light and its propagation. Its more original contributions deal with the phenomena of light interference and of diffraction. Grimaldi was the first to identify and study the diffraction of light. Newton quotes him several times in his *Opticks* when talking about this optical phenomenon.

Ignace Gaston Pardies (1636–1673), professor at the college Clermont at Paris, author of a popular book of geometry, *Éléments de Géometrie* (Elements of geometry, 1671), left unpublished a comprehensive work on optics *Traité complet d'optique* (Complete work on optics).[38] Pardies moved away from the views of Grimaldi, who had considered light as a fluid, and advanced the consideration of its undulatory nature. This idea was later developed by Huygens, who had access to Pardies' manuscript, in his famous treatise on light. Pardies was influenced by Descartes' physics, though he criticized other aspects of Cartesian philosophy, such as the mechanical nature of animals. Pardies wrote a criticism of Newton's theory of light and colors referring to it as an "extraordinary hypothesis". Newton, annoyed by the term "hypothesis" used by Pardies, answered him and explained his experiments with the prism and the decomposition of the colors of light. Pardies was satisfied with this answer and commented that he understood now what he had missed before. Pardies' objections and Newton's answer were published in the *Philosophical Transactions* (1672, No. 84 and 85) of the Royal Society. The English Jesuit Francis Line (1594–1675), professor at the English College of Liège, continued this debate between Pardies and Newton. Line also questioned the results of Newton's experiments with the prism in a letter to Henry Oldenburg, Secretary of the Royal Society, to which Newton answered explaining his experiments. A year later, after Line's death, his successor at the college Anthony Lucas (1633–1669) repeated experiments with the prism and insisted with the same critiques to which Newton patiently answered again (*Philosophical Transactions* 1675–1776, No. 110, 123 and 128). This is a sign of Newton's appreciation for the Jesuits' works on optics, in spite of the opposition they encountered in England. Pierre Ango (1640–1694), professor at the college of la Flèche Partes published parts of Pardies' work on optics in his book *Optique* (Optics, 1682). Huygens commented that Ango should have limited himself to transcribing the exact text of Pardies, since his commentaries did not help to make the text clearer. Pardies also wrote a text about mechanics, *La statique ou la*

[37] Carmelo Oñate Guillén, Primer texto de la difracción de la luz. *Revista Española de Física*, 7 (1993), 54–57.

[38] August Ziggelaar, *Le physicien Ignace-Gaston Pardies S. J. (1636–1673).* Copenhague (1971).

science des forces mouvantes, (Statics or the science of moving forces, 1673). Pardies' scientific work was cut short by his early death while attending the sick.

Jesuit scientists also shared at this time in the study of magnetism.[39] The first was Leonardo Garzoni (1567–1592), professor at the college of Venice, who in about 1580 wrote the work *Trattati della calamita* (Treatise about the loadstone) which was not published.[40] In this work, Garzoni treated of the double polarity of magnets, their attraction of iron, and the orientation of magnetic needles toward the Earth's poles. Garzoni's manuscript was used by Giovanni Battista della Porta and Paolo Sarpi in their work about magnetism. Through their work, Garzoni's ideas influenced William Gilbert, the founder of modern magnetism with his work *De magnete* (About the magnet, 1600), where the Earth is considered for the first time as a large magnet. Cabeo used Garzoni's manuscript to write up his *Philosophia magnetica* (Magnetic philosophy, 1629). Cabeo presented all that was known in his time about magnetism and added his own observations and experiments. He explained for the first time magnetic and electric attraction. He did not accept Gilbert's ideas about the origin of the Earth's magnetic field and erroneously maintained that magnetic declination was constant. Jacques Grandami (1588–1672), a disciple of Biancani, wrote *Nova demonstratio immobilitatis terrea petita ex virtute magnetica* (New demonstration of the stability of the earth from magnetic force, 1645), where he tried to prove the stability of the earth and the geocentric system from the existence of the earth's magnetic field. Gautruche also used the same argument to refute the motion of the earth. De Chales, professor at the college of Lyon, in his comprehensive work *Cursus seu mundus mathematicus* (Course or world of mathematics, 1674) included a very complete treatment of magnetism. In 1727, Nicolas Sarrabat (1698–1739) published a book about the variations of the terrestrial magnetic field (*Nouvelle hypothèse sur les variations de l'aiguille aimantée*, New hypothesis about the variations of the magnetic needle) which received an award from the *Académie des Sciences*. Hell during his journey to Lapland in 1769 carried out a series observations of the magnetic declination several times each day at the island of Vardø and along the Norwegian coast during his return journey to Copenhagen. These were the first magnetic observations made at high latitudes. Hell observed the relation of strong magnetic time variations and the occurrence of the aurora borealis.[41] We will find out about Kircher's ideas on magnetism in the next chapter and about the magnetic observations made by Jesuits in China in Chap. 4. Together with magnetism Jesuits became interested in the phenomena of static electricity especially after 1740 with the discovery of the Leyden jar. Electricity was then considered as

[39] Pierre de Vregille, Les jésuites et l'étude du magnétisme terrestre. *Études*, 104 (1905), 495–511.

[40] Leonardo Garzoni, *Trattati della Calamita a cura di Monica Ugalia* (M. Ugalia: Introduzione 7–83). Milan: FrancoAngeli (2005).

[41] Truls L. Hansen and Per Pippin Aspaas, *Maximilian Hell's geomagnetic observations in Norway 1769*. Tronsø: University of Tronsø (2005).

a fluid produced by some types of bodies.[42] Among the Jesuits interested on electricity we find Cabeo, Kaspar Schott and Francesco Lana-Terzi, and they will be mentioned in the next section.

2.7 Introduction of Modern Physics

About the beginning of the eighteenth century, an evolution took place in the teaching of physics in Jesuit schools with the complete introduction of modern physics, not only by professors of mathematics but also by those of natural philosophy.[43] We have seen that professors of mathematics had been using mathematical formalism and the recourse to observations and experiments in their classes of astronomy, optics, and mechanics, and had tried to explain the new phenomena of magnetism and electricity. They now began to enter the traditional field of natural philosophers in order to explain the nature of physical phenomena. Thus the barrier created between applied mathematics and *physica*, began to be erased. Already, from the middle of the seventeenth century, Jesuits could not ignore the contributions to experimental physics made by Torricelli, Pascal, Otto von Guericke and Robert Boyle, which questioned many elements of the Aristotelian physics, in such areas as the compression and expansion of gases, atmospheric pressure and the existence of vacuum. In Jesuit schools these questions had to be handled, not only in the classes of mathematics but also in those of natural philosophy, so that a bridge was needed between the disciplines of mixed mathematics and what was still called *physica*.

A representative figure in this process was the German Kaspar Schott (1608–1666), disciple and assistant of Kircher in Rome and afterwards professor at Mainz and Würzburg.[44] He wrote his *Cursus mathematicus* (Course of mathematics, 1661) in 28 books in which he covers a great variety of subjects, such as astronomy, optics, hydrostatics and cosmography, and another book entitled *Physica curiosa* (Curious physics, 1662), where he presented many strange and curious experiments of physics. For Schott what he called "experimental natural philosophy" was the bridge between mathematics and *physica*. He had been very impressed with the experiments of von Guericke with the vacuum pump, especially those carried out in 1654 at Regensburg in the presence of the Emperor Ferdinand III, where he had demonstrated the existence of empty space. Schott repeated these experiments and confirmed the interesting phenomena which were produced. However, he did not accept the existence of the vacuum, influenced as he was by the Aristotelian doctrine which denied its possibility. It seems that Schott and not Kircher was the

[42] John L. Heilbron, *Electricity in the 17th and 18th centuries. A study in early modern physics*. Mineola (N.Y.): Dover (1999) (The Jesuit school, 180–192).

[43] Marcus Hellyer, *Catholic Physics*, 165–180.

[44] Hans-Joachim Vollrath, *Wunderbar berechenbar: die Welt des Würzburger Mathematiker Kaspar Schott, 1608–1666*. Würzburg: Echter Verlag (2007); Julius Oswald, Leben und Werk des würzburger mathematiker Kaspar Schott S.J., *Archivum Historicum S. I.* 78 (2008), 417–441.

source of inspiration for Father Caspar Wanderdrossel in Umberto Eco's novel, *The island of the day before*. Acceptance of the existence of vacuum or empty space constituted a real problem among Jesuits because of its supposed impossibility in the Aristotelian *physica* that they were obliged to follow. Francesco Lana-Terzi (1631–1687), a disciple and assistant of both Kircher and Schott at the Roman College and later professor at Brescia and Ferrara, accepted the existence of the vacuum and the modern concepts of air density and weight, and the principles of Archimedes and Torriceli. Based on these ideas, he proposed an aerial ship suspended from four metallic balloons, a vacuum having been created inside each of them. This was an early notion of the hot air balloons in which Joseph Michel and Jacques Étienne Montgolfier made their first ascent in 1782. Naturally, Lana-Terzi's proposal was never carried out. He presented a large number of other physical experiments in: *Magisterium naturae et artis, opus physico-mathematicum* (Mastership of nature and art, physical-mathematical work, 1684).

Another problem presented by modern science concerned atomism, already introduced by Galileo and followed by most modern authors. The mechanicism of Descartes was added to this problem, as he reduced all interactions between bodies to those of a mechanical nature.[45] At that time there were two types of atomism: one suggested by Descartes in which atoms were the last elements of a continuous matter identified with space itself, and the other followed by Galileo and Newton in which atoms were solid and impenetrable particles separated by empty space. In the Middle Ages, the ancient Greek atomism of Leucippus and Democritus was identified with their atheistic philosophy and condemned. Aristotle had already rejected it and had proposed his hylomorphic theory (composition of matter and form) which was the accepted theory in the Christian West from the thirteenth century. In 1649 Pierre Gassendi suggested the compatibility of atomism and Christian thought with the consideration that atoms were created by God. This was also maintained by other authors like Newton. The acceptance by Jesuits of atomism was made difficult by their Aristotelian philosophy and by the already mentioned theological problem of interpreting the presence of Christ in the Eucharist in terms of transubstantiation, which was explained according to that philosophy. Jesuits moved away from the hylomorphic theory in physics, but in order to avoid theological problems avoided the word "atoms" and used instead "particles" or "corpuscles". The word "atom" had the connotation of the condemned classical atomism of Leucippus and Democritus revived by modern authors, but corpuscles could be understood as compatible with Aristotelian physics. Honoré Fabri (1608–1688), professor at Lyon between 1640 and 1647, wrote in Rome his work *Physica* (Physics, 1670), where, pretending to follow Aristotle, he made an attempt to come to an agreement with the physics of corpuscles. His position was close to that of Gassendi, though he pretended not to accept atomism. Fabri's modern tendencies had been shown in his disagreement with the proofs presented by Riccioli against the motion of the Earth. He held that no man ought to scruple to follow the hypothesis of the Earth's motion,

[45] Hellyer, *Catholic physics*, 217–227.

but only forbear to maintain it in public, till the prohibition be lifted. He was also the author of a popular textbook of geometry (*Symnopsis geometriae*, Compendium of geometry, 1669) where he presents an ingenious method for the quadrature of the cycloid. Maximus Mangold (1722–1797), a professor at Mainz, went further and accepted completely Gassendi's atomism, defending that natural bodies are made up of homogeneous atoms or ultimately indivisible particles. Atomism in one form or another was widely accepted by Jesuit physicists about the middle of the eighteenth century. In the next chapter we will see the peculiar atomic theory of matter presented by Boscovich .

In spite of the prohibitions, there were among Jesuits, especially French, some followers of Descartes' natural philosophy.[46] This implied rejection of Aristotelian natural philosophy and Superiors took measures against this tendency, removing professors who followed these views from their posts. René Rapin (1620–1687) was among Jesuit sympathizers of Descartes. He found Descartes' doctrine better organized than that of Galileo and Newton. George Fournier (1595–1652) quoted Descartes very often in his book on hydrography. Another follower of Descartes was Noël Regnault (1683–1762), professor of mathematics in the College Louis-le-Grand in Paris, who published, *Entretiens physiques* (1734), which contributed to popularize the interest for physics in France. Later, Aimé-Henri Paulian (1722–1801) tried to make a synthesis between Descartes and Newton, which he called *le système newtocartesien*, in his book *Traité de paix entre Descartes et Newton* (Treaty of peace between Descartes and Newton, 1763) . Newtonian physics was spreading through continental Europe by the middle of the eighteenth century and its acceptance among Jesuits became inevitable. At first, Jesuits took an eclectic position with a mixture of principles of scholastic natural philosophy and elements of Cartesian and Newtonian physics, but finally Newton's physics was totally incorporated into their teaching. This process, in which Boscovich had a great influence, too place about 1750, and included the acceptance of the heliocentric system and the theory of universal gravitation. However, there were still problems, as with Carlo Benvenutti (1716–1797), a disciple of Boscovich who was removed from his teaching post at the Roman College for his book *Sypnosis physicae generalis* (Summary of general physics), with its Newtonian and experimental contents, because he did not include the thesis of Aristotelian natural philosophy.

Leopold Gottlieb Biwald (1731–1805), professor of physics at the University of Graz from 1764 where he continued teaching after the suppression of the Society, and was rector in 1786 and 1798, is an example of the total acceptance of Newtonian physics and astronomy. He was the author of two textbooks of physics *Physica generalis* (General physics, 1766) and *Physica particularis* (Particular physics, 1766) in several editions, which contributed to the spread of Newtonian physics in Austria. The first covered the principles and applications of mechanics and astronomy and the second fluids, light, electricity and magnetism. Biwald presented the physics of Newton, whom he called the first father of the true physics, and also the atomic

[46] Oneill and Dominguez, *Diccionario Histórico*, vol. II (Filosofía, V. Cartesianismo), 1443–1447.

theory of Boscovich about the composition of matter. He defended fully heliocentric Newtonian astronomy and rejected the objections against it, especially those from a literal interpretation of the texts of the Bible. The German Jacob von Zallinger (1753–1813), professor at Munich and Dillingen, is another example,[47] with his *Interpretatio naturae, seu philosophia Newtoniana metodo exposita* (Interpretation of nature, or Newtonian philosophy presented methodically, 1773–1775). His presentation is totally mathematical and could be considered even then as a modern textbook of physics. The same can be said of the work of the Hungarian Ker János Horváth (1732–1799), professor at the Jesuit university of Nagyszombat (today Trnav, Slovakia). This university had been founded in 1635 by the Jesuit Cardinal Peter Pázmány and was moved to Pest in 1784 (today the state Eötvös Loránd University of Budapest). Horváth was professor of physics from 1770, the year he published his textbook *Physica Generalis – Physica Particularis* (General physics-Particular physics, 1770). In this book he follows the physics of Newton and was also influenced by Boscovich's ideas. Newtonian physics was generally followed in Jesuit colleges from the middle of the eighteenth century, but the expulsion of the Jesuits from Portugal, France and Spain between 1759 and 1767, and the suppression of the order in 1773, cut short this development. It is difficult to envision what would have been the further development of Jesuits' scientific work had these events not taken place.

[47] Hellyer, *Catholic physics*, 231–233.

Chapter 3
Kircher and Boscovich, Two Leading Figures

3.1 Athanasius Kircher, His Work and Museum

Athanasius Kircher (1602–1680), the last of nine children, was born in the small town of Geisa, near Fulda, where his father, Johann Kircher, administered the affairs of the Prince-Abbot of Fulda, during the difficult times of the Protestant reform.[1] Kircher studied in the Jesuit college of Fulda and in 1616 joined the Jesuits at the novitiate of Paderborn (Fig. 3.1). These were difficult times due to the Thirty Years war between Catholics and Protestants in Germany. Because of this, Kircher had to leave Paderborn in 1622 and to travel to Münster and then to Cologne and Koblenz, where he finished his studies of philosophy. During these studies, Kircher began to get interested in languages and mathematics, two subjects that would become his lifelong passion. In mathematics he was more interested in the practical applications, especially, in the construction of elaborate optical and mechanical machines. In 1624 he traveled to Heiligenstadt where he taught mathematics and Hebrew and Syriac languages. His curious machines so impressed the Archbishop-Elector of Mainz on his visit to the college that he invited Kircher to his court at Aschaffenburg. There in only 3 months he carried out a survey of the principality. In 1630 he returned to Mainz to study theology and after finishing these studies, he went to teach mathematics and ancient languages in the Jesuit College at Würzburg. At this

[1] Litterature about Kircher is very abundant, for example: Conor Reilly, *Athanasius Kircher S. J., Master of a hundred arts (1602–1680)*. Rome: Edizioni del Mondo (1974); John Fletcher (ed.) *Athanasius Kircher und seiner Beziehungen zum gelehrten Europa seiner Zeit.* Wiesbaden: Otto Harrassowitz (1988); Ignacio Gómez de Liaño, *Athanasius Kircher, itinerario del éxtasis o las imagines de un saber universal.* Madrid: Siruela, (2001); Paula Findlen (ed.), *Athanasius Kircher, The last man who knew everything.* London: Routledge (2004); Joscelyn Godwin, *Athanasius Kircher's theatre of the world. The life and work of the last man to search for universal knowledge.* Rochester (Vermont): Inner Traditions (2009). John E Fletcher, *A study of the life and Works of Athanasius Kircher 'Germanus Incredibilis'* (Edited for publication by Elizabeth Fletcher). Leiden: Brill (2011).

A. Udías, *Jesuit Contribution to Science: A History*,
DOI 10.1007/978-3-319-08365-0_3

Fig. 3.1 Athanasius Kircher (1602–1680), the man who knew everything (Arch. Romanum S.I.)

time his interest in magnetism began and he met Kaspar Schott, a young Jesuit, who became his disciple and faithful companion and friend. A year later Kircher and Schott had to escape from Würzburg to Avignon on the arrival of the army of the King of Sweden Gustav Adolf, who was helping the Protestant German princes. At Avignon, Kircher installed an astronomic observatory and an elaborate planetarium which showed the position of the Sun, Moon and planets by means of a complex system of mirrors. The planetarium became a popular attraction in the city. In 1635 Kircher described the system in a book about mirrors, *Primitiae gnomicae catoptricae* (Gnomic and catoptric first-results). At this time, because of his teaching of ancient languages, his interest was kindled for the interpretation of Egyptian hieroglyphs.

In 1632 Kircher met the wealthy collector and patron of scholarship, Nicolaus Claude Fabri de Peiresc, who had been a student at the Jesuit colleges of Tournon and Avignon and was also interested in Egyptian hieroglyphs. Peiresc had a good library with many Coptic and Arabic manuscripts and he corresponded with many scientists. Through him, Kircher came into contact with Pierre Gassendi and Marin Mersenne, key figures together with Descartes in the development of modern science in France. Peiresc was impressed with the ingenuity and inventiveness of the young professor and thought that Kircher could help him to interpret Egyptian hieroglyphs. Kircher discovered the connection between the Coptic language and that of ancient Egypt and wrote the first Coptic grammar. Kircher had a universal curiosity and great capacity for work and could pass with ease from one subject to another. He had a especial interest for antiquity and exotic subjects and expected to find in ancient documents traces of a lost pristine wisdom. He boasted to Peiresc that he had a rare Arabic manuscript with the key to the interpretation of Egyptian hieroglyphs and also a sunflower that marked the hours of the day, neither of which was confirmed. Kircher had also a passion for designing curious mechanical and optical instruments and the ability to construct them. His desire to be sent to the mission of China was not heeded by his superiors who thought that he was most needed in Europe. He was appointed professor of mathematics at the prestigious Roman College where he arrive in 1633, the year of Galileo's condemnation, and where he spent the rest of his life.

Kircher occupied the chair of mathematics at the Roman College for only 8 years, after which he was freed from all teaching duties so that he could dedicate himself completely to his research and his enormous work as a writer. At Rome, Kircher had access to the information that arrived from Jesuits in different parts of the world, especially from Asia and America, that he could use in his writings. In 1636 he accompanied the German Prince, Friedrich of Hesse, who was visiting Rome, in a journey to Sicily and Malta. During this journey Kircher saw the eruptions of the volcanoes of Etna and Stromboli and experienced an earthquake. In Naples he found that the Vesuvius was also active; he climbed the volcano and had himself lowered into the crater to observe the phenomenon close up. These experiences aroused his curiosity and interest for the terrestrial phenomena that he would study in his monumental work about the interior of the Earth. After his return to Rome he never left the city again. At Rome, in 1652–1654, Kircher enjoyed the company of Schott who proof-read and edited some of his books. Kircher's activity spread over a great variety of subjects to which he dedicated his universal curiosity.

In Kircher's work one can see two main interests. The first is the search in ancient texts, especially those from Egypt, for a pristine and original wisdom (*prisca sapientia*), which would provide the key to all knowledge.[2] In this wisdom he expected to find also a pristine theology (*prisca theologia*), which would form the basis of all religions. In this Kircher was influenced by the Hermetic currents of Italian humanists like Marsilio Ficino and Domenico Pico della Mirandola. In this current there were many syncretistic elements, especially in the writings known as the *Corpus Hermeticum*, attributed to a supposed Egyptian sage contemporary of Moses (*Hermes Trimegistus*), but which were in reality the work of several authors of neoplatonic and gnostic tendencies from the first to the third century. Kircher, who believed Hermes was an ancient Egyptian sage, claimed that traces of the origin of all knowledge are to be found in Egyptian culture. He also thought that in the newly discovered cultures of Asia and America one could find traces of this pristine wisdom and that a parallelism between West and East existed. Knowledge about these cultures was being transmitted at that time by Jesuit missionaries. The second great interest concerned magnetism as the force which pervades the entire universe and constitutes the ultimate cause of all phenomena. This led him to begin with the study of the properties of the loadstone and the characteristics of the Earth's magnetism. In his studies Kircher used a combination of scholastic and Aristotelian physics with mathematical applications and a continuous recourse to

[2] Joscelyn Godwin, *Athanasius Kircher, la búsqueda del saber de la antigüedad*. El Escorial: Swan, Avantos y Hakeldama (1986).

experiments as demonstrative arguments, following the lines of the new experimental sciences. His constant use of observations and experiments, which he called "the new method of philosophy", took often him away from Aristotelian doctrine. His insistence on the novelty of his method, his recourse to observations and experiments and his search for a natural explanations of phenomena, place him to some extent within the currents of the "new science", although other characteristics of his work take him outside its limits. Kircher's universal curiosity often led him to accept uncritically unbelievable things and events, such as, for example, the existence of dragons and mermaids. In this sense, he appears more in the realm of the Baroque than as a modern scientist. Moreover, for him there was always an ultimate end in his investigation of the works of nature, viz. their function as means to reveal the mysteries of the Creator.

In 1651, Kircher was assigned to organize the collections of objects of different types which belonged to the Roman College. He established a museum in one of the rooms of the college which was soon known as the *Museum Kircherianum.* The collection contained archeological objects of Roman and Etruscan origin, given in great part by the Italian aristocrat Alfonso Donnino. It included also archeological pieces from ancient Egypt, such as obelisks and mummies and curiosities brought by Jesuit missionaries from all parts of the world. Another section was formed of minerals, rocks, fossils and strange stuffed animals and skeletons. The museum also contained an art gallery with paintings and statues. Kircher added a collection of 19 machines and instruments built by himself for his experiments and described in his books. This formed the most remarkable part of the museum. Among them there were an automatic organ which imitated the songs of birds, a magic lantern (*Laterna magica*) to project images and other instruments used for hydraulics, optical and acoustic experiments. Some instruments were of statues, with spiral tubes in their interior that seemed to talk. One of these acoustic tubes connected the museum with the porter's lodge to announce visitors. These curious machines were designed in order to attract the attention of visitors, similar to what is found in modern science museums. The museum presented an image of universal science, as understood by Kircher. Emphasis was placed on the marvelous and the museum was conceived as a room of wonders and a theater of the world. The museum also presented a model of the cosmos with all its different aspects on a small scale. Soon Kircher's museum was something no visitor to Rome could miss. Kircher himself dedicated a lot of effort to showing the museum to the most distinguished visitors, and for many of them, amazed by the machines and his explanations the time spent with Kircher at the museum was the memory they most cherished of their visit to the eternal city. Among his visitors was Christina, Queen of Sweden, to whom Kircher presented a small obelisk with praises of her in hieroglyphic form. For Kircher this was also a work he liked to do, as mentioned in some of his books.[3]

[3] Paula Findlen, Introduction. The last man who knew everything... or did he? Athanasius Kircher S. J. (1602–1680) and his world. In P. Findlen (ed.) *A. K. The last man*, 1–48.

Kircher stands out above all as a prolific writer with 32 books and some 23 manuscripts of works that were not published. His books covered a great variety of subjects, such as, Egyptian hieroglyphs and culture, magnetism, sundials, optics, acoustics, music, astronomy, philology, logic, arithmetic, mathematical tables, geology, geography, Chinese culture and religion, Roman antiquities, the medical problems of plague, Noah's ark and the tower of Babel.[4] It is difficult to classify Kircher. He is best considered as a universal scholar and polymath and justly deserves his description as the man who knew everything. In 1661 he came into contact with the Dutch publisher Johann Jansson van Waesberghe (Janssonius), who in view of the success of Kircher's books saw in his writings a rich mine and asked him continuously for new publications. This forced Kircher into continuous activity to satisfy his eager publisher. Janssonius became the main publisher of Kircher's works: fifteen of his books some running into several editions. He also bought the rights for central Europe of the books, already published in Rome. Janssonius's editions of Kircher books were of large format (folio) with elaborate frontispieces and title pages and beautiful engravings by the best artists. The books are still today true bibliophile's jewels.

Outstanding among the six books dedicated to philological subjects are the four volumes dedicated to Egyptian hieroglyphs, *Oedipus Aegyptiacus* (The Egyptian Oedipus, 1652–1654). Kircher suspected that the present Coptic language had traces of the antique Egyptian language and that it held the key for the interpretation of hieroglyphs. He thought incorrectly that he could interpret them as simple symbols. Kircher dedicated four more books to this subject: *Prodomus coptus sive aegyptiacus* (Introduction to the Coptic or Egyptian language, 1636), *Lingua aegyptiaca restituta* (Egyptian language recovered, 1643), *Obeliscus Pamphilius* (The Panphilian obelisk, 1650) and *Obelisci aegyptiaci interpretatio* (Interpretation of an Egyptian obelisk, 1666). Kircher also contributed to spreading in Europe knowledge about China and its culture, using the information sent by Jesuit missionaries. His work *China illustrata* (Illustrated China, 1667), was very popular and was translated into French and Dutch. In this book he presented for the first time in Europe many curiosities about this hitherto great unknown country. In 1665 Kircher turned his attention to the plague that raged in Rome and based on his microscopic observations assigned the transmission of the sickness to fine invisible worms through the air. He published this in *Scutrinium pestis* (Investigation of the plague, 1658). As the culmination of his work, Kircher tried with his *Ars magna sciendi* (The great art of knowing, 1669) to present a universal science, following on the ideas of Raymond Lull's combinatory science. Between 1665 and 1700 some ten reviews and references of Kircher's publications appeared in the *Philosophical Transactions* of the Royal Society.

An important aspect of Kircher's work was his enormous correspondence. Among some 760 addressees of his letters are to be found four popes, two emperors,

[4] John Fletcher, "Athanasius Kircher: a man under pressure" and "Kircher's works" and "Bibliography". In J. Fletcher (ed.) *A. K. Beziehungen*, 1–15 and 179–195.

the kings of France and Spain, the Queen of Sweden, cardinals, bishops, members of the European nobility, scholars and Jesuit companions. Among the scientists one finds Cassini, Mersenne, Huygens, Leibniz and Gassendi. His correspondence played a similar role in the Germanic countries as that of Mersenne in France. A special section in the correspondence is that devoted to the more than 200 Jesuits spread through Europe, Asia and America, from whom he received the information he later included in his books.[5] Special attention needs to be given to some of his scientific books, that is, those dedicated to magnetism, geology, optics, acoustics and astronomy.

3.2 The Cosmic Force of Magnetism

Kircher's interest in magnetism can be traced back to his teaching at Heiligenstadt and the publication in 1631 of his first book, *Ars Magnesia* (Magnetic art). Magnetism became from then on his preferred subject. Ten years later in 1641, he published his definitive work on the subject, *Magnes sive de arte magnetica* (The magnet or the magnetic art) dedicated to the emperor Ferdinand III, reedited in 1643 and 1654.[6] Magnetism reappears in 1664 in parts of *Mundus Subterraneus* (Subterranean world) and in 1667 in *Magneticum naturae regnum* (The magnetic kingdom of nature). In the preface to *Magnes*, Kircher insisted on the universality of the magnetic phenomenon, which for him is present in all kinds of attractions and repulsions, sympathies and antipathies which occur in nature. He was aware that others had studied what he called "magnetic philosophy", but he preferred to talk of "magnetic art", that is, he wanted to join theory and practice and apply rules and experiments. Kircher claimed to use a simple style, as was becoming to German simplicity (*germano candore*), not one that was affected or rhetorical, even though the subject of magnetism is very abstruse. He maintained that his study was one of experimental philosophy, so that in his explanations he would not allow any supernatural or satanic elements and he explained everything only from a natural point of view (*sola vi naturae*). He quoted above all Gilbert's work *De magnete* (On the magnet, 1600), "the first who has written on the nature of magnetism" and the Jesuits Garzoni and Cabeo, whose works were mentioned in Chap. 2. Kircher's work is divided into three books. The first deals with the properties of magnets, the Earth's magnetism and the use of the compass. The second treats of different applications of magnetism, its relation to the heavenly bodies and the distribution and origin of the magnetic declination and inclination. The third explores the cosmic influence of magnetism and its unifying force which pervades the whole universe.[7]

---

[5] John Fletcher, Athanasius Kircher and his correspondence. In J. Fletcher (ed.) *A. K. Beziehungen*, 139–178.

[6] *Magnes sive de arte magnetica, Opus Tripartitum*. Rome: Ludovici Grignani (1641).

[7] William Hine, Athanasius Kircher and magnetism. In J. Fletcher (ed.) *A. K. Beziehungen*, 79–97.

It should be pointed out that Kircher did not agree with Gilbert about the Earth's magnetism, though he often quoted his work. Kircher did not accept that the Earth itself was a great magnet, as proposed by Gilbert, and defended the view that terrestrial magnetism was produced by veins of magnetic material on the Earth's surface oriented toward the north and south poles. Since Kircher did not accept the Earth's rotation, for him the north and south poles refer to the celestial poles of the rotation of the sphere of the fixed stars. According to him it is precisely the influence of the celestial poles that has oriented the Earth's magnetism in their direction. For Gilbert terrestrial magnetism was related to the Earth's rotation. Kircher also rejected Kepler's idea that the revolution of the planets around the Sun was due to a magnetic force emitted by the Sun. According to Kircher, Kepler was a good mathematician but he failed as a physicist. Kircher maintained, at least in public, the stable position of the Earth in the center of the universe in accordance with Aristotelian doctrine.

An interesting aspect of Kircher's work is his treatment of magnetic declination (the angle between the direction of the compass and the geographic north) and magnetic inclination or dip (the angle of the magnetic needle and the horizontal) and their geographical distribution, which he called "magnetic geography". He affirmed that it is necessary to reformulate geography from the point of view of terrestrial magnetism. Connected with this, he proposed that the Society of Jesus should establish what he called a "geographical council" (*Consilium geographicum*), in charge of processing magnetic observations carried out by Jesuits around the world. Thus he describes methods for the measurement of the magnetic declination and inclination or dip and their relation with geographic latitude and longitude. Kircher recognized that it was not possible to measure geographic longitude from the magnetic declination, as was proposed at that time. The accurate determination of geographical longitude in long sea journeys was an important problem. Since 1667 the Royal Society had been considering this problem and had tried to use the magnetic declination for this purpose, though finally it gave up. The problem was not solved until the development of the first marine chronometers by John Harrison in 1773. Kircher in his book presents a list of 65 observations of magnetic declination made in different parts of the world, of which 39 were made by Jesuits, among them, Scheiner, Cabeo, Biancani and Saint-Vicent. Kircher did not accept the explanation given by Gilbert about the cause of magnetic declination and its variations from one place to another. Instead he proposed that both were due to irregularities in the magnetic veins or dikes in the Earth, which he claimed were the cause of the main magnetic field. He suggested the possibility of drawing a magnetic map (*Mapam magneticam*) that would present the values of declination on a world map, but he did not do so himself – a pity, because this would have been the first magnetic map, instead of the one published by Edmond Halley in England in 1700. Kircher accepted that magnetic declination varies with time, as pointed out by Henry Gellibrand in 1635, but he could not find an explanation for this.

In the third part of the book entitled , *Mundus magneticus seu catena magnetica* (Magnetic world or magnetic chain), Kircher took the bold approach of considering magnetism as a cosmic force that pervades everything and that finally has its origin

Fig. 3.2 Kircher's world magnetic chain (*Magnes sive de magnetica arte*, 1641) (Biblioteca Histórica Universidad Complutense Madrid, BH-DER 4558)

in God (Fig. 3.2). He proposed that all elementary and hidden qualities, all the motions of sympathy and antipathy present in nature are manifestations of the cosmic and spiritual force of magnetism. For these considerations he based himself on the idea of the "Sefirot", which according to the Jewish Cabala are the ten emanations or attributes of the divinity, which as celestial influxes govern everything in creation. Kircher identified these influences with magnetism. He saw them as links or concatenations, which through the cosmic force of magnetism unite everything. He found this similar to the "tree of life" of the Sefirot, which he had discussed in *Oedipus Aegyptiacus*. In this universal concatenation Kircher included the magnetism of the planets, the meteors, the Sun and the Moon and their influence on the sea, plants and animals. He stated that a certain magnetic force is internally present in all elements and is the origin of all the processes which take place in the world. Magnetic forces are present in plants (*phytomagnetismus*) and animals (*zoomagnetismus*), and they are the cause of the strange and curious phenomena, which he often uncritically accepted. Finally, Kircher ended identifying magnetism with love (*De magnetismo amoris*), which is for him the force and origin of all the movements of attraction which keep everything in order. All these movements have their ultimate origin in God, the magnet of the whole of nature (*Deus totius naturae magnes*). Kircher went even further and linked this unifying motion with the mystery of the Trinity, in which everything is attracted in God the Father through the Eternal Truth (the Son) by the unifying force of the Spirit, constituting the "uni-triune magnetic principle" (*Unitrino principio magnetico,*). In this form, he finally concluded that all creation is attracted by the force of love toward God, center and magnet of everything.

3.3 The Mysteries of the Geocosmos

Kircher's experiences on his journey to Sicily – which included the eruptions of Etna and Stromboli, the earthquake and finally the activity of the Vesuvius, into whose crater he descended – led Kircher to become interested in the mysteries of the Earth or of the "geocosmos", as he called it. The result was the publication in 1664 of *Mundus subterraneus*, which was one of his greatest successes with two further editions in 1665 and in 1678.[8] Many of the ideas presented in this book influenced the development of the Earth sciences which were beginning at that time.[9] The book is of large format and divided into two volumes containing many engravings; the first volume was dedicated to Pope Alexander VII and the second to Emperor Leopold I. The novelty of the subject was expressed, with little modesty, in the preface: "I dare to open a path hitherto not tried, I believe, by anybody, one that no human mind has attempted; but I do this not of my own will, but inspired by the Divine Numen". *Mundus subterraneus* is divided into twelve books. Books 1–7 in the first volume are about general subjects, considering the Earth as the center of the universe, the structure of its interior and the origin of the elements which form it. The second volume, Books 8–12, contains many curious facts about what is found in the Earth's interior, such as, bones (fossils), animals, men and demons, the art of metals, a long disquisition about alchemy and the problem of what he call the principle of the "panspermia".

Kircher took for granted that the Earth is located at the center of the universe, and explained the fall of bodies in accordance with Aristotelian doctrine. He quotes the opinions of modern authors like Galileo, Mersenne, Torricelli and Gassendi and tries to refute them, using experiments to measure the rate of falling bodies. We can see that he did not understand completely the mathematical abstraction of the laws proposed by Galileo for the motion of falling bodies and for the parabolic path of projectiles. In discussing the motion of pendulums he pointed out that this could be the foundation for a new type of clock (*novum horologii genus*). This idea had been already proposed by Galileo, but it was only carried out in practice by Huygens in 1673.

---

[8] *Mundus subterraneus in XII libros digestus.* Amsterdam: Joannem Janssonium Waesberghe (1664).

[9] Eduardo Sierra, El geocosmos de Kircher. Una cosmovisi6n cientffica del siglo XVII. *Geocrítica*, 33–34 (1981), 5–19. Frank D. Adams, *The birth and development of the geological sciences.* New York: Dover (1954), 433–439.

The most original part his book is that dealing with the structure of the Earth's interior (Books 2–4). Kircher proposed the existence of three systems of conduits or channels through which air (*aerofilacios*), water (*hidrofilacios*) and fire (*pyrofilacios*) flow. The fire conduits or *pyrofilacios* are related with the volcanoes and connect with a permanent fire in the center of the Earth. Their interaction with the *aerofilacios* is the cause of earthquakes. The *hidrofilacios* connect the water of the oceans and seas with masses of water in the interior and with the lakes and the sources of the rivers. In this way all the water on the Earth is connected by underground channels. Two main water vortices are located at the Earth's poles, with water going in through the north poles and out through the South Pole. The interaction of the *pyrofilacios* with the *hidrofilacios* explains the existence of hot springs. The circulation of water in oceans and river is represented in several interesting geographic maps. Kircher considered these three types of conduits in the Earth's interior as analogous to the veins in the human body. The comparison continues with the claim that the mountain chains in the Earth are similar to the bones in man. Thus a complete analogy is established between the geocosmos (Earth) and the micro cosmos (human body). These ideas influenced the "organicist" currents popular at that time. This analogy allows Kircher to speak of the "very fecund womb of the geocosmos" where minerals, plants and animals are generated.

In the context of the geocosmos, Kircher also presents his idea of the *panspermia rerum* or "primeval seeds" from which everything originates.[10] Quoting Saint Basil (*Hexaemeron,* s. IV) Kircher proposed that God had created first a chaotic mass in which everything was present as in the form of seeds and from which all beings were later formed. He tried to find the nature of these seeds and finally proposed "a certain spirituous vapor of a sulphurous-saline-mercurial nature". A similar idea had been already proposed by Paralcelsus, whose work was known to Kircher, though he did not agree with him about magic and alchemy. Kircher explained that he was not referring to the elements of sulphur, salt and mercury, but to their forces (*vires*). These forces constitute for him the universal primeval seeds of the potentialities of matter, forming a primitive chaotic feminine mass from which everything, mineral, plants and animals, will be brought out in the geocosmos. Kircher insisted that he had arrived at the knowledge of this triple principle (sulphur-salt-mercury) by physical experiment and not from metaphysical speculation. In the discussion about the origin of metals (Book 9), Kicher suggests the influence of the pervading magnetic forces on the primeval seeds (*panspermia*). He has taken for granted in Book 8 the existence of the cosmic magnetic force, that we saw in his book about magnetism (*Arte magnetica*), which make similar things repel each other and different ones attract each other and which is present in minerals, plants and animals. For him, this cosmic magnetic force is precisely what gives form and figure to all things through a mysterious entity he called the "architectonic spirit". It may surprise that in this discussion Kircher never mentioned atomism which was being resuscitated in his

[10] Ingrid D. Rowland, Athanasius Kircher, Giordano Bruno and the Panspermia of the infinite universe. In P. Findlen (ed.) *A. K. The last man*, 191–205.

time by authors like Galileo and Gassendi. However, he mentioned that the vapors and exhalations which fill the geocosmos are formed by "faint, subtle and insensible corpuscles which are the cause of all changes and mutations in the terrestrial world". These ideas may seem strange to us and mere speculations without any empirical basis, but we should consider them as an attempt, although in great part mistaken, in Kircher's time, to search for a natural unifying and dynamic explanation of all terrestrial phenomena.

Many other subjects appear in the different books of *Mundus subterraneus*, some of a practical character, such as mining, the art of metals, chemistry, the preparation of medicines, the composition of glasses and gunpowder and the pyrotechnic art. There are also sections with descriptions of insects and plants. A long section is dedicated to alchemy (Book 11); Kircher denied the possibility of producing gold from other metals and the existence of the so called "philosopher's stone" or "elixir of life" which could be used for this purpose. In some areas, Kircher shows at times little critical judgment, accepting the existence of marvelous animals like dragons, and of strange fossils, stones with human and animal figures, underground men and demons, though he did not accept the existence of giants in antiquity. This work can be thought of as a great encyclopedia full of lights and shadows, some interesting insights and a lot of fantasy, where Kircher attempted to find a natural explanation for the phenomena he presented. There is a continuous recourse to experiments, although it is difficult to believe that many of them were ever carried out, and to the information uncritically accepted as provided by Jesuit missionaries from exotic lands. No wonder the book was a best seller in the society of that time eager for all kinds of novelties.

3.4 Light, Sound and a Journey Through the Cosmos

Kircher dealt with the nature of light and sound, which he considered to be related, in two different treatises, both with a predominantly practical approach to the construction of curious instruments. Light is the subject matter of *Ars magna lucis et umbrae* (The great art of light and shadow) published in 1646 and 1671.[11] The book begins, following Euclid's text, with 17 definitions and 8 axioms. Its approach is basically Aristotelian, far from the modern treatises on light, such as the one which Huygens will publish in 1690. For example, Kircher's treatment of reflection and refraction does not add anything new. He gives much space to explain the construction of different types of sundials and the related astronomical problems. The most original part is found in book 10 entitled *Magia lucis et umbrae* (Magic of light and shadow). In the preface to this book Kircher clarifies that he uses the term "magic" to refer to astonishing effects produced by purely natural means. He presents many

[11] *Ars magna lucis et umbrae*. Rome: H. Scheus (1646); Amsterdam: Jansson van Waesberghe (1671).

types of projections of optical images using the reflection and refraction of light. Among them he describes the construction of the "camera obscura" and the magic lantern. Although the invention of the latter, used for projecting images by means of the light of a lamp and a lens, is often attributed to him, it was already in use from about 1550. Kircher was, however, the first to describe with detail its construction in the 1671 edition. The book deals also with the phenomena produced by reflection from different types of mirrors, plane, spherical and cylindrical, and the refraction of light through water and glass. Kircher examines the possibility of burning ships with solar rays concentrated by mirrors, as was attributed to Archimedes in the siege of Syracuse by the Roman navy. He concludes that it was not possible. In the part devoted to sundials Kircher show his true obsession with this subject presenting a variety of types. He includes a sundial which shows the time at different parts of the world and another built from a sun-flower. He mentions also a moon-dial, to show the time during the night, but we do not know if he ever made one.

Kircher discussed sound in two books *Phonurgia nova* (New sound-making, 1673)[12] about sound in general and *Musurgia universalis* (Universal music-making, 1650)[13] about music. The first is about the nature of sound and begins with 19 definitions, 10 axioms and 4 postulates, in the form of the beginning of Euclid's geometry. He recognizes that sound is similar to light, but still following Aristotelian ideas he defines it as a "quality" (*qualitas passibilis*) which affects the ear. He was far from the mechanistic theory of modern science that already considered sound and light as waves. Kircher put the question as to whether sound can be transmitted through a vacuum. This led him to consider the existence of the vacuum which, faithful to Aristotelian doctrine, he denied. He experimented with echo and the transmission of sound through large acoustic tubes at great distances producing marvelous effects. Referring to prodigies connected with sounds he mentions those of bells which toll by themselves, the fall of the walls of Jericho, and surprisingly the legend of the Pied Piper of Hamelin, accepting their veracity. In *Mussurgia* Kircher expounded all aspects of music, its different kinds and instruments. He includes many types of music scores and engravings of instruments; thus the book forms an interesting treatise on the theory and practice of music. In the part of the book that he calls the magic of consonant and dissonant harmony (*magia consoni et disoni*) Kircher mentions the influence of music on sicknesses and moods. In book X, he deals with the harmony of the universe as a work of God, whom he calls "the eternal first musician" (*Archimusicus ille aeternus*). He compares the structure of the world with that of an organ (*mundani organi fabrica*). Using this analogy he compares the ten registers of an organ with the "symphony" of the world which begins with the minerals, plants, animals and men, then reaching the choirs of angels and the divine music. This vision of a universal harmony or symphony extends from the non-living, the living, the human body, the heavenly bodies, human society, and the choirs

[12] *Phonurgia nova sive conjugium mechanico-physicum artis et naturae paranymphia phonoscophia consignatum*. Kempten: Rudolphus Dreher (1673).

[13] *Musurgia universalis sive ars magna consoni et disoni in X libri digesta*. Rome: F. Corbelleti (1650).

of angels to end with the music of the Trinitarian mystery of God (*De musica Dei triniti*). For him the mystery of the Trinity is present in the whole harmony of the universe, where we can find many Trinitarian parallels (for example: vegetal-animal-rational; intelligence-memory-will; father-mother-child, etc.). However, he rejects Kepler's celestial harmony based on the values of the periods of the rotation of the planets around the Sun.

Kircher's cosmological vision is found in his work *Itinerarium exstaticum* (Ecstatic journey, 1656), with a second part about the subterranean and submarine world, *Iter exstaticum II* (Ecstatic journey II, 1657). Schott reedited these Works with the title *Iter exstaticum coeleste* (Celestial ecstatic journey, 1660 and 1671).[14] The origin of this work is presented as a long dream by Kircher in 1656, when he was exhausted with the work on his two books about Egyptian hieroglyphs and the subterranean world. The book has the form a dialogue between Kircher, under the name of Theodidacto (taught by God), and a celestial companion, Cosmiel, "minister of God the Most High and genius of the world", in a journey from the Earth to the stars passing through the Moon, Sun and the planets. Kircher sets aside Aristotelian cosmology and admits changes in the celestial world, which he suggests is formed of the same materials as the Earth. Ecclesiastic censure prevented Kircher from accepting the Copernican system and he opts for the system of Tycho Brahe. We saw in the previous chapter that this was the compromise solution taken by many Jesuit astronomers. Kircher quotes Riccioli's astronomical work where he proposed a modified Tychonian system. We are not sure up to what point Kircher was convinced of this view. Peiresc in a letter to Gasendi wrote that "the good father Athanasius" has admitted to us that he is pressed and obliged to write following Aristotle's common suppositions and that the same father Scheiner follows them only by force and obedience.[15]

However, Kircher uses his celestial journey to introduce many elements of the modern image of the universe. For example, he eliminates the existence of the heavenly spheres and affirms that the heavens are fluid; he describes the sunspots and says that the Sun itself is made of fire and that the Moon like the earth has mountains and valleys and an atmosphere. He presents a map of the surface of the Moon based on observations made between 1636 and 1650. He accepts that Venus rotates around the Sun and has phases like the Moon and is formed by earth and water; that Jupiter has an atmosphere and a rough surface and four satellites and that Saturn has an elliptical form with two satellites adjoining it (Huygens correctly explained Saturn's rings for the first time in 1656). Kircher maintained that the stars are not fixed on a sphere, but are distributed at different distances in a fluid region. The dimensions of the universe are very large, but not infinite, and it is filled with a light material, the ether. The second book (*Iter exstaticum II*) is divided into three dialogues. The first dialogue deals with the submarine world, and Kircher's companion here is called Hidriel.

[14] *Itinerarium exstaticum.* Rome: Mascardi (1656); *Iter exstaticum II.* Rome: Mascardi (1657). Edited by Kaspar Schott, *Iter exstaticum coeleste*. Würzburg: Endter (1660 and 1671).

[15] John Fletcher, *A. K. Beziehung*, 4.

The other two dialogues are about the earth and the subterranean world, where Kircher is accompanied again by Cosmiel. In this part, he repeats many of the ideas already proposed in *Mundus subterraneus*.

To finish this brief note on the figure of Kircher, which still fascinates many of us, the verses dedicated to him by Wolfgang Goethe are still relevant:

> So before you know it, Father Kircher is back again.
>
> However, I won't be ashamed of this:
> we are always fumbling after problems.[16]

3.5 Roger Boscovich and Modern Science

Roger Joseph Boscovich (Rudjer Josif Bošković) (1711–1787) was born in Dubrovnik (previously Ragusa), Croatia, the eighth of nine children of Nikola Boscovich and Pavica Bettera (Fig. 3.3).[17] Dubrovnik was at that time an independent city-state with close relations with Venice. Boscovich studied at the Jesuit

Fig. 3.3 Roger Boscovich (1711–1787), physicist and astronomer (Arch. Romanum S.I.)

[16] Godwin, *A. Kircher's theatre*, 134.

[17] Henry V. Gill, *Roger Boscovich S. J. (1711–1787). Forerunner of modern physical theories.* Dublin: Gill and Sons (1941); Lancelot. L. Whyte (ed.) *Roger Joseph Boscovich, S. J., F. R. S., 1711–1787: Studies of his life and work on the 20th anniversary of his birth.* London: George Allen and Unwin (1961); Elizabeth Hill, Roger Boscovich: A biographical essay. In Whyte (ed.) *Roger Joseph Boscovich,*(1961), 16–101; Zeijko Marcović, Bošcović, Rudjer J., In C. C. Gillespie (ed.), *Dictionary of scientific biography*, New York (1970–1980), 2, 326–332. Carmelo Oñate, Un sabio del siglo XVIII: R. J. Boscovich. *Letras de Deusto,* 18 (1988), 199–209; Juan Casanovas, Per il secondo centenario della morte del P. Rugiero Boscovich. *La Civiltà Cattolica*, IV (1988) 531–544.

college of his city until, at age fourteen, he traveled to Rome to enter the Jesuit novitiate. He then studied humanities and philosophy at the Roman College. There he was a student of Orazio Borgondio, professor of mathematics, who introduced him to the study of this science. At that time he began the study of Newton's works, *Principia mathematica* and *Opticks* which will become the foundations of all his works. We have seen how the General Congregation 16 (1731) had opened the door to the teaching of natural philosophy so that Newton's physics was taught in Jesuit colleges. Boscovich was for 5 years a professor of humanities, but any private time he had was employed in the study of mathematics and experimental physics. In 1737 he began his studies of theology and before completing these, he succeeded Borgondio in 1740 in the chair of mathematics at the Roman College, in the same year that Pope Benedict XIV, who had a progressive mentality, was elected to the papacy. In 1757 the Pope lifted the prohibition of the Index against all books which defended the motion of the Earth, the main obstacle against the Copernican system. Boscovich, who had a good relationship with the Pope and was a close friend of his secretary Cardinal Silvio Valenti Gonzaga, seems to have influenced this decision. Progress in astronomy and the general acceptance of Newton's theory of gravitation made it impossible to continue rejecting the heliocentric system. Boscovich oriented his interests very early to the natural sciences. In 1736 he had already carried out some observations of the transit of Mercury across the solar disc and in 1746 he determined the orbit followed by a planet according to Newtonian astronomy. He can be counted among the first in continental Europe to have fully accepted Newton's theories.

In 1750 Boscovich received the assignment from the Pope to measure the length of the degree along the meridian and to revise the map of the Pontifical States, as will be mentioned below. Other Papal assignments followed. Boscovich was called to inspect the cracks that had appeared in the dome of Saint Peter's Basilica. He proposed the installation of iron braces, a solution also given by other architects, and that was carried out. Other papal assignments involved hydrology, such as the drying out of the Pontine marches and the redirection of the course of the river Tiber. He also studied the flow of other rivers, the Adile, Panaro and Tidone and of the waterfalls of Caina and Nistore. Papal authorities asked Boscovich's opinion on other matters, such as the modifications of the ports of Rimini, Savona, Magnavaca and Fiumicino. He also had a great interest in archeology and participated in the excavations of an antique Roman villa near Frascati and made a study of the Augustan obelisk of the Campo Marzio.

In 1757, the calm and peace of Boscovich's scientific work in Rome began to be disturbed by a number of journeys and other activities. A year before this, his good friend Cardinal Gonzaga had died and 2 years later Pope Benedict XIV, who had shown him constant support and backing, also died. Boscovich found that the atmosphere in Rome was becoming stifling and oppressive. His ideas, which had moved away from traditional Aristotelian natural philosophy and embraced modern scientific theories, especially those of Newton, were under suspicion and even met hostility. Thus, he took any opportunity he could to get away from Rome. In 1757 he travelled to Vienna in a diplomatic mission to defend the rights over the waters of the Lake Bientina of his native Republic of Lucca, against the claims of Florence. The rights of Florence were defended by another Jesuit, Leonardo Ximenes, who, as we have seen, had created an astronomical observatory in that city. After successfully accomplishing his mission Boscovich remained in Vienna and was commissioned by the Empress Maria Theresa to study the repairs needed in the city library. Boscovich suggested to the Empress the convenience of measuring the length of the degree of the meridian in Austria, Moravia and Hungary. He proposed for this work Joseph Liesganig, the assistant of Maximilian Hell in the Vienna observatory. Other diplomatic issues in favor of his native city took him to France to accompany the Marquis Michelangelo Romagnoli. The scientific circles of Paris gave Boscovich a warm reception. He was made a correspondent member of the *Académie des Sciences* in 1748. In Paris he strengthened his friendship with Joseph Jérôme de Lalande, professor of astronomy at the *Collège de France* and director of the astronomical observatory. After his long visit in 1760 he travelled from Paris to London invited by the president of Royal Society. On January 15 1761 Boscovich was elected a fellow of that Society. During his stay in London he defended the rights of his native city, Dubrovnik, against the accusations of collaborating with France by letting French warships into its port against the interest of England. In London Boscovich published *De solis et lunae defectibus* (About the defects of the Sun and the Moon, 1760) in Latin verse dedicated to the Royal Society. In 1761 he travelled to Istanbul to observe the transit of Venus, but arrived too late to make the observations. There he was detained for 6 more months due to sickness. His return journey was a long one through the eastern countries of Bulgaria, Moldavia and Poland where he became sick once more, and remained for 6 months as the guest of the noble Poniatowski family. Finally he arrived in Rome via Vienna in 1763. Boscovich wrote a lengthy account of his journey which was published in 1783 (*Giornale di un viaggio da Constantinopoli in Polonia*).

After his long absence Boscovich found the atmosphere in Rome very changed, especially after the death of Benedict XIV. We saw that in 1744 he had made the plans for an observatory in the Roman College that were not carried out. Now, the rector of the college of Brera in Milan, asked him to take charge of the construction of an astronomical observatory. Not wishing to stay in Rome, he accepted in 1764 the offer of a chair of mathematics and astronomy at the University of Pavía, but as he did not find there a favorable situation, he dedicated most of his time to the observatory of Brera. In 1768 he left the chair in Pavia and accepted a chair of optics and astronomy at the *Scuole Palatina* in Milan. Difficulties in understanding, in part due to his strong character, led to conflicts with the director of the Brera Observatory, Louis Lagrange, and he left the observatory. Also finally in 1772 he left his post in Milan. Boscovich moved to Venice at the time of the suppression of the Society of Jesus. His French colleagues, specially his friend the astronomer Lalande, urged him to move to Paris where the government of Louis XV offered him the direction of the Laboratory of Optics at the Naval School. He was not too happy there either. His candidacy to the *Académie de Sciences*, though counting with the support of several of its members, especially Lalande, was blocked by the influential Jean d'Alembert, who was harsh in his opposition. In 1782, worn out with this situation, Boscovich received permission from the French government for a long absence in Italy to work on the publication of his works on optics and astronomy. He found this work very demanding on his health and prolonged his stay in Italy. In Milan he was allowed to work in the Brera observatory now a state institution. His health suffered and after a long illness he died in 1787. Zeljko Markovic ends his short biography saying: "Sharp in thought, bold in spirit, independent in judgment, zealous to be exact, Boscovich was a man of the eighteenth century European science in some respects and far ahead of his time in others".[18] His production was very abundant with some 80 publications about a variety of scientific topics. Some of the more important ones deserve to be presented in detail.

3.6 A Dynamic Atomic Theory

The most important work of Boscovich, where he presents his dynamic atomic theory reducing all forces acting in nature to one, *Theoría philosophiae naturalis* (Theory of natural philosophy) was published for the first time in 1758 in Vienna under the care of Karl Scherffer (Fig. 3.4). A second edition was published in 1763, with the supervision and revision of the author.[19] In this work Boscovich abandoned totally Aristotelian natural philosophy and the hylomorphic doctrine which were still in theory, but not in practice, the obligatory teaching among Jesuits. His starting point was based on the ideas of Leibniz and Newton; from that base he put forward new ideas about the atomic composition of matter and the forces acting in nature. At the end of his work on optics (*Opticks*, 1703) Newton had presented his atomic theory in which atoms are endowed with active forces of gravitation, cohesion and fermentation. Boscovich took a bold step forward and reduced all forces to just one which had an attractive and repulsive nature depending on distance. He, thus, in fact proposed the unification of all the forces of nature.[20] This idea can be traced back to an early work *De viribus vivis* (About living forces, 1745), and appeared later in *De lumine* (About light, 1748) and especially in *De continuitatis lege* (About the law of continuity, 1754) and *De lege virium in natura existentium* (About the law of the forces which exist in nature, 1755).

---

[18] Marcović, "Bošcović", 332.

[19] *Philosophia naturalis theoria redacta ad unicam legem virium in natura existentium*. Vienna (1758). *Theoria philosophia naturalis redacta ad unicam legem virium in natura existentium, nun ab ipso perpolita et aucta, ipso auctore praesente et corrigente*. Venice: Typographia Remondiniana (1763). English translation: *A theory of natural philosophy* (biography by B. Petronievich, introduction by J.M. Child). Cambridge (Mass.): MIT Press (1966).

[20] Mariano Colubi, *Boskovich y la vision mecánica de la naturaleza* (1740–1785). Madrid: Universidad Autónoma de Madrid (1999).

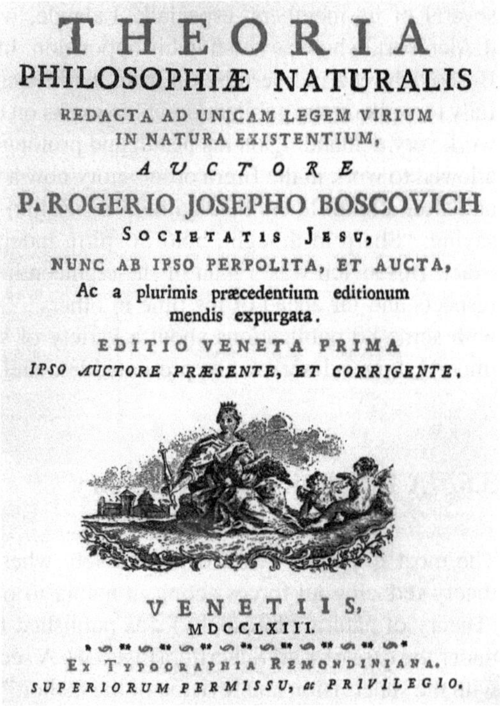
THEORIA
PHILOSOPHIÆ NATURALIS
REDACTA AD UNICAM LEGEM VIRIUM
IN NATURA EXISTENTIUM,
AUCTORE
P. ROGERIO JOSEPHO BOSCOVICH
SOCIETATIS JESU,
NUNC AB IPSO PERPOLITA, ET AUCTA,
Ac a plurimis præcedentium editionum
mendis expurgata.
EDITIO VENETA PRIMA
IPSO AUCTORE PRÆSENTE, ET CORRIGENTE.

VENETIIS,
MDCCLXIII.
EX TYPOGRAPHIA REMONDINIANA.
SUPERIORUM PERMISSU, ac PRIVILEGIO.

Fig. 3.4 *Theoria Philosophiae Naturalis* (1763) of Roger Boscovich (Arch. Romanum S.I.)

Philosophia naturalis is divided into three parts: the first presents the basic notions of the theory, the second its applications to mechanics, and the third its applications to other parts of physics. Boscovich begins with the principle of continuity (a concept taken from Leibniz but with roots in scholastic philosophy) which states that in nature all processes are gradual and without jumps, and that there is a principle of impenetrability, that is, two particles cannot occupy the same place. His fundamental idea is that matter is formed by simple points, indivisible and without extension, that is, atoms without mass similar to geometric points, located in an infinite empty space, but finite in number. These atoms or points possess inertia and are the center of a unique active force which is repulsive at short distances and attractive at long distances in accordance with Newtonian gravitation. When the points are very near, the repulsive force increases to infinity, so that two atoms can never join together. At intermediate distances the force changes alternatively from repulsive to attractive several times. The curve representing the intensity of the force is continuous with several changes of signs and points of zero value. At long distances the force is attractive and follows the law of the inverse square of the distance according to Newtonian law of gravitation. Boscovich, however, states that Newton's law is not followed exactly, but only very approximately. Material bodies are then formed by arrangements of this type of atoms and their properties depend only on their spatial distribution. Mass for him was, then, not a property of each atom, as held by Newton and the classic atomists, but a consequence of the number and distribution of the atoms in a body. The extension of matter is explained as a dynamic configuration of a finite number of atoms or centers of interaction. Difference between bodies depends only on the number and relative position and configuration of the atoms which form them. Thus, as the atoms are the ultimate elements of nature and are all equal, the principles of homogeneity and simplicity are fully respected. Since atoms do not have mass, the force with which they interact among each other, attracting and repelling according to their distance, is only related to their relative acceleration. In this way, the dynamic interactions among atoms are reduced to purely kinematic relations. In his theory Boscovich is only interested in the relative positions and motion of the atoms according to the forces that attract and repel them.

In the second part dedicated to the applications of the theory to the problems of mechanics, Boscovich begins by considering the centers of oscillation and of gravity of bodies, which for him are only assemblies of atoms without mass. Since he does not allow any contact between atoms, he has to explain how collisions between elastic bodies are produced. Collisions between atoms with mass was the central point of the mechanical theory of the school of Descartes, which did not accept the existence of forces acting at a distance, as in Newtonian gravitation. Boscovich tried to show with geometrical demonstrations that all mechanical interactions between bodies can be explained by his single force between atoms. In the third part he faced a greater problem trying to explain all properties of bodies, such as extension, mass, density, inertia, gravity and cohesion, in accordance with his theory. The difficulty became even greater in explaining the differences between solids and liquids, chemical reactions, light, electricity and magnetism. Newton's law of universal gravitation is supposed in the theory of Boscovich as the force acting between bodies at large distances. Cohesion between atoms inside bodies is explained by the limit between attracting and repelling forces at short distances and the difference between solids and liquids by the different amount of cohesion between their atoms. In liquids a lesser degree of cohesion results in a greater mobility of the atoms that form them. Fire and light are explained in terms of very rapid moving particles. Boscovich followed Newton's corpuscular theory of light and rejected Huygens's wave theory, claiming that it does not fit the observations. He followed the same line of argument to explain the transmission of sound through the air by the interaction of its particles. Boscovich also tried to explain electrical and magnetic phenomena in terms of the different forms of attraction of the particles, according to the single law he was proposing. Although his explanations cannot be accepted today, they were a consequence of his efforts to reduce all natural forces to one.

Boscovich ended his work with a metaphysical appendix on the soul and God and with four supplements. He held that the soul, as the principle of intelligence and will, is of a different nature from matter. He considered the relation between soul and body and denied that the soul is located in any particular part of the body, as Descartes had done with reference to the pineal gland. In this respect, he moved away from both Descartes and Leibniz. He also rejected the materialistic view that the universe originated by pure hazard and defended the opinion that the order in nature, shown by his law of a universal force, implies that it is the work of a Creator of infinite power, wisdom and providence. Boscovich ends this appendix saying that we should acknowledge with gratitude that the infinite wisdom and freedom of God has chosen adequate means for the purpose he had in his creation. The first two supplements deal with the problem of space and time. He considers them to be the forms of existence of physical bodies, while empty space and time are only imaginary entities. The last two supplements present an analogical version of his law and a refutation of the existence of attractive forces at very short distances.

Boscovich affirmed that his theory was a middle road between those of Leibniz and Newton. It has many things in common with both, but also differs strongly from them. It is more simple, since it can explain all natural properties from a single force. He rejected Descartes' mechanistic ideas which reduced all physical interaction to collisions between bodies. His Newtonian views forced him to accept that forces act at a distance, but he reduced them to only one type. His effort in this respect may be seen as part of the unification process present in the whole history of physics up to present times. Boscovich was highly esteemed in England being one of the first continental scientists to accept Newton's ideas. In the footsteps of Newton, he always showed a preference for geometry rather than mathematical analysis in his demonstrations, while his contemporaries, like Leonhard Euler and Joseph Lagrange, were imposing mathematical analysis to simplify demonstrations.

3.7 Mathematics, Astronomy, Optics and Geodesy

Boschovich had a wide range of interests in the field of mathematical physics. For him mathematics was always the central point, following his two masters Newton and Leibniz. His most important work on mathematics, based on his teaching experience from 1737, is contained in the three volumes for students of *Elementorum universae matheseos* (Elements of complete mathematics, 1757) that he had promised to publish in 1747. The first volume deals with the geometry of plane and solid figures, arithmetic and trigonometry; the second with algebra and the solution and properties of equations up to fourth grade; and the third, considered the most original part, with conic sections (ellipse, parabola and hyperbola). His friend, the astronomer Lalande, wrote in the *Journal des Sçavans*, that Boscovich's genius appears, reaching sublimity, especially in the presentation of conic sections. Other books of his on mathematics are *Trigonometriae sphaericae constructio* (Construction of spherical trigonometry, 1737) and De *natura et usu infinitorum et infinite parvorum* (On the nature and use of infinites and infinitesimals, 1741). In the preface of the first volume of *Elementorum*, Boscovich promised a fourth volume where he would deal with the geometrical and algebraic treatment of infinitesimals and the foundations of the differential and integral calculus. He also promised that after dealing with pure mathematics he would write about applied mathematics, covering optics, astronomy, chronology, architecture and music. He never found the time to fulfill these promises.

Boscovich dedicated much of his energy to astronomy. We have seen how he prepared the blueprints for an observatory at the Roman College on the roof of the church of Saint Ignatius, a project that would not be carried out until many years later. He directed the construction of the observatory of the College of Brera in Milan, considered one of the best of those times. In this observatory Boscovich dedicated much of his time to the calibration of the instruments and to astronomical observations. Some of his earliest observations were those of the transit of Mercury across the solar disk in 1736 (*De Mercurio novissimo infra solem transitu*, On the new transit of Mercury under the Sun, 1737). We saw how different circumstances did not allow him to observe the transits of Venus of 1761 and 1769.

Another of his works in astronomy was about the interaction between Jupiter and Saturn (*De inaequalitatibus Saturnus et Jupiter sibi mutuo videntur inducere*, On the inequalities that Saturn and Jupiter seem to induce in each other, 1756). Also interesting is the long Latin poem of 5128 verses on the Sun and Moon , *De Solis et Lunae defectibus libri V* (Five books on the defects of the Sun and Moon, 1760), published during his stay in London and dedicated to the Royal Society. In this he describes Newton's astronomy and theory of light. He proposed a method to determine the orbit of a comet with only three observations (*De cometis*, 1746) and of that of a planet (*De determinanda orbita planetae,* 1749). In other works Boscovich joined astronomy and optics is his treatment of astronomical instrumentation and the design of new types of telescopes, *De novo telescopii usu ad objecta coelestia determinanda*, (On the new use of a telescope to determine celestial objects, 1739) and *Memorie sulli cannocchiale diottrici*, (Report on dioptric telescopes, 1771). He also wrote several works on the nature of light, such as *Dissertatio de lumine* (Dissertation on light, 1748). Boscovich proposed an interesting experiment to decide between Huygens' wave theory of light and Newton's corpuscular theory, based on astronomical observations made using a telescope full of water, but the method was not conclusive. Boscovich's works on optics and astronomy were published together at the end of his life in *Opera pertinentia ad opticam et astronomiam maxima ex parte nova et omnia hucusque inedita in V tomos distributa* (Works of optics and astronomy in great part new and unedited distributed in five volumes, 1785).

As a consequence of his studies of Newton's works, Boscovich soon became interested in the problem of the figure of the Earth. In 1739 he published *Dissertatio de Telluris figura* (Dissertation on the shape of the Earth) and in 1740 *De inaequalitate gravitatis in diversis terrae locis* (On the inequalities of gravity in different places of the Earth). At that time France and England were divided by a controversy about the shape of the Earth. In England scientists held that the Earth was flattened at the poles, following Newton's proposal that its form must be one of equilibrium of a gravitating body in rotation. In France, on the contrary, the general opinion was that it was elongated at the poles and flattened at the equator. This had been proposed by Descartes and followed from the measurements made by Jean Picard and Giovanni D. Cassini of a degree of latitude in France. This problem was not solved until Pierre Maupertuis and Claude Clairaut published, in 1738 and 1743, the results from the two expeditions organized by the *Académie de Sciences* of Paris to measure a degree of latitude near the poles and near the equator; these confirmed Newton's prediction.

Boscovich became interested in this problem and offered to John V of Portugal to make this type of measurement in Brazil. The Holy See intervened and instead proposed to Boscovich in 1750 to make these measurements in the Papal States along the meridian between Rome and Rimini. Since this petition came from his friend Cardinal Gonzaga he could not refuse. The assignment included also a revision of the map of all the Papal States. Boscovich interrupted his classes at the Roman College to make these measurements with the help of Christopher Maire (1697–1767), professor of mathematics and later rector at the English College in Rome. Boscovich and Maire began the measurements in July 1750 making triangulations along a total length of 240 km. They began with a careful measurement of a base line near Rome and then made triangulations from eleven vertices with three measurements from each. The final result for the length of the degree of the meridian arch was 56,979 *toesas* (111,252 m), a fairly exact value. In these calculations Boscovich used an original error analysis, which can be considered an advance on Laplace's method of the least squares. As he was aware of possible errors in the measurements and of the irregularities in the Earth's composition, Boscovich concluded that the shape of the Earth is most probably flattened at the poles, but that its exact shape is still uncertain. He determined that its flatness may vary from 1/230 to 1/255 (today the accepted value is 1/298.25). The results from these measurements were published in a long report of more than 500 pages, *De litteraria expeditione.. ad dimidiendos duos meridiani gradus* (Expedition to measure two degrees of the meridian, 1755). This report was published in French in 1770.[21] The work included also the cartography of a new detailed and corrected map of the Papal States, published in the French edition. Pope Benedict XIV was very pleased with the results and always had a favorable relation with Boscovich. The French geodesists, especially, de la Condomine were also very pleased with this work. During his visit to London in 1760, Boscovich suggested to the Royal Society that similar measurements be made in the British colonies of North America, reminding the members that such measurements have been already made in France. Some years later the Royal Society informed him that Charles Dixon and Jeremiah Mason made this type of measurements in Maryland between 1763 and 1767 as part of wide program of geodesic measurements.

Boscovich, though he never taught natural philosophy is remembered above all as an exponent of Newtonian natural philosophy. Through the chemist Joseph Priestly, the discoverer of oxygen, the theories of Boscovich about the atomic

[21] *De litteraria expeditione per pontificiam ditionem ad dimidiendos duos meridiani gradus et corrigendam mappam geographicam, iussu et auspicii Benedicto XIV, suscepto a Patribus Societatis Iesu Chrsitophoro Maire et Rogerio Josepho Boscovich.* Rome: Nicolaus e Marcus Pelearini (1755). French translation : *Voyage astronomique et géographique dans l'état de l'Église entrepise par l'ordre et sous les auspices du pape Benoit XIV, pour mesurer deux degrés du méridien et corriger la carte dans l'état ecclésiastique par les PP. Maire et Boscovich traduit de latin.* Paris: N. M. Tilliard (1770).

constitution of matter influenced some of the physicists of nineteenth-century England, such as William Thomson (Lord Kelvin), Michael Faraday and James C. Maxwell, pioneers of the theory of heat, electricity and magnetism respectively. Boscovich remained a scientist of his century and we should not see in him, as some try to do, a precursor of modern quantum-relativistic physics. For example, some have tried to see in his rejection of the absolute character of Newton's concepts of space and time a preview of Einstein's ideas His contributions must be always seen in the context of the physics of the eighteenth century. Boscovich is an important figure in the history of physics, although he remains today somewhat forgotten, except in his native Croatia, where his image appears on mail stamps and the bills of 10,000 dinars. Dmitri Mendeleev, the creator of the periodic table of elements, said that Boscovich together with Copernicus are the pride of the Slavic people, and that he can be thought of as one of the founding fathers of modern atomism.

Chapter 4
Jesuit Astronomers in China and India

4.1 Matteo Ricci, from Macerata to Beijing

The Jesuit interest for astronomy, that we have seen present in the colleges in Europe, also had an important role to play in mission countries, especially, in China. Saint Francis Xavier (1506–1552), the first Jesuit missionary to the Far East, wrote to Rome after his second journey to Japan, asking for some Jesuits with knowledge of astronomy to be sent out. He argued that the Japanese were very curious about natural sciences and particularly interested in the motion of the heavenly bodies, eclipses of Sun and Moon, phases of the Moon, the origin of rain, snow, thunder, lightning and other phenomena.[1] However, it was in China, and not in Japan, that for Jesuit missionaries astronomy and mathematics would be a great help in introducing the Christian faith. It was, in fact, owing to the Chinese interest in astronomy that the first Jesuit missionaries could enter China, which under the Ming dynasty had been closed to all foreign influence.

As a country of millenary culture, China had been interested for a long time in astronomical observations, and astronomy had developed early on, especially during the Western Han dynasty (206 B.C. to 9 A.D.), when it acquired an official character. A Board of Astronomy was created and put in charge of the preparation of the calendar. This official calendar, which the Emperor promulgated every year, was obligatory for the whole empire and for any vassal states. In the fourth century, the Chinese possessed good catalogues of stars and long lists of astronomical observations, such as those for appearances of comets and sunspots. In the fifth century, the mathematician Zu Chongzi wrote important works that were in use centuries later and he carried out a reform of the calendar. During the Tang dynasty (618–907), through the influence of Indian astronomy, Chinese astronomers progressed in methods for carrying out astronomical calculations. In fact, Chinese

[1] *Cartas y escritos de San Francisco Javier (*Annotated by Felix Zubilaga, 3rd ed.). Madrid: Biblioteca de Autores Cristianos, (1979), (Letter to Saint Ignatius, 9 April 1552) 449.

A. Udías, *Jesuit Contribution to Science: A History*,
DOI 10.1007/978-3-319-08365-0_4

astronomers were ahead of Westerners in conceiving the heavenly bodies as floating in infinite empty space, and in the use of a polar-equatorial reference system of celestial coordinates, a system first introduced in Europe by Tycho Brahe in 1585. They also used for their studies gnomes and armillary and celestial spheres. In some of their theories, like the concept of a vacuum between heavenly bodies and that of the infinity of the universe, the Chinese were more advanced than the Greeks, whose ideas were dominant in Europe during the whole of the Middle Ages. From early centuries, Chinese astronomers carefully observed and recorded celestial phenomena, such as the eclipses of the Sun and Moon, planetary conjunctions, appearance of comets, sunspots, nova and supernova stars. Chinese astronomers were more interested in the unusual and rare celestial phenomena than in the regular motion of the celestial bodies. For example, they did not have a geometrical model of the planetary system. During the Yuan dynasty (1279–1368), which followed the Mongol invasion of China after the conquests of Genghis Khan, there was collaboration between Chinese astronomers and those from Persia and Arabia, which resulted in the establishment in China of an Arabian school of astronomy. During this dynasty this school took charge of the Board of Astronomy and the preparation of the calendar. Guo Shoujing stands out among the astronomers of this school and under his direction the observatories of Nanking and Beijing were established. After the expulsion of the Mongols and the establishment of the Ming dynasty (1368–1644), the traditional Chinese astronomy school regained power, but the Arabic school maintained a department in the Board of Astronomy. Under the Ming dynasty, China practically closed itself off from all foreign influence and the sciences suffered a general decline. Suspicion and opposition to everything foreign was expressed by the word *xiyu*, which from the Ming dynasty onwards was applied indiscriminately to anything coming from outside China. During this time astronomy and mathematics, in particular, underwent a strong decline with no new contributions.[2] For example, the prediction of eclipses was very unreliable with major errors that affected the trustworthiness of the calendar and caused social discomfort. This situation would allow Jesuits to show the superiority of European astronomy and lead to their being put in charge of the reform of the calendar and taking over the directorship of the Imperial Observatory.[3]

The epic story of the Jesuits in China and its relation with astronomy and mathematics begins with Matteo Ricci (1552–1610) (Fig. 4.1).[4] Ricci was born in Macerata, Italy, and studied in the Jesuit school of that city until 1568 when he was

---

[2] J. Needham, *Science and civilization in China.* (vol. III, Mathematics and the sciences of heavens and earth. The time of the Jesuits). Cambridge: Cambridge University Press, (1959) 437–461; Ho Peng Yoke, Li Qi and Shu: *An introduction to science and civilization in China* (Chap. VII. The decline of Chinese mathematics and the coming of the Jesuits). Hong Kong: Hong Kong University Press (1985), 10–109.

[3] Agustín Udías, Jesuit astronomers in Beijing, 1601–1805. *Quarterly Journal Royal Astronomical Society,* 34 (1994), 463–478.

[4] Vincent Cronin, *The wise man from the west.* London: Rupert Hart-Davies, 1959; Michela Fontana, *Matteo Ricci. Un gesuita alla corte di Ming*. Milan: Mondadori (2005); R. Po-Chia Hsia, *A Jesuit in the forbidden city: Matteo Ricchi , 1552–1610.* Oxford: Oxford University Press (2010).

Fig. 4.1 Matteo Ricci (1552–1610) and Xu Guangqi (1562–1633) (Kircher, *China illustrata*, 1667) (Biblioteca Histórica Universidad Complutense Madrid, BH-DER 10718)

sent by his father to Rome to study law. In Rome in 1571 he joined the Jesuits in the novitiate of San Andrea on the Quirinal. In the Roman College he studied philosophy which included courses in mathematics and astronomy. He took the courses offered by Clavius and it is possible that he belonged to the academy of mathematics together with the most promising students. Ricci may not have been aware of the importance these studies would have for his future work in China.[5] In 1576 Ricci asked to be sent to the Far Eastern mission and 2 years later in Lisbon he boarded the ship São Luis for his long journey to India. Six months later he arrived in Goa, the main Portuguese settlement in India. Later he moved to Cochin, in the south, where he studied theology and was ordained to the priesthood. He returned to Goa in 1580 to teach Latin and Greek in the college of Saint Paul. In 1573 the Jesuits appointed the Italian Alessandro Valignano (1539–1606) as general overseer of the missions in India and the Far East. Valignano, a man of far reaching vision, understood the need to accommodate to Eastern cultures and tried to put this conviction into practice. Accordingly, he sent Michele Ruggieri (1543–1607), who had traveled with the same ship as Ricci, to Macao, the only Portuguese commercial post

[5] Juan Casanovas, Il P. C. Clavio professore di matematica del P. M. Ricci nel Collegio Romano. *Atti del Primo Convegno Internazionale di Studi Ricciani*, Macerata (1984), 229–239.

allowed by the Chinese authorities, where the Jesuits had a residence from 1563. In 1579, Ruggieri arrived to Macao with the exclusive aim of learning the Chinese language and culture in preparation for entry into mainland China. In 1582 Ricci joined Ruggieri in Macao and also began to learn the Chinese language making rapid progress thanks to his extraordinary memory.

In 1583, after several attempts and despite the great difficulties imposed by the Chinese authorities on all foreigners, Ricci and Ruggieri finally obtained permission to settle in Zhaoqing, a city not far from Macao, with the protection of the Prefect Wang Pan. Soon Ricci's knowledge of mathematics and astronomy attracted the attention of the learned men of the city. This was the beginning of the Jesuits' marvelous adventure, using their knowledge of mathematics and astronomy, to bring the Christian faith to the closed empire of the late Ming dynasty.[6] In 1595 Ricci, who had tried first with Ruggieri the grey robes of Buddhist monks, adopted now the long blue silk robe and tall four-cornered hat of Chinese Confucian scholars and literati. He presented himself as a Western scholar (*xiru*) eager to establish contact with Chinese wisdom and he took the Chinese name of Li Madou. He soon became aware of the keen interest among Chinese literati for mathematics and astronomy and their relative backwardness in these subjects. He found the classes of Clavius, that he had followed at the Roman College, very useful. His aim was the imperial court of Beijing and he tried all means to get there. The first stage of his journey took him to the city of Nanchang, where he arrived in 1595 and stayed for a year. There Ricci wrote his "Treatise on friendship" (*Jiaoyou lun*), the first literary work written in Chinese by a European. The treatise based on Latin and Greek classical authors was an instant success, since it recalled to its readers similar ideas about friendship found in Confucius. In 1598 he arrived to Nanjing, the southern capital, on the second stage of his journey. There Ricci could admire the astronomical instruments of the observatory, gnomes, armillary spheres and celestial globes, all very finely constructed. These instruments had been made by the astronomer Guo Shoujing in the fourteenth century under the Mongol emperor Khublai Khan, about whom Marco Polo had spoken in his account of his journeys. Ricci, who had taken measurements of the latitude on his journeys, realized that the orientation of the instruments was not correct for the latitude of Nanjing. This confirmed his view about the shortcomings of the contemporary Chinese astronomical knowledge. The instruments had been designed for and installed first at a different location, 4° of latitude to the north, and later moved to Nanjing without changing their orientations.

Ricci arrived in Beijing on January 24, after a very arduous journey, some of it along the Imperial canal, where he had to bear many hardships due to the Court eunuchs with whom he made part of the journey. Accompanied by the Spanish

[6]Arnold H. Rowbotham, *Missionary and mandarin. The Jesuits at the court of China.* Berkeley (Cal.): University of California Press (1942); George H. Dunne, *Generation of giants. The story of the Jesuits in China in the last decades of the Ming dynasty.* London: Burns and Oates (1962); Liam M. Brockey, *Journey to the East. The Jesuit mission to China, 1572–1724.* Cambridge (Mass.): Harvard University Press (2007); Florence C. Hsia, *Sojourners in a strange land. Jesuits and their scientific missions in late imperial China.* Chicago: University of Chicago Press (2009).

Jesuit Diego de Pantoja (1571–1618), Ricci presented himself as a scholar well versed in astronomy and mathematics, who had come from very far, to place himself at the service of the emperor. He brought several gifts, among them two clocks, a clavichord, a geographic atlas and two crystal prisms. The clocks and the crystal prisms especially impressed the Chinese of the Court. However, Ricci, in spite of all his efforts, was never received personally by the Emperor Wanli. In the only audience granted him Ricci had to make a reverence in front of an empty throne. However, the Jesuits had access to the "forbidden city" and to members of the Court and they obtained permission to live in Beijing. The maintenance of the clocks and the lessons given to Court members on how to play the clavichord provided them with frequent access to the imperial palace. Pantoja, who was a good musician, took charge of the teaching. Ricci's charisma and his knowledge of astronomy and mathematics attracted many scholars, literati and high dignitaries of the imperial Court. Among them were Xu Guanqui and Li Zhizao, members of the Board of Rites, and Li Yingshi, a high military officer, three of the first and most important converts to Christianity. In 1603 Ricci published his Chinese catechism with the title "True doctrine of the Lord of Heavens" (*Tianzhu shiyi*), where he used terms drawn from Chinese thought to present the Christian message. In particular, he used the Chinese term "Lord of Heaven" (*Tianzhu*), to refer to God. This term would create problems for the Jesuits in the future. In his book Ricci tried to show not only that Christianity was compatible with the doctrine of Confucius, but that it agreed with the most authentic elements of Confucianism, before the latter was contaminated by ideas drawn from Buddhism and Taoism. The Catechism had a wide circulation and thus Ricci successfully concluded his great work of cultural adaptation.

Ricci used the interest of Chinese scholars for astronomy and mathematics and their lack of knowledge in these subjects to attract them toward the Christian faith. He thus began the publication of several books of mathematics in Chinese. Outstanding among them was the translation into Chinese, with the collaboration of Xu Guanqui, of the first six books of Euclid's *Elements of geometry*, using Clavius' edition and commentary.[7] Published in 1607 with the title *Yihe Yaunben* (Origin of quantity), this was the first edition in China of this fundamental book of geometry. The translation used terms from classical Chinese, but new terms were introduced for some concepts for which there were no Chinese equivalents. Xu Guangqi wrote a preface where he stated that mathematics was a complement to Confucius' doctrine. The first six books form a unity, so that it was common to publish them apart. Xu Guangqi wanted to translate the other books, but Ricci thought it would be better to wait and see the reception given to the first books before translating the rest. In fact, the other books were never translated by Ricci or by any of his Jesuit successors. They were first translated in 1865 by the Chinese mathematician Li Shanlan along with the British protestant missionary Alexander Wylie. In 1611, after Ricci's death, Xu Guangqi published a second edition with Ricci's corrections

[7] Peter Engelfriet, *Euclides in China. The genesis of the first translation of Euclid's Elements in 1607 and its reception up to 1723*. Leiden: Brill (1998).

and also those of Pantoja and Sabatino de Ursis. In 1626 this second edition was included in the encyclopedia of mathematics and astronomy *Tianxue chuhan,* published by Li Zhizao. Toward the end of the eighteenth century this book was included in the Imperial Collection known as *Siku quanshu.*

In addition, Ricci published a Chinese translation of Clavius, *Epitome Arithmeticae Practicae,* which together with Euclid's *Elements* became the two basic texts of arithmetic and geometry used to introduce the Chinese to elementary mathematics. Ricci also wrote some more elementary books of geometry in Chinese. In astronomy he published a translation of Clavius' *Gnomonica* alongside his own work a "Treatise on Heavens and Earth" (*Quiankun tiyi*), an elementary book on astronomy and geography. In this work he gave an explanation of the celestial and terrestrial spheres, since the Chinese had no clear idea at that time about the shape of the Earth. The book also contained a catalogue of stars with their celestial coordinates. In 1584, during his stay in Zhaoqing, Ricci had published the first world map in modern European style showing the correct relationship between China and other countries. It indicated to the Chinese for the first time the American continent. This map, as we will see in next chapter, soon became very popular, being reprinted many times during his lifetime. It contrasted with the traditional Chinese maps, which showed only the Chinese Empire at the center surrounded by small countries without any real sense of scale. Ricci, recalling the travel stories of his compatriot Marco Polo, realized that the kingdom of Cathay was really China. In 1610, the year of his death, Ricci wrote to his superiors in Rome asking for some books of mathematics and, especially, for a good mathematician and astronomer who could continue the work that he had begun "with his weak forces and poor knowledge". After his death, the emperor Wanli decreed that Ricci should be buried with the highest honors.

4.2 The Reform of the Calendar

We can hardly understand today in the West the importance that the calendar had in the old society of China.[8] In line with the teaching of Confucius, which pervaded the whole of Chinese culture at that time, the ideal of human life consisted in harmony between humanity and nature. Thus, human events must follow the rhythm of nature, specially, that of the heavenly bodies. Social celebrations and private lives must adjust to particular astronomical phenomena, such as, eclipses and the conjunction of planets. In accord with such phenomena, some days were propitious for certain feasts or ceremonies, civil or religious, and other were not. This affected even the small details of private lives. Since the Qin dynasty (210 B.C.), the emperor carried the title of *Tianzi,* "Son of Heavens", and he represented the link between the order of the heavens and the lives of human beings. Thus it was very important to

[8] Needham, *Science and civilization*, (Calendrical and planetary astronomy), 390–408.

have a calendar which predicted accurately the position of the heavenly bodies and the occurrence of events such as eclipses of the Sun and Moon, planetary conjunctions and the appearance of comets. The occurrence of an unexpected astronomical event could lead to serious problems in public and private life. From the time of the Western Han dynasty (206 B.C.–9 A.D.), the Directorate or Bureau of Astronomy (*Qintianjian*) and the Imperial Observatory (*Guanxiangtai*), both depending on the Ministry of Rites (*Libu*), were in charge of drawing up the calendar that regulated the life of Chinese society and was promulgated by the emperor on the first day of each year. The Ministry of Rites was in fact one of the ministries with the greatest influence in the imperial government. The traditional Chinese calendar was a luni-solar calendar; it had the duration of the solar year divided into 12 lunar months of 29 or 30 days which began always with the new Moon. To adjust the solar and lunar cycles, 7 times during each 19 years an additional month was added to the year, and those years had 13 months. The year always began with the second new moon after the winter solstice. Chinese astronomers tried for a long time to synchronize the moon cycle and the length of the solar year, but they had not always succeeded. The calendar also contained the ephemerides of the Sun, Moon and planets, and predictions of the weather and of the occurrence of extraordinary phenomena, such as eclipses, the conjunction of planets and the appearance of comets. According to both the position of the heavenly bodies and other astrological considerations, days were established as propitious or not for certain types of events and ceremonies.

Due to the shortcomings of the very old astronomical tables being used, the calendar was not very exact in its predictions in Ricci's time and there was a general recognition of the need for its reform. The errors were creating social concern in establishing the right days for certain activities. In 1597 an influential mandarin in Nanjing, impressed by Ricci's knowledge of mathematics and astronomy, had proposed to him the possibility of being put in charge of the reform of the calendar. In 1605 Ricci, aware of the prestige this could bring to the Jesuits, wrote to Rome asking for new astronomical tables and a good astronomer to be sent to China who could do this work. He insisted in his letter on the great reputation this would bring to the Jesuits and how it would open a door for missionary work and make it more stable. In 1610 the Chinese astronomers made a gross error in the prediction of a solar eclipse, and this emphasized the need for reform. Sabatino de Ursis (1575–1620), an Italian Jesuit astronomer who had arrived in China in 1603 and came to Beijing in 1607 to help Ricci, had correctly predicted the eclipse. He was the first Jesuit to predict accurately an eclipse in China, thus showing the superiority of European astronomy. Xu-Guangqi, who was a member of the Ministry of Rites, took advantage of this prediction to convince the other members of the Ministry and finally the emperor to entrust the Jesuits with the reform of the calendar. Although lacking any formal official authorization, de Ursis and Pantoja, with the help of some Chinese scholars, among them Xu Guangqi and Li Zhizao, began to work. They translated into Chinese a treatise on the motion of the planets and determined the geographical coordinates of some important cities. That same year 1610, the Portuguese Manuel Dias (1574–1606), who had arrived to China in 1605, published in Chinese an astronomical treatise on the celestial sphere, which was very well received. In this work Dias presented in a very clear form traditional geocentric astronomy, adding Galileo's recent discoveries of the phases of Venus and of the satellites of Jupiter, and a description of the telescope, which was still unknown in China. In 1616 Shen Ch'üe, a Chinese astronomer envious of the Jesuits, provoked a persecution against them which cut short these promising developments. De Ursis was expelled to Macao where he died a few years later. This was the end of this first phase.

In 1620, the emperor Wanli died and after the short reign of Tianqi, in 1627 Chongzhen became emperor. He appointed Xu Guangqi vice-president of the Ministry of Rites and one of his counselors. Xu Guangqi became the most important figure in the Chinese Christian community. Using his influential position, he renewed the attempt to entrust the Jesuits with the reform of the calendar. In 1613 Nicolas Trigault (1577–1628), who had arrived to China in 1610, returned to Rome in order to secure aid and new personnel for the China mission. He was the first missionary to return to Europe from China. Trigault traveled through Europe and obtained valuable aid, especially from the German princes. During his stay in Europe he published a Latin translation of Ricci's Italian account of his work in China, *De Christiana expeditione apud Sinas*, (On the Christian expedition to China, 1615). This book caused a great impression with its first description of a China unknown in Europe and many young Jesuits offered themselves for this mission. Trigault made sure that among those selected were some with a good ability and knowledge of mathematics and astronomy. Twenty two young Jesuits were selected and boarded the ships with Trigault in Lisbon on his journey back to China with a good number of scientific books and instruments. Only eight of them arrived in China in 1618, among them Johann Schreck, Giacomo Rho and Johann Adam Schall, who would play an important role in the reform of the calendar.

Johann Schreck (better known by his Latinized name Terrentius or Terrenz, 1576–1630), had been born in Switzerland and been a student of Clavius in Rome. Before joining the Jesuits he had already had a promising scientific career. He studied medicine at Altdorf near Nuremberg and in Padua, where he knew Galileo. In 1611 he was elected the seventh member after Galileo of the *Academia dei Lincei*. The Academy asked Schreck to translate into Latin the book of Francisco Hernández about the fauna and flora of Mexico. The same year he joined the Jesuits and offered himself to the China mission after listening to Trigault and accompanying him in his journeys through Europe. During his journey to China, with a long stop in India, he collected specimens of plants for a future book on botany which he never wrote. Schreck took with him the first telescope to arrive in China. After his arrival he began his scientific work and in 1627 helped Wang Chen, a Chinese mathematician, with the publication of a book on problems of mechanics. Because of the importance in China of the correct prediction of eclipses, Schreck wrote to Galileo and Kepler, whom he knew personally, asking for help in this matter. He did not receive any answer from Galileo but Kepler sent him a short work on this subject which he later published in Europe. Schreck together with Nicolò Longobardi (1565–1655) published several scientific works in Chinese, some of them on astronomy where they described the telescope and the sunspots – recently discovered in Europe, but known from old in China. They also compared the systems to locate the heavenly bodies, equatorial (used in China) and ecliptic (traditional in Europe). In 1629 the official Chinese astronomers of both the traditional Chinese and the Arabic schools were not able to predict correctly a solar eclipse. Schreck succeeded in predicting accurately the time of the eclipse and this increased his prestige. He stated that the error of the Chinese astronomers was not caused by personal negligence but that it was due to a failure of their astronomical system and he pointed to the need to accept European astronomy. Xu Guangqi took advantage of this situation and convinced the emperor Chongzhen to entrust the reform of the calendar to the Jesuits, and the emperor approved. Chongzhen did not issue the official decree until much later, in 1638, 5 years after Xu Guangqi's death. However, Schreck, and Longobardi, with the help of Xu Guangqi, had begun to work in earnest on the reform. Giacomo Rho (1592–1638), who had arrived with Schreck in China, already had a good knowledge of mathematics and astronomy.[9] He published in Chinese several books in which he explained the motion of the Moon and the planets according to European astronomy. He also described a mathematical method for rapid calculation that had been published by John Napier. Napier was the first mathematician who developed logarithms in a publication in 1614. Logarithms were introduced into China by the Polish Jesuit, Mikolaj Smogulecki (1610–1666). Schreck's premature death halted for a time the work on the reform of the calendar.

4.3 Johann Schall and Ferdinad Verbiest

Johann Adam Schall von Bell (1592–1666) was born from a noble family of Cologne in Germany; he became a student at the Jesuit German College in Rome and possibly attended some of the courses given by Clavius (Fig. 4.2).[10] In 1611 he may have been present at the reception given to Galileo at the Roman College. In 1619 he arrived in China with Trigault and Schreck, and in 1623 moved to Beijing, where he began his astronomical observations, such as the two lunar eclipses of 1623 and 1624. He published his observations together with a treatise on lunar eclipses, and also established by astronomical means a precise determination for the difference in geographical longitude between Beijing and Rome. In 1627 Schall moved to Singafu, capital of Shensi province, where he continued his astronomical observations. In 1630, after Schreck's death, Xu Guangqi called Schall to Beijing to take charge, together with Rho, of the reform of the calendar. Schall had a great working capacity and soon began his analysis of the calendar. At the same time, he published in Chinese several treatises on mathematics and astronomy. When Xu Guangqi died in 1633, he was succeeded by Li Tianjing, also a Christian, as head of the Directorate of Astronomy,

[9] José Antonio Cervera, *Las varillas de Napier. Giacomo Rho S. J. (1592–1638) y su trabajo como matemático y astrónomo en Beijing*. Mexico: El Colegio de México (2011).

[10] Alfons Väth, *Johann Adam von Bell S.J. Missionar in China, kaiserlicher Astronom und Ratgeber am Hofe von Peking 1592–1666. Ein Lebens und Zeitbild.* (new edition) Nettetal: Steyler (1991).

Fig. 4.2 Adam Schall (1592–1666), astronomer and mandarin (Kircher, *China illustrata*, 1667) (Biblioteca Histórica Universidad Complutense Madrid, BH-DER 10718)

and he confirmed Schall in his work, but without any official recognition. Rho died in the same year and Schall had to carry out the work alone. In 1641, with the emperor participating in the observations, he made a new prediction of a solar eclipse which convinced the emperor of the superiority of European astronomy. Just when Schall was ready to introduce the new calendar, the political situation deteriorated with revolts spreading and cases of treason. The Manchu army, attacking from the north, began to invade the Chinese empire. In 1644 Chongzhen, the last emperor of the Ming dynasty, aware of the lost situation committed suicide. In a short time the invading Manchu army occupied the capital and the rest of the empire. The new Manchu emperor, Shunzhi, inaugurated the new Qing dynasty. Shunzhi was only 6 years old and it was the powerful regent Dorgin Amawang who governed in his name. The new Manchu government soon became aware of the advantages of European science and technology and maintained good relations with the Jesuits. Schall kept working on the calendar and a new opportunity presented itself with a solar eclipse on the first of September 1644. The Chinese astronomer made an error of 1 h in its prediction, while Schall predicted it correctly. This increased Schall's prestige and convinced the new government of the superiority of European astronomy. At the end of that year, Schall was appointed director of the Imperial Observatory with the dignity of Mandarin of fifth rank. His calendar for 1645 was finally approved and officially proclaimed.

Schall in his new post as director of the observatory reduced the five calendars which were in use during the Ming dynasty to only two. The first was of a more technical nature and was for the imperial administration, containing the ephemerides of the Sun, Moon and planets and other astronomical phenomena. The second was for the general public and it gave simpler information, including weather forecasts, and established the propitious days for events and celebrations. In this line, Schall reorganized the Directorate of Astronomy reducing the five departments to two. He tried to add a western school of astronomy to the other two, Chinese and Arabic, but he gave up in face of the strong opposition. Schall decided not to use the term "western astronomy", but spoke instead of the "new astronomy", because of the suspicion among Chinese dignitaries against anything foreign. Talking about the new astronomy, he did not present it as coming from outside China. Schall's influence in the Court kept increasing. The young emperor Sunzhi, who had assumed power in 1651, became more and more attracted to him. His visits to the Court were continuous; he began to give the young emperor instruction about western science and culture. Sunzhi addressed Schall as "venerable father" and awarded him the title of "Research master of the mysteries of heaven" and the dignity of "Mandarin of the first class", an honor reserved to princes and ministers. Schall took part in important state decisions, even in the choice of the crown prince. In the whole history of China, probably no Westerner ever had the influence that Schall had during the life of emperor Sunzhi.

Besides preparing each year's calendar Schall was prodigiously active, teaching astronomy and mathematics to Chinese astronomers, designing new instruments, and publishing more than 30 volumes in Chinese on diverse scientific topics. These included the use of the telescope, the theory of eclipses, catalogues of stars, and a summary of European astronomy, including the latest works of Copernicus, Tycho Brahe, Galileo and Kepler. His most important book was an encyclopedia of astronomy in 150 volumes, divided into three parts: Theoretical and Practical Astronomy, Astronomical Tables and Auxiliary Sciences. He made a great number of instruments for astronomy, optics and hydrology. He introduced in China the keeping of astronomical observations through the use of the telescope and made a gift of one to the emperor. Schall counted on the help of Johann Gruber (1623–1680), who was also a consummate painter, and Ferdinand Verbiest, his assistant and successor, who will be discussed later, along with some 150 Chinese government employees. At Schall's insistence, Gruber made a journey to Rome by land, as will be seen in the next chapter. One of the greatest advantages Jesuit astronomers had over their Chinese colleagues was the use of much better astronomical tables. Chinese astronomers used very old Arabic tables and were ignorant of geometric models for the position of the heavenly bodies. Ricci brought the traditional Alphonsine tables, but afterwards Jesuits brought newer tables. In 1646 Schall received Kepler's tables (*Tabulae Rudolphinae*) published in 1627, that Kepler had calculated using the system of Copernicus, and Tycho Brahe's observations. Schall's astronomical work had an ultimate missionary purpose. He was convinced that the progress of Christianity in

China was linked with astronomy. However, Schall's work at the imperial Court took many forms. For example, he took charge of the construction of 20 cannons for the defense of the northern frontier. He accepted this job from the emperor when faced with the alternative that aid would be asked from the Dutch Protestants in detriment to the mission. Although Schall had no experience in this type of work, his cannons were very effective.

Schall's familiarity with the emperor and his great influence in the imperial Court was the source of envy and animosity, not only from the Chinese but also from some of his fellow Jesuits. Not all Jesuits agreed with Schall's work on the calendar, with his dignity as a mandarin, and with his influence at Court. Many considered such things to be unacceptable in a religious. Ludovico Buglio (1606–1682) and Gabriel Magalhães (1610–1682), two Jesuits who had arrived to China in 1636 and 1640 and worked together in the province of Chengdun, were specially opposed to Schall's position at Court and wrote to Rome denouncing the existence of superstitious elements in the calendar, such as the designation of propitious and unpropitious days, as well as other astrological aspects. They wrote that Schall's mandarin dignity was incompatible with his condition as a religious. Schall defended himself, making clear that he was only responsible for the astronomical aspects of the calendar and that, reluctantly on his part he had been forced by the emperor to accept the mandarin dignity. In 1655, a commission of professors of the Roman College was established to examine the case, and gave a ruling favorable to Schall. This was confirmed in 1664 by a second commission and given verbal approval by Pope Alexander VII. The approval referred to Schall's position as director of the Astronomical Observatory, to his work in making the calendar, and to his mandarin dignity.

The envy roused among some Court members, especially the astronomers Yang Guanxian of the Chinese school and Wu Mingxuan of the Arabic school, had far worse consequences. These two astronomers resented having been removed by Schall from their previous posts. In 1661, the opportunity for revenge came with the premature death of emperor Shunzhi at the age of 23. During the minority of his successor Kamxi, who was only 7 years old, a Regency Counsel took power. The Counsel accepted the accusations presented against Schall and the Jesuits. There were three of these: namely, that the Christians conspired against the empire; that Christian doctrine was pernicious and dangerous for Chinese society; and that European astronomy was full of errors. Schall was accused personally of having changed the burial day of a baby son of the emperor to a non-propitious day and thus, according to the accusers caused the death of his mother. He was even accused of having caused the emperor's death. In 1664, Schall, Verbiest, Magalhães and Buglio and their Chinese co-workers were put in prison. A long trial began that would last 7 months. Since one of the accusations involved European astronomy, the judges decided to make a test with the prediction of the time of the solar eclipse of January 16, 1665. The judges organized a solemn public act that day at which members of the court and a throng of onlookers were present. A herald announced the times of the predictions. Yang Guanxian and Wu Mingxuan failed in their predictions by three quarters of an hour and half an hour respectively. The Jesuits correctly predicted the right time at 3 h 26 min in the afternoon. Verbiest then proposed a further test for the Chinese astronomers; this consisted in predicting the length of the shadow of a vertical bar at noon on an appointed day. The accusers were unable to do this and refused to accept the test. Verbiest, who was acting as defendant because of Schall's serious illness, insisted that in astronomy the verification of predictions was the decisive test and not ancient doctrines. In spite of all this, the judges condemned Schall to death and the other three Jesuits to exile. Thirteen of the Chinese assistants in the observatory were also condemned to death. The following day a large earthquake shocked and caused severe damage in the capital. According to traditional Chinese ideas this was a sign that a great injustice had been committed and the trial was revised. The judges released Schall and the other Jesuits but ordered the execution of five of the Chinese assistants who were Christians. Schall died in the following year, his health having deteriorated because of all his sufferings. His innocence, however, was officially recognized only in 1669 by a decree of Kangxi, the new emperor.

Ferdinand Verbiest (1623–1688) was born in the small Flemish village of Pitthem. He arrived to Beijing in 1660, took the second post in the observatory under Schall's direction, and became his most faithful collaborator.[11] In the difficult time of the trial, Verbiest shared his prison, took care of him, and made all the calculations for the time of the eclipse. In 1665, during the stay of the Jesuits in prison, Yang Guanxian was appointed director of the observatory and put in charge of the calendar. The shortcomings of the new calendar were soon evident. In 1667 Kangxi, though only 13 years old, personally assumed power. He would be one of the most powerful Chinese emperors with a long reign of 52 years. Two years later, he consolidated his power and put Oboi, who had been the most powerful regent, in prison. Toward the end of 1668, Verbiest made clear in front of the emperor and his Court the errors of the calendar proposed by Yang Guanxian and Wu Mingxuan for 1669. He again challenged both astronomers to predict the length of the shadow of a vertical bar at noon of a particular day, but they would not accept. Verbiest successfully made the prediction in front of the emperor, as well as other astronomical observations. All these proofs convinced Kangxi of the superiority of European astronomy and he appointed Verbiest director of the Imperial Observatory and ordered him to revise the proposed calendar. Jesuits held this post from that date till the suppression of the Society of Jesus in 1773, although a formal decree was only published in 1725.[12]

Verbiest gradually gained the esteem of the new emperor who received from him lessons in mathematics and astronomy. In 1674 Kangxi granted him the dignity of mandarin of the second class and in 1682 a higher dignity, equivalent to that of a vice-minister. Verbiest became one of the closest advisors of the emperor for all

[11] Henri Bosmans, Ferdinand Verbiest, directeur de l'Observatoire de Peking (1623–1688). *Revue des Questions Scientifiques*, 71 (1912), 195–273, 375–464; John W. Witek (ed.), *Ferdinand Verbiest (1623–1688) Jesuit missionary, scientist, engineer and diplomat*. Nettetal: Steyler Verlag (1994); Noël Golvers, *Ferdinand Verbiest, S. J. (1623–1688) and the Chinese Heaven*, Louvain: Leuven University Press (2003).

[12] Minghui Hu, Provenance in contest: Searching for the origins of Jesuit astronomy in early Qing China. *International History Review*, 24 (2002), 1–36.

kinds of affairs. Kangxi's high esteem for him was shown in 1675, at the time of the controversy about the Chinese rites. In a personal visit he paid to the Jesuits' house, he gave Verbiest a large sheet of paper with the personal hand-written message, "I revere the heavens" (*Jing tian*), in two large characters. This poster was fixed at the doors of Jesuit churches and later became the cause for a new controversy. Verbiest's influence with the emperor was a great help for the missionaries' work, an activity that was officially allowed. Some who had been exiled to Macao were also allowed to return to the mainland.

Verbiest was a tireless worker. He wrote in Chinese more than 20 astronomical books in which he presented the Tychonic geo-heliocentric system, as was the custom among many Jesuit astronomers of that time. Among his books, *Astronomia perpetua Imperatoris Kam Hi* (Perpetual astronomy of the emperor Kam Hi, 1683), stands out; this was dedicated to the emperor and contains the ephemerides of the Sun, Moon and planets and tables of solar and lunar eclipses for 2,000 years. His most popular book was *Liber organicus astronomiae europaeae* (Organic book of European astronomy, 1668, 1687) written first in Chinese and then published in Europe in Latin, where it had a wide circulation.[13] Although Schall had constructed some astronomical instruments, Verbiest, beginning in 1669, completely renewed the old instruments on the platform of the observatory. The six new instruments were finely worked with decorative figures by Chinese artists. They were an ecliptic armillary sphere resting on four dragon heads; an equatorial armillary sphere resting on the back of a dragon; a large celestial globe of nearly two meters diameter with 1888 stars represented; a horizontal circle for azimuth measurements and a large-size quadrant and sextant. The celestial globe was accompanied by a catalog of stars with their ecliptic and equatorial coordinates.To construct these instruments Verbiest used as models those found in Tycho Brahe's *Astronomiae instauratae mechanica* (Mechanics of the established astronomy, 1598), which were already obsolete in Europe. These instruments were installed to demonstrate the superiority of European astronomy and they were accompanied by their description, published in Chinese in 1673 with detailed images. Actually the observations were made through telescopes which were not installed in a permanent way. The instruments can be seen today in the same place where they were installed originally. In 1900 during the Boxer war, German and French troops took some of them and brought them to Europe. They were only returned after the end of the First World War. Verbiest also introduced into China meteorological instruments such as thermometers and hygrometers, and he wrote a book about them in Chinese.

Following Ricci's example, Verbiest also worked in geography and with the help of Buglio and Magalhães published a large map of the world divided into two hemispheres (Eurasia and Africa in one and America in the other) of a meter and a half in diameter, in a stereographic projection following Mercator's model. The map was reedited in several sizes and accompanied by several treatises on geography

[13] Noël Golvers, *The Astronomia Europea of Ferdinand Verbiest , S.J. (Dillingen, 1687). Text, translation notes and commentaries*. Nettetal: Steyler, (1993).

which gave information about the different parts of the world. Verbiest had a great skill as an engineer for all types of constructions and not only in his astronomical instruments. He made other kinds of devices for the enjoyment of the emperor. An interesting one was a small car moved under the impulse of steam which turned its wheels. This can be considered a kind of precursor of the steam engine. Verbiest directed the construction of more than 500 cannons of various types for the Machu army, an activity unusual for a Jesuit but as seen also with Schall, although then on a lesser scale. Verbiest also wrote a long treatise on this subject in Chinese. In many other important affairs he acted as an advisor to the emperor. One of them was during the diplomatic negotiations between China and Russia about the frontiers between the two countries. In 1676, he acted as interpreter for the Russian embassy of Nikolai Spathary in Beijing using Latin as the common language. On this occasion Verbiest made a map of the regions in conflict. He had a particular interest in the success of these negotiations, because he envisaged a possible land route for future missionaries to China through Russia, a project that was never realized.

In spite of his continuous and diverse activities Verbiest never forgot his primary interest as a missionary. His scientific work was always a means to spreading the Christian message. This close relationship between astronomy and the spread of Christianity in China was clearly presented in the last paragraph of his work *Liber organicus*:

> Hence, the Christian Religion in China is justly represented as a most August Queen who appears publicly with her arm leaning on *Astronomy*, while all the other Mathematical Sciences crowd around the Virgin at her side as her most charming companions; this is because she was first introduced into China through *Astronomy*, because she was left untouched thanks to *Astronomy*, and because, after having been banished several times, she was called back each time and successfully restored to her former dignity by *Astronomy*.[14]

4.4 Jesuit Directors of the Imperial Observatory

After Verbiest's death, the Jesuits continued in charge of the Imperial Observatory until the suppression of the Society of Jesus in 1773.[15] The number of Jesuits in China at that time was relatively small, between 40 and 60 at any one time, with a maximum of about 90 in 1700 and of these, only four or five worked at the observatory. The first successors of Verbiest were Filippo Grimaldi (1638–1712) and Tomas Pereira (1645–1708), who alternated as directors until 1712. They had arrived to China in 1666 and had been Verbiest's collaborators. With them Antoine Thomas (1644–1709), who had arrived to China in 1684, and became the closest collaborator of Verbiest during the last years of his life, also worked at the observatory. Thomas' main interest was in cartography and he drew up several maps of the north

[14] Golvers, *The Astronomia Europea*, 132, 447.

[15] Udías, "Jesuit astronomers", 473–474.

of China and Mongolia. He was the first in China to measure the length of a degree of a meridian arc, as we will see in next chapter.

About this time began the famous controversy about the Chinese rites which also affected the Jesuit astronomers. Without going into details here about this well-known controversy, the core of it turned on the use of the term introduced by Ricci, "Lord of Heaven" (*Tianzhu*), to refer to God, and to allow Christians to participate in the ceremonies of veneration of the forefathers and of Confucius. Such practices had been observed by Jesuit missionaries from the time of Ricci. With the arrival of non-Jesuit missionaries, complaints were sent to Rome about these practices, which were considered to be superstitious and prohibited to Christians. The Jesuits, however, held that the term *Tianzhu* was perfectly adequate and that the practices were merely ceremonies of a civil character and should not be prohibited. The debate in Rome was not favorable to the Jesuits, especially, after the visit to China of the Papal Legate, Cardinal Charles Thomas Maillart de Tournon, who had absolutely no understanding of the situation. He prohibited the use of the term "Lord of Heaven", claiming that it referred to the veneration of Heaven and not of God, and that the practices were idolatrous. Pope Clement XI in 1707 and 1715 and again Pope Benedict XIV in 1742, condemned all these practices. The Emperor Kangxi was angry with this decision, since he had supported the Jesuits' position, and he prohibited the teaching of Christian doctrine in China, rescinding the edict of tolerance he had issued in 1692. The Jesuit astronomers, however, were maintained in their posts. These were very difficult times for the missionaries and only the scientific work of the Jesuits ensured their presence in the imperial court.

The last four Jesuit directors of the Observatory were from Germany. Kaspar Kastner (1665–1709) occupied the post for only two years. He had good relations with the emperor, who put him in charge of the education of some of the princes. Kastner took an active part in the rites controversy defending the Jesuit position while on a journey to Rome. Kilian Stumpf (1655–1720) held the post between 1712 and 1720. He designed new astronomical instruments, among them an altazimuth quadrant that he added to Verbiest's instruments on the platform of the Observatory. He also participated in the rites controversy with the work, *Informatio pro veritate* (Information in favor of truth) which was condemned in Rome. Ignaz Kögler (1680–1746), who had been professor of mathematics at the University of Ingolstadt, held the post from 1720 until his death. Kangxi held him in great esteem and his successor, the emperor Yongzhen, appointed him officially director of the observatory with the rank of Mandarin of Second Class, Member of the Ministry of Rites, and tutor of the crown prince. Kögler contributed a newly elaborated equatorial armillary sphere to the instruments on the platform and published several texts in Chinese on logarithmic tables, eclipses, the satellites of Jupiter, and a catalog of more than 3,000 stars giving their coordinates and brightness. The last Jesuit director was Augustin von Hallerstein (1703–1774), who had arrived to China 1739 and assumed the direction of the Observatory in 1742. He carried out a program of astronomical observations and introduced new tables for the calculation of ephemerides and eclipses based on Newtonian astronomy. He published in Chinese a collection of astronomical observations in 30 volumes, including his own work and that of his collaborators and predecessors between 1717 and 1752. In 1768, Hell in Vienna published some of these observations. Hallerstein also dedicated much of his time to experiments on electricity. He maintained an active correspondence with the Royal Society of London and the Academy of Science of Saint Petersburg. He also undertook other official tasks and was the first to determine the population of China: according to his calculations, 196 837 977 inhabitants in 1760. News of the suppression of the Society of Jesus, which arrived to China in 1774, caused his death from a stroke. Four Portuguese former Jesuits, Felix da Rocha (1713–1781), José da Espinha (1722–1788), André Rodrigues (1729–1796) and José Bernardo de Almeida (1728–1805), continued as directors of the Observatory until 1805. With these four priests, the long tradition of Jesuits in charge of the Imperial Observatory of Beijing, begun by Schall in 1644 and continued for more than 150 years, came to an end.

4.5 The French Mission

In 1678 Verbiest wrote a letter to the Jesuits in Europe asking for material and personal aid for the China mission, especially, in its scientific work. The French Jesuit François de la Chaise (1624–1709), personal confessor to Louis XIV, convinced the king of the advantages of financing an expedition of Jesuit astronomers and mathematicians. The director of the Paris Observatory, Domenico Cassini, Riccioli's former student and friend, supported this idea. He saw in the Jesuits' long sea journey an opportunity to test his method for determining geographic longitude by observing the satellites of Jupiter. With the King's approval, a group was formed of seven Jesuits, who received the title "Mathematicians of the King of France", and were linked to the *Académie des Sciences* by special privilege. On March 1686 the expedition left Brest taking with them a good quantity of scientific instruments, such as, telescopes, sextants, pendulums, thermometers and barometers. In order not to interfere with the patronage (*padroado*) of the King of Portugal on all the missions to India and the Far East, the French Jesuits insisted that they were not going as missionaries but as mathematicians. Jean Baptiste Fontenay (1653–1715), professor of mathematics at the *Collège Louis-le-Grand* of Paris, was the head of the expedition. He had good experience in astronomy and had made frequent observations, for example, those of the solar eclipse of 1678. The expedition made a stop at Cape Town (South Africa) where they were well received by the leaders of the Dutch East India Company, in spite of some Protestant antipathy for Jesuits. There they tried out Cassini's method based on the occultations of Jupiter's satellites, and found a difference of longitude with the meridian of Paris of 18°, an error of about 2°. Later the expedition was detained for more than a year in Siam (Thailand), and arrived to China in 1687. The ministry of Rites, always suspicious of foreigners, at first forbade their entrance, but thanks to Verbiest's mediation with the emperor, they were allowed to enter and proceed to Beijing.

The French Jesuits established a house in Beijing with a good library of scientific books, where they installed a small astronomical and meteorological observatory.[16] They soon became very active at Court and enjoyed the favor of the emperor. This created a conflict with the Jesuits of the Imperial Observatory, who also depended on the emperor's favor. The Jesuits of the French mission, under Fontenay's leadership, kept a certain independence from the rest of the Jesuits, who from the beginning of the China mission were under the patronage of the King of Portugal. Joachim Bouvet (1656–1730) and Jean François Gerbillon (1654–1707) learned the Manchu language, used in the court during the Qing dynasty, and between 1689 and 1692 succeeded in being appointed to teach mathematics to the emperor. They introduced the more modern geometry of Pardies to complement the text of Euclid translated by Ricci and Xu Guangqi. The latest developments in mathematical analysis and calculus were translated into Chinese and later included in the new imperial collection of mathematics, *Shuli jingyun* (Essence of mathematical principles), published between 1709 and 1723. The French Jesuits also participated in forming the collection of maps known as the *Atlas of Kangxi*, published between 1690 and 1720. These maps included the new territories of the Chinese empire to the north and west. Gerbillon took part with Pereira as mediators in the negotiations between Russia and China that led to the Nerchinsk treaty in 1689. As we saw above, Verbiest had also been interested in the negotiations between China and Russia. Pereira, however, was opposed to them, since he thought that they were detrimental to the interests of Portugal. Kangxi's satisfaction with the good services of the Jesuits in these negotiations influenced his decree of 1692 ensuring tolerance for the Christian religion.

Bouvet was always welcome at Court and wrote a biography of the emperor which was published in France. He maintained correspondence with Leibniz on mathematical problems, and thought he had discovered in some old Chinese text what could be considered binary mathematical formulation. Leibniz was very interested in China and based on the information obtained from the Jesuits, he wrote in 1697 *Novissima Sinica* (News about China). Bouvet tried to link astronomy and Christianity by using the Chinese term *tianxue* (learning from heaven) to refer to the knowledge of both. In a long treatise in Chinese entitled, *Essence of the doctrine about heaven*, he tried to show how the astronomy brought by the Jesuits agreed with the ancient texts of Chinese scholars and Confucian orthodoxy. The veneration of heaven found in these texts, according to Bouvet, referred really to its Creator. With great erudition he tried to show that there is, in classical Chinese texts, an implicit recognition of God, who rules the motion of the heavenly bodies and the destinies of men.

French Jesuits continued to arrive to Beijing to form part of the French Mission. Antoine Gaubil (1689–1759), had studied astronomy with Cassini in Paris and worked on the history of Chinese astronomy. In 1732 he published in France a pioneer book on this subject (*Histoire de l'atronomie chinoise,* History of Chinese astronomy),

[16] Hsia, *Sojourners in a strange land*, 91–126.

highly praised by Joseph Needham, the historian of Chinese culture. Alexander von Humboldt considered him to be the wisest among the Jesuit missionaries in China. Gaubil restarted the astronomical observations that had been practically abandoned after Fontenay's death. Michel Benoist (1715–1774) was entrusted with different tasks by the emperor among them a new geographical atlas of China. He also published a new map of the world, which updated with modern methods those of Ricci and Verbiest. In 1761 Benoist was the first to formally introduce Chinese astronomers to Copernican astronomy. In 1692 the *Académie des Sciences* in Paris published the astronomical observations made by French Jesuits in China together with others made in India, prepared and annotated by Thomas Gouye (1650–1725), a professor of mathematics at the *Collège Louis-le-Grand* of Paris. In 1699 Luis XIV appointed Gouye an honorary member of the Academy. He acted as president between 1711 and 1725 and appointed Bouvet and Gerbillon as corresponding members. In time, however, the relations between the Academy and the Jesuits began to deteriorate. In 1729 Étienne Souciet (1671–1744), also professor and librarian of the Paris college, published another collection of astronomical observations together with those of temperature and atmospheric pressure made by the French Jesuits in China, adding also observations from old Chinese texts (*Observations mathématiques, geographiques et physiques,* Mathematical, geographic and physical observations). Jean Baptiste du Halde (1674–1743) published in Paris a monumental work about China using all the information available supplied by the French Mission (*Description...de l'empire de la Chine*, 4 vols., Description of the Chinese empire, 1734). Du Halde was the editor between 1711 and 1743 of the Jesuit periodical publication, *Lettres édifiantes et curieuses* (Edifying and curious letters, vols. 9 to 26), where abundant information was presented about China.

4.6 The French Expedition to Siam

While the French Jesuit astronomers were on their journey to China, they made a stop in Siam for almost 1 year. King Phra Narai welcomed them and invited them to Louvo (Lop Buri), where he had his palace. They received great help from the Prime Minister, Constancio Phaulkon, a Greek who had arrived in Siam in 1675 as an employee of the British East Indian Company, and had converted to the Catholic Church. In Louvo the Jesuits installed their astronomical instruments and began making observations. On December 11, 1685, a lunar eclipse took place which the King could observe through a telescope; he was very much impressed. From the observation of the satellites of Jupiter, and using Cassini's method, the Jesuits determined the difference in geographical longitude from Paris as $98^\circ\ 31'$ (an error of only 1°). One of the Jesuits, Guy Tachard (1648–1712), remained in Siam, while the rest continued their journey to China. King Phra Narai was very interested in European science and thought of creating, with the aid of the Jesuits, an academy of science and an observatory similar to those in Paris and Beijing. For this purpose the king sent Tachard back to France accompanied by a royal delegation.

He was supposed to return with a group of Jesuits astronomers and mathematicians to carry out the king's wishes.

Tachard and the Siamese delegation arrived to Brest in June 1686. King Louis XIV welcomed the Siamese ambassadors and promised his support to the project. During his stay in Paris, Tachard published a narrative of his journey, *Voyage de Siam* (Journey to Siam, 1686), which was circulated widely.[17] Using his influence, Tachard succeeded in convincing the Jesuit superiors to send 14 Jesuits competent in mathematics and astronomy to Siam. As in the case of the Chinese expedition, the Jesuits sent to Siam were designated "Mathematicians of the King of France". The *Académie des Sciences* also offered them its support, entitling them extraordinary members and supplying them with all kinds of scientific instruments, such as, telescopes, quadrants and pendulums. The expedition led by Tachard sailed from Brest in March 1687 with a fleet of six warships. In addition to the Jesuits and the Siamese delegation, there were on board Simon de la Loubère, appointed extraordinary ambassador to the King of Siam and a detachment of soldiers. During the long journey, the Jesuits observed the solar eclipse of May 11, 1687, to the great admiration of the Siamese. After their arrival in Bangkok the King welcomed the expedition with great enthusiasm and promised the Jesuits to begin building an observatory, a house and a church. All predicted a brilliant future.

Tachard returned to France in January of the following year with the returning fleet of de la Loubère, together with a group of 12 young noble Siamese sent to study in the Jesuit college of Paris. In April the Jesuits organized the observation of another solar eclipse in which the King participated and was very impressed. Work on the observatory began near the royal palace in Louvo. A month later, taking advantage of the King's serious illness, the mandarin Petracha led a palace conspiracy and succeeded in taking power. Phaulkon, the Prime Minister, tried to suppress the rebellion with the aid of the French soldiers who had remained in Siam, but was defeated. Phaulkon was taken prisoner and beheaded together with some members of the King's family. The King Phra Narai died some time later and Petracha occupied the throne. The Jesuits did not suffer any harm, but their whole project collapsed and they had to leave the country. They boarded a ship in Bangkok to India, except for one who remained in Siam as a missioner along with the Portuguese Jesuits. Two of them died during the journey to India and four others in a shipwreck 4 years later. Gouye prepared and commented on the observations made during the journey and stay of the Jesuits in Siam, and these were published in France in 1688 by the *Académie des Sciences.* Along with these observations were those that determined the geographical longitude made with Cassinis' method, and observations of eclipses and measurements of the period of the oscillations of a pendulum.

---

[17] Florence C. Hsia, Jesuits, Jupiter's satellites, and the Académie Royale des Sciences. In J.W. O'Malley, G.A. Bailey, S.T. Harris and F Kennedy (eds.), *The Jesuits. Cultures, science and the arts*. Toronto: University of Toronto Press (1999), 241–257; Hsia, *Sojourners in a strange land*, 51–109.

4.7 Jesuit Astronomers in India

The work of Jesuit astronomers in India is less known than that in China, but was also very important.[18] In 1542 Saint Francis Xavier established the first Jesuit mission in India in Goa, where in 1510 Alfonso de Albuquerque had founded a Portuguese settlement. In Goa the Jesuits established their first school and from there they spread mainly through the south of India. In 1580, the Mogul Emperor Akbar invited the Jesuits for the first time to the north of India, where they founded a school in Agra and houses in Lahore and Delhi, the new capital of the Mogul Empire. India had known an early development of mathematics with important achievements such as the decimal system, and applications to astronomy. The mathematicians and astronomers, Aryabhata in the sixth century and Brahmagupta in the seventh century, were important early figures. Brahmagupta was director of the observatory of Ujjain, which became the main astronomical center. In the twelfth century Bhaskara was director of Ujjain; his astronomical works were still in use when the Jesuits arrived to India. Antonio Rubino (1578–1643), who had come to India in 1602, is one of the first Jesuits there with an interest in astronomy. Although he was more interested in geography and cartography, in 1613 he wrote to Grienberger asking for some books with information about the newly discovered telescope. He is actually credited with the introduction of the telescope into India. In 1638 Rubino left India for Macao and in 1642 entered Japan, from where all missionaries had been expelled. He was soon arrested and executed.

The French Jesuits established a mission in Pondicherry in the south of India. They had brought with them astronomical instruments and made observations between 1688 and 1735. In 1689, two survivors of the Siam expedition, Jean Venant Bouchet (1655–1732) and Jean Richaud (1633–1693), settled in Pondicherry, where they made observations with telescopes. In 1689, Richaud, while he was observing a comet, discovered that Alpha-Centauri, the star nearest to the Sun, is a double star. This was many years before the French astronomer, Nicolas Louis de Lacaille, to whom this discovery is often attributed made his announcement in 1752. Bouchet traveled along the coast of Coromandel making astronomical observations, determining the latitude, and preparing maps of the region. Pierre François Xavier Duchamp (1693–1740) dedicated his energy to the study of astronomy in ancient Indian texts and he published a book about this. He observed a lunar eclipse in 1732 and detected discrepancies between his observations and the astronomical tables used by Indian astronomers. Duchamp was very critical of Indian astronomers whom he criticized for their lack of interest in learning new techniques. He maintained a correspondence with Gaubil, who was studying at that time old Chinese astronomy.

[18] R. K. Kochhar, Secondary tools of empire: Jesuit men of science in India. In T. R. de Souza, *Discoveries, missionary expansion and Asian cultures*. New Delhi: Concept Pub. Co. (1994), 175–183; Virendra N. Sharma and Lila Hubbert, Jesuit astronomers in eighteenth century India. *Archives Internationales d'Histoire des Sciences*, 34 (1984), 99–107.

Between 1730 and 1740 a group of French Jesuits carried out astronomical observations in Chandernagore near Calcutta (Kolkata) under Claude Stanislas Boudier's (1686–1757) direction. They used several telescopes (between 2 and 5 m long), sextants, clocks and other instruments brought from France. For his observations and calculations Boudier used the new astronomical tables published by Philippe de la Hire in Paris between 1682 and 1702. He carried out all kinds of astronomical observation, e.g., solar and lunar eclipses, sunspots, right ascension and declination of stars and occultations of Jupiter satellites. Duchamp in Pondicherry made some of these observations simultaneously with those of Chandernagore. Boudier was a very diligent observer, but in some cases his observations lacked the exactitude of those of his colleagues in France. He blamed these deficiencies on the lack of proper instruments and kept asking for better instruments to be sent to him from France.

An important chapter of the astronomical work of the Jesuits in India was their collaboration with the Rajah of Amber and Jaipur, Sawai Jai Singh.[19] This Indian Prince had a great interest in astronomy and built five observatories in the north of India. The most important was in Jaipur, where he had installed his new capital city. These observatories consisted of curious large buildings made of stone, which simulated large quadrants and sextants and other elements for the observation of the heavenly bodies. Work was finished in 1728 and some of them have been renovated and can still be seen today, an extraordinary sight. Indian astronomy in the eighteenth century was a mixture of traditional ancient doctrine and contributions from Arabic astronomy. The traditional Indian astronomy was based on old texts, such as the *Surya Siddhanta* and the works of Aryabhata, Brahmagupta and Baskhara. Sawai Jai Singh was aware of the shortcomings of this astronomy and wanted to have contacts with modern European astronomy. For example, he was worried by the lack of accuracy of Indian astronomical tables. To solve this problem in 1728 he asked Manuel Figueredo (1689–?), Rector of the Jesuit school of Agra, to bring from Europe some of the newest tables. Two years later Jesuits brought the astronomical tables of de la Hire, new books of astronomy, telescopes, clocks and other instruments (Fig. 4.3). The Rajah ordered his astronomers to use the new tables and to resolve the inconsistencies between these and the traditional Indian tables and he invited Boudier to come to Jaipur. Boudier accompanied by Jean François Pons (1698–1752) made the long journey of more than 1,200 km between Chandernagore and Jaipur, carrying a large telescope, a quadrant and other instruments. On their way, the Jesuits measured the geographical latitudes and longitudes of the cities visited during the journey using observations of stars and the satellites of Jupiter. In all they made 50 determinations between Chandernagore and Adra and 10 between Adra and Jaipur. Longitudes were determined with reference to the Paris meridian. In

[19] George M. Moraes, Astronomical missions to the court of Jaipur, 1730–1743. *Journal Bombay Branch of Royal Asiatic Society*, 27 (1951–1952), 61–65, 85; Virendra N. Sharma, The impact of the eighteenth century Jesuits astronomers on the astronomy of India and China. *Indian Journal of History of Science* 17 (1982), 345–352.

Fig. 4.3 Manuel Figueredo and the Rajah Sawai Jai Singh (Udías 2003)

Delhi they stopped at the observatory which was similar to that of Jaipur. Here they observed the solar eclipse of May 3rd, 1734. Boudier compared the times of his observations with those derived from the tables of de la Hire and found some discrepancies. Boudier and Pons remained at Jaipur for 2 years helping the Rajah and his astronomers in their observations. Boudier maintained a correspondence with Gaubil in China sending him his observations.

Sawai Jai wanted to solve all the problems of differences between the European and Indian astronomical tables and asked for other Jesuit astronomers to work on these. Two German Jesuits, Anton Gabelsberger (1701–1741) and Andreas Strobl (1703–1770), left from Europe and arrived at Jaipur in 1740. There they helped the Indian astronomers with their observations and calculations, especially on the positions of the Moon. Gabelsberger died a year later and Strobl continued with the work in Jaipur. In 1743 Sawai Jai died and the work at the observatory ceased. The observatory itself was abandoned and with time was reduced to ruins. It was only restored and renovated in 1901. In 1743, Joseph Tieffenthaler (1710–1770), a Jesuit from Tyrol arrived in India with the intention of working at the Jaipur observatory. He was not aware that the Rajah Jai Singh had died only a few months before and that work at the observatory had ceased. Tieffenthaler traveled to Delhi, where there was an observatory similar to that of Jaipur. There he met Strobl, who had moved there after the death of the Rajah and for some time they made astronomical observations together. Tieffenthaler , a tireless traveler, moved to Ujjain, another observatory built by Sawai Jai, where the center of traditional Indian astronomy had been. However, he decided to dedicate his efforts to geography and cartography, as we will see in the next chapter.

4.8 Scientific Exchange Between West and East

Jesuit astronomers were the first to bring the new science that was beginning in Europe to India and China. Joseph Needham, the historian of science in China, opens his chapter on the Jesuits saying: "In the history of intercourse between civilizations there seems no parallel to the arrival in China in the seventeenth century of a group of Europeans so inspired by religious fervor as were the Jesuits, and, at the same time, so expert in most of those sciences which had developed with the Renaissance and the rise of capitalism".[20] Leibniz had already said that the mission of Jesuits in China constituted the greatest enterprise of his time. To a lesser degree, the Jesuits also brought to India the first contacts with modern European science. This transfer was not only from West to East, but through their letters the Jesuits made Eastern science culture and civilization known in Europe, for the first time. This information came to satisfy the interest and curiosity awakening in Europe at that time about the culture and characteristics of those far-away lands. Manuscript copies of the missionaries' letters were widely distributed in Europe. In France a collection of such letters was published between 1704 and 1776 with the title *Lettres édifiantes et curieuses* (Edifying and curious letters) in 34 volumes, which were translated into Spanish, Italian, English and German. These letters became very popular and were the main source of information in Europe about the lands and peoples of Asia, Africa and America. For example, Leibniz became very interested in Chinese culture thanks to the information received from Jesuit missionaries, and he valued the exchange between the two civilizations. He recognized that Europe was more advanced in science, philosophy and technology, but admitted that China was ahead in the development of a civil ethics and politics and in the establishment of a peaceful social order.

There is no doubt that Jesuits brought to China many of the new European developments in mathematics and astronomy that were unknown in China. Thus, in mathematics they introduced the geometry of Euclid with their translation of the first six books into Chinese. They also introduced spherical geometry and trigonometry necessary for astronomy; logarithms; stereographic projection for cartography; and new developments in algebra, analytic geometry, calculus and computational methods. In astronomy they brought, among other novelties, the geometrical description of the motion of planets, the division of the Earth by meridians and parallels and methods for the prediction of eclipses. They introduced the use of the telescope, first brought by Schreck in 1618. Also Jesuits carried out an extensive

[20] Needham, *Science in China*, III, 437.

program of astronomical observations during the 150 years they were in charge of the Imperial Observatory. These observations included catalogs of the positions of stars, solar and lunar eclipses, observations of planets and the satellites of Jupiter, and the appearances of comets. These observations were published in China in a large number of volumes in Chinese, and some of them also in Europe.

An important negative aspect in the Jesuit transmission of modern astronomy was their failure to introduce the heliocentric system. We have already considered the difficulties encountered by Jesuit astronomers in accepting the Copernican system after the ecclesiastical censure of 1616 and the condemnation of Galileo in 1633. According to Needham, before this condemnation some Jesuits, like the Bohemian Wenzel Kirwitzer (1588–1626), who arrived to China with Schall and Schreck, but died very young, and Smogulecki, who was Polish and had studied in Cracow, were probably Copernican. In fact the geocentric cosmological theory presented by the Jesuits, was already obsolete in Europe, where heliocentrism was widely accepted toward the end of the seventeenth century. The book by Dias published in 1610, which was very popular among Chinese astronomers was totally Ptolemaic, presenting a finite universe limited by the sphere of the fixed stars and maintaining the existence of the crystalline spheres for the planets. After Schall, as was the case with Jesuit astronomers in Europe, the Ptolemaic system was discarded and replaced by the geo-heliocentric theory of Tycho Brahe. They also agreed on the fluid nature of planetary space. However, although the Jesuits in China did not present, at least in public, the Copernican system as representative of the real situation, they did expound it as a hypothesis, as they did in Europe. We know that they used Kepler's astronomic tables, which were calculated with the Copernican system. As we saw earlier, it was not until 1761 that Benoist formally presented in China the Copernican system. However, Kögler and Hallerstein used Newtonian astronomy to calculate the positions of the Moon. As happened in Europe, Jesuit astronomers avoided cosmological problems and centered their work on observational astronomy. In China, during the reign of emperor Kangxi, a current of strong nationalism, inherited from the Ming dynasty, spread the idea that Jesuit astronomy had been already known in ancient China. Thus the Jesuits were merely reminding the Chinese of what they had already known in antiquity, but had been lost. Moreover, they suggested that science had its origin in China and was transmitted to Europe. Contrary to this view, some Jesuits proposed that ancient Chinese thought had its origin in the Jewish patriarchs and had passed to China. They claimed that this thought had been corrupted later under Taoist and Buddhist influence.[21]

The scientific exchange in India was of less importance because the Jesuits did not hold official positions, unlike in China where they were involved with the direction of the Imperial observatory and the preparation of the calendar. Relations with Indian astronomers occurred only in Jaipur and Delhi during the time of Sawai Jai Singh. There Jesuits introduced the use of telescopes and modern astronomical tables, but once again one has to acknowledge their failure to introduce the heliocentric system,

[21] Minghui, "Provenance" (2002), 22–25.

although de la Hire's tables were calculated using that system. In general, the influence of Jesuit astronomers in India was of shorter duration and lesser importance than that in China. Furthermore Indian astronomers showed a certain resistance to accepting the European astronomy brought by the Jesuits.

In any case, in China as well as in India, astronomy and mathematics were for the Jesuits not an end in themselves but a means to propagate the Christian message.[22] However, this does not diminish its scientific value. Jesuit missionaries were aware of the value of science in itself, as it had formed an important part of their training. They came from Europe where at that time modern science was experiencing its beginning with continuous new developments. The Jesuits found in China and India a situation in which mathematics and astronomy had flourished in the past, but were at that time stagnant with nothing new developed. The Jesuits made a great effort to translate, particularly into Chinese, the new contributions of European mathematics and astronomy, and to arouse the Chinese scholars' interest for these sciences. Later on scientists in Europe took advantage of the Jesuits' numerous observations and their cartographic work in China and India. The Jesuits also learned to appreciate the achievements of these two ancient civilizations and communicated them to Europe in their letters and publications. They created in Europe a new interest for these two cultures. During the eighteenth century China and India adapted themselves to modern science, a process in which the Jesuits played an important role. There can be no doubt, therefore, that the Jesuits for almost two centuries contributed to the scientific exchange between West and East which benefited both sides. Needham, referring to the Jesuits' work in China, concludes: "All in all, the contribution of the Jesuits, checkered though it was, had qualities of noble adventure. If the bringing of the science and mathematics of Europe was for them a means to an end, it stands for all time nevertheless as an example of cultural relations at the highest level between two civilizations theretofore sundered".[23]

[22] Antonella Romano, Les jésuites entre apostolate missionaire et activité scientifique (XVIe–XVIIIe siécles). *Archivum Historicum S.I.*, 74 (2005), 213–236.

[23] Needham, *Science in China*, III, 457.

Chapter 5
Naturalists, Geographers and Explorers

5.1 Acosta and the Nature of American Lands

The Society of Jesus, a few years after its founding, sent missionaries to Asia, Africa and the newly discovered American continent. During the life of Saint Ignatius, the first Jesuit missionaries were sent to India, Japan, Ethiopia and Brazil. At first, missions in Spanish America were reserved by royal decree to the traditional religious orders, so that Jesuits could not send missionaries to Florida until 1566, under Father General Francis Borja. In 1567 Jesuits arrived in Peru and in 1572 in Mexico. In the previous chapter we saw the great interest in Europe for news from Jesuit missionaries. This interest was not only for the customs and culture of the peoples but extended also to the nature of the land, plants and animals of those far-away and unknown regions. A letter from Saint Ignatius in 1554 to Gaspar Berce, a Jesuit missionary in India, shows how important it was for Jesuits to satisfy this interest:

> Some important persons of this city (Rome) read with great interest the letters from India and expressed their desire to read about the cosmography of those regions where ours live; for example, how long are the days in summer and winter, when the summer begins, if the shadows fall to the left or to the right. Finally, they wish to know if there is anything else that seems to be extraordinary, as it may be about animals and plants not known in Europe or not with their size.[1]

Following these recommendations, the Jesuit missionaries' correspondence became an important source about the nature of the lands, fauna and flora of those unknown regions, together with the customs of their inhabitants. These letters also served as effective means of propaganda about Jesuit works and a source for their financing. We must also take into account that natural history (with the description of animals, plants and minerals, as well as the conditions of nature) was included in the studies of natural philosophy, which formed a part of the philosophical training of every Jesuit. For these studies, besides the works of Aristotle on natural philosophy,

---

[1] Saint Ignatius of Loyola, *Obras* (Letter to P. Gaspar Berce, Roma 24 February 1554), 985.

A. Udías, *Jesuit Contribution to Science: A History*,
DOI 10.1007/978-3-319-08365-0_5

Pliny's book, *Historia naturalis* (Natural history), was an important reference. Jesuit missionaries often referred to Pliny's work when they described the new animals and plants found in the lands they visited.

Many Jesuit missionaries became naturalists, moved by this desire to satisfy the interest aroused in Europe for the unusual features of these far-off lands and by their own intellectual curiosity, to an extent not found in the missionaries of other religious orders. In their letters and books they described the geographic and climatic conditions, the fauna and flora of the lands they visited, often being the first Europeans to do so. Their mathematical training allowed them to make astronomical observations and to draw up maps, often the first, of those regions, with measurements of geographic latitude and longitude of localities and of distances between different points and along the course of rivers. Conditions in Africa and Asia were already known, in part, in Europe, but the newly discovered lands of America were completely unknown and offered much new information for the Jesuits to communicate. From the first narratives in the form of letters, the Jesuit proceeded to write really scientific works, which described in detail the geography and climatology, the occurrence of earthquakes and volcanic eruptions, the flora and fauna, as well as, the customs and culture of the people of those lands. The study of plants found in America also had a practical aspect for their medicinal use. In this area, the Jesuits made use of the knowledge of medicine men, witch doctors and shamans, although they did not always trust them. Thus, Jesuits became, especially in America, the first explorers, geographers, cartographers, naturalists and anthropologists.[2] Naturally, they were not the only ones, since at that time there were many others who wrote about these matters. Also one should not forget that the main aim of the Jesuits was to serve as instruments to preach the Gospel, and not simply to satisfy the curiosity of their readers in Europe.

José de Acosta (1540–1600), stands out among the first Jesuits to describe the natural conditions in America.[3] He was born in Medina del Campo (Valladolid), joined the Jesuits in 1553 and traveled to America in 1572, in the third Jesuit expedition to those regions. He remained in America for 15 years making frequent journeys which took him to regions that today form the countries of Peru, Bolivia, Chile and Mexico. During his travels he took notes about his observations of the aspects of the sky, temperature, rainfall and winds, the occurrence of earthquakes and volcanic eruptions, the great variety of plants, animals and minerals, unknown in Europe, as well as of the customs and history of their inhabitants. Having returned to Spain in 1587 Acosta wrote his *Historia natural y moral de las Indias* (History of the nature and customs of the Indies, 1590), which became very popular with

---

[2] Luis Millones Figueroa and Domingo Ledezma (eds.) *El saber de los jesuitas, historias naturales del Nuevo Mundo*. Madrid: Iberoamericana (2005); Steven J. Harris, Jesuit scientific activity in the overseas missions, 1540–1773. *Isis*, 96 (2005), 71–79; Andrés I. Prieto, *Missionary scientists. Jesuit science in Spanish South America, 1570–1810*. Nashville (Tennessee): Vanderbilt University Press (2011).

[3] José Rodriguez Carracido, *El P. José de Acosta y su importancia en la literatura científica española*. Madrid: Sucesores de Rivadeneira (1899); Claudio M. Burgaleta, *José de Acosta (1540–1600) his life and thought*. Chicago: Loyola Press (1999).

four editions in Spain in less than 20 years and was immediately translated into Latin (1590), Italian (1596), French (1598), German (1598), and English (1604).[4] Modern interest for this work can be seen in recent editions and translations.[5] *Historia Natural* is divided into seven books: the first three deal with general topics of geography, climate, earthquakes and volcanic eruptions; the fourth with minerals, plants and animals and the last three with the customs, religion and political organization of the Indians of Peru and Mexico. The first two books are a translation of a book Acosta wrote in Latin and published in 1588 in Peru, *De natura novi orbis* (Nature of the new world). In his preface to *Historia Natural* Acosta admitted that other authors had already written about the new lands of America, but they had limited themselves to describing the unusual and strange things found in nature, without trying to find their causes. Thus he writes: "Till now I have not seen an author who dares to find the causes and reasons of the new and strange things of nature". This is very important for him since many of the observations cannot be explained by the traditional philosophy (Aristotelian). For example, when he talks about the climate of the American torrid or tropical zone, he points out the abundance of water and vegetation against the Aristotelian opinion that it should be extremely dry and hot. Thus he adds: "Aristotle was of this opinion and though he was a great philosopher he was wrong on this point". In his book Acosta often remarks that his observations cannot be explained by the traditional natural philosophy taught at the universities "because they are natural things which fall outside the generally accepted philosophy". Acosta's emphasis on observation as superior to the authority of Aristotle puts him in line with empirical modern science. He tried to find explanations for these new natural phenomena and objects found in America, admitting that they contradicted the accepted opinions of scholastic philosophy. Jerónimo Feijoo, one of the pioneers of modern science in Spain in the eighteenth century, called him the "Pliny of the new world", both because of his descriptions of natural phenomena in those regions and the breadth of subjects in his book. Alexander von Humboldt considered him to be one of the founders of modern physical geography and the first to attempt a systematic study of the geography and natural history of the New World. Details of the first four books on cosmography, geography, geophysics, mineralogy, botany and zoology are worth considering.

Among the subjects of geophysical interest discussed by Acosta in the first three books are magnetic declination, tides and currents in the oceans, earthquakes, volcanoes and climate, especially, the types and causes of winds.[6] Climatic and meteorological phenomena occupy books II and III, perhaps the most interesting part of his work

[4] José de Acosta, *Obras. Historia natural y moral de las Indias.* (Estudio preliminar y edición de Francisco Mateos). Biblioteca de Autores Españoles 73. Madrid: Rivadeneira (1954).

[5] Jose de Acosta, *Natural and Moral History of the Indies.* (edited by J. E. Mangan, introduction by W.D. Mignolo, translation Frances López-Morillo) Durham (NC): Duke University Press (2002); Jose de Acosta, *Historia natural y moral de las Indias.* Mexico: Fondo de Cultura Económica (2006); Madrid: Consejo Superior de Investigaciones Científicas (2008).

[6] Agustín Udías, José de Acosta (1539–1600): A pioneer of geophysics. *Eos, Transactions of the American Geophysical Union,* 67 (1986) 461–462.

for their novelty. The first problem raised concerns the climate of tropical regions, which according to Aristotelian doctrine should be dry and hot, but in America is humid and temperate. Acosta mentions the different climates found in those regions depending on their altitude and proximity to the coast, the prevalence of winds and rain, and the properties of the land. He was the first to discuss altitude sickness, which he had experienced crossing the Andes, though he mistakenly attributed it to the winds. He describes the regularity in force and direction of the trade winds and of the winds in extra-tropical latitudes, and he assigns their cause to the influence of the rotation of the heavenly spheres which drag the air. Acosta did not accept the rotation of the Earth, proposed by Copernicus in 1540, and he found no other convincing mechanism to explain the trade winds. He shows how the constant direction of these winds was used to chart the sea route of Spanish ships across the Atlantic between Spain and America and across the Pacific between Mexico and the Philippines.

Acosta dedicates considerable space to the problem of variations in magnetic declination across the Atlantic Ocean and refers to places where the declination is null; he recognized the difficulty in explaining this phenomenon. He points out that one of these locations was near the Azores, where the declination changes from east to west. Regarding ocean tides he questions if this was a global or a local phenomenon and if tides have the same motion on both sides of the same ocean. Basing himself on observations made by Spanish sailors in the straits of Magellan, Acosta concludes that this is a global phenomenon that affects the whole ocean. He also describes the periodicities of the tides and their relation with the phases of the Moon. In his account of the fountains and rivers he gives one of the first descriptions of the great Amazon River. He closes book III with a discussion of volcanoes and earthquakes (on earthquakes see Sect. 8.1).

In the 42 chapters of book V Acosta deals with minerals, plants and animals found in the newly discovered lands of America. His special interest is in those which are not known in Europe and he describes them in great detail giving curious facts about them. Talking about the minerals he stresses the abundance of gold, silver and mercury and describes the methods used in the mines of Potosí in Peru. He also mentions the abundance of emeralds and pearls and the methods used to obtain them. The chapters dedicated to the plants found in America and not known in Europe have even greater interest. He acknowledges the work of Francisco Hernández, a doctor appointed by the Spanish King Philip II in 1575 to study the flora of Mexico, which covered up to 1,200 new plants. Acosta begins with the plants which form the staple food of the Indians, especially corn (maize) from which they make bread and drinks, and tubers, such as, potatoes, cassaba and yucca. He gives a long list follows of plants, fruits and flowers proper to the American lands with their characteristics, and in some cases their nutritional value and medicinal properties. Thus he mentions chili used as a condiment, cacao highly appreciated as a drink and coca which Indians chew to gain strength. After the plants, Acosta describes the animals proper to America, such as vicunas, guanacos and llamas, and the many different types of monkeys. In his descriptions he appears as a keen observer and gives a wealth of information about plants and animals.

Acosta dedicates the last three books to the "moral history": the customs, religion, and social and political organization of the Indians. The last book is dedicated to the history of Mexico before the arrival of the Spaniards and ends with the conquest by Hernán Cortés and the death of Montezuma. By spreading knowledge of the land and inhabitants of the New World, Acosta hoped to assist in the preaching of Christian doctrine among the Indians. The first two books of *Historia natural,* were first written in Latin as an introduction to his work, *De procuranda Indorum salute* (On how to gain the salvation of the Indians, 1588). In this work he dealt with the methods and problems found in preaching the Christian faith to the American Indians. Acosta's work was very influential among Jesuit naturalists in America, just as that of Clavius had been among Jesuit mathematicians.

5.2 Jesuit Naturalists of the New World and Asia

The path opened up by Acosta was followed by many Jesuits who reported their observations of the newly discovered American lands – or the "New World" as they called it – and thus became the first naturalists and geographers. Cristóbal de Acuña (1598–1670), who arrived to America in 1622 and worked mainly in Chile and Ecuador, was one of the first. In October 1637, a Portuguese expedition led by Pedro de Texeira had left the Portuguese settlement of Pará (today Belem, Brazil), at the mouth of the Amazon River, and going upstream arrived a year later in Quito. Francisco de Orellana, the first to travel along the river in 1542, had given it this name, taken from Greek mythology of female warriors (*amazonas*), because he believed there were *amazonas* among the Indian tribes. Luis Jerónimo Fernández de Cabrera, Count of Chinchón, the Viceroy of Peru, decided that two Spaniards should accompany the Portuguese expedition on the return journey and appointed the Jesuits, Acuña and Andrés de Artieda (1605–1651). The journey began in February 1639 at the Napo River in Ecuador and the expedition arrived in Pará in December. During the journey Acuña made observations and took notes as he was crossing the different regions. A year later he returned to Spain and published *Nuevo descubrimiento del gran Río de las Amazonas* (New discovery of the great Amazon River, 1641).[7] The book was an instant success with a second edition published in 1656, a translation into French in 1682 and into English (from the French edition) in 1698; further editions were published in German and Portuguese. The book was dedicated to Gaspar de Guzmán, Count-Duke de Olivares, First Minister of Philip IV, then King of Spain. In the inscription Acuña wrote: "with great care I tried to find out everything I could about the regions we were crossing and to put it down in a few pages when it was worth many volumes". In the Preface he stressed the novelty of the contents of his book in which he wrote "about a new world, new nations,

---

[7] Cristóbal de Acuña, *Nuevo descubrimiento del gran río de las Amazonas*. Madrid: Imprenta del Reyno (1641).

new kingdoms, new works, new ways of living … a large river of fresh water as we sailed to its mouth, filled with things not seen before". He insisted that he was not making things up and commented: "I only wrote the things I heard with my ears and investigated with great care, after having taken notes with particular care about everything which was worth being recorded, such as the names of the rivers and of the different nations".

The text follows the course of his journey, down the Napo River in Ecuador to where this river joins the Amazon near Iquitos (Peru), and then along the Amazon to its mouth near Pará. Acuña located the source of the Amazon River near the mountains of Guanamá and Pulcá and estimated its total length to be 1356 Castilian leagues (7,557 km, modern estimations give 7,020 km) and its width varying from one fourth of a league in the narrowest parts to several leagues in the widest. He made measurements of the latitude along the course of the river and found that they were never more than 4° to the south of the Equator. His geographical descriptions are very detailed, giving the names of the tributaries and mentioning the existence of lakes. Acuña did not draw a map of the Amazon River and its tributaries, but one based on his descriptions was given in the French edition. He wrote about the climate, calling it temperate and healthy, and he compared it with that of Paradise. Plants are described in detail beginning with the trees in great number, diversity and size: "the trees are numberless and so high that they reach the clouds and so wide as to cause astonishment". He talks of cedars with more than 30 palms (6 m) of circumference. He describes the great variety of plants and fruits, especially, those that are used as food by the people of those lands. In particular Acuña commented on the medicinal plants of which there is a great variety with many different effects. He highlights from a practical point of view the abundance of wood, cacao, tobacco and sweet cane. With regard to the fauna, Acuña mentions especially the fish, giving an exceptional description of what he calls the *pege buey* (manatee or sea cow), and he explains the varied forms of fishing used by the Indians. The different tribes found along the journey are also special objects of his interest. Basing himself, he claims, on trustworthy testimony, he admits the existence of amazons, supposed to be a tribe formed only by warrior women (*mujeres varoniles*). He relates that once a year the amazons received a visit of men in order to have children, but afterwards kept only the girls.

In 1596 Bernabé Cobo (1580–1657), a young man from Jaén (Spain), arrived on the island of Hispaniola, as a member of a failed expedition organized to search for the mythical El Dorado. He moved to Peru and after some years studying in Lima he joined the Jesuits in 1601. In 1616, he finished his training as a Jesuit in Lima and Cuzco, where he learned Quechua and Aymara, the two main Indian languages, and he began his travels throughout the region, which aroused his interest for the study of the nature and history of those lands. As he says in the Preface of his book, he was moved by natural inclination to search and discover the secrets of these lands. In 1629 Cobo made a journey to the territories of Central America (Nicaragua, Guatemala and Mexico) on what he considered a study mission and in 1642 returned to Peru where he completed his great scientific and historical work,

Historia del Nuevo Mundo (History of the New World).[8] He had worked on it during 40 years and considered it finished in 1653. In this work he used all the information he had been able to gather from native Indians and early Spanish settlers. As he wrote: "the first motive I had, which led me to write this work, was the diversity of opinions found in the different accounts written about this New World and the desire to find out the truth about them". The work is divided into 43 books which deal with the geography, climate, minerals, plants and animals of the regions he had visited, together with a long political and ecclesiastical history. It was never published during the author's life, probably because Cobo could not find financial support. Only 17 of the 43 books are preserved. Of these the first ten are on natural history: books 1 and 2 on geography, book 3 on minerals, books 4–9 on plants and animals proper to America and book 10 on plants and animal brought by the Spaniards. In 1804 Antonio José Cabanilles, director of the Botanical Garden of Madrid, was the first to recognize the importance of this work, which was published for the first time only in 1890–1893.

In the first book Cobo presents some basic notions of cosmography, following traditional geocentric Aristotelian doctrine. The only novelty is the chapter on the aspect of the southern sky, where he says that very few have studied it. He mentions that the Southern Cross constellation has its largest star at 30° from the Antarctic Pole. He also, like Acosta, considers why the tropical region is inhabitable, being humid and with cool winds, contrary to what was predicted by Aristotelian doctrine. He describes in detail the cordillera of the Andes and the high Andean plateau (*altiplano*). He estimated that the highest mountains were about one league and a half above sea level (8,400 m, in fact 6,962 m). Cobo divided the lands of Peru into six levels between the sea coast and the high plateau, and he gives characteristic features of each. He describes the high altitude sickness, which he attributes to "very dry, thin and light air", and says that it causes a sudden illness which ceases when one comes down to lower heights. He claims that this feeling is not comparable to what is felt in the Alps. On the Peruvian lowlands (*llanos*), he notices that they have very odd and strange properties, especially a total absence of rainfall: this, he says, is one of the great mysteries of nature for which we cannot find the cause. He describes the eruption of a volcano in 1586 near Quito (Ecuador) and another in 1600 near Arequipa (Peru). He mentions that Peru is the land of all America with most frequent earthquakes and describes several of them: in Arequipa (1582), Lima (1586) and Arica (1604). This last had a long duration and was followed by a tsunami with three consecutive sea waves that affected Lima, Callao and Pisco.

The third book has for its subject what he calls the "perfectly mixed" or minerals that according to him are generated inside the earth and lack all life. The book has 30 chapters about different types of minerals and gems found in America, mentioning some of their medicinal properties. Metals receive special attention, in particular,

[8] Bernabé Cobo, *Obras. Historia del nuevo mundo.* (Estudio preliminar y edición de Francisco Mateos). Biblioteca de Autores Españoles 91 and 92. Madrid: Rivadeneira (1959).

gold, silver, mercury, copper, iron, tin and lead. He mentions the main mines and the methods used in them, in particular in those for gold, silver and mercury. With regard to gemstones he mentions the great abundance of emeralds.

The most important part of the *Historia* is that in seven books dealing with plants and animals. Cobo distinguishes between plants that are proper to the New World and those brought by the Spaniards. He was aware of the difficulty in making this distinction, because the plants brought by the Spaniards have spread so much throughout the land that they appear as if they were native to the land. Even today we find it difficult to realize that plants like bananas, sugar cane, mango and coffee are not original to America. To identify the plants Cobo usually gives a description of their leaves and fruits along with their names in the languages of Peru, Mexico, Central America and the Antilles. His descriptions are more detailed than those of Acosta and he gives more information about their nutritional value and medicinal use. In total Cobo describes 108 species of plants original to America and 87 species that are also found in Europe. He describes 129 species of trees which he divides into five classes, separating fruit trees from those used only for building and fire wood. He divides the animals between those which live in water, including fishes, cetaceans, amphibians and mollusks (55 species). He then lists birds (59 species) and terrestrial animals, including mammals, reptiles, insects and spiders (71 species). He dedicates the tenth book, the last on natural history, to the plants and animals brought by the Spaniards, and gives some reasons why they have prospered so widely and rapidly. Cobo's descriptions, compared to those of Acosta and Acuña, are more complete and detailed, specifying a greater number of species from tropical and extra-tropical regions. He was critical of some aspects of Acosta's work. It is regrettable that Cobo's work was not published in its time, that some parts have been lost, and that it remained practically unknown until the end of the nineteenth century.

With time Jesuit naturalist studies became more numerous, but each was limited to a particular region. Only those of Acosta and Cobo are of a general character. In 1549 the first Jesuits arrived in Brazil. The Portuguese Fernão Cardim (1549–1625) was one of the first to write about the nature of this land in his *Do Clima e terra de Brasil* (About the climate and land of Brazil, 1585). The English travel writer Samuel Purchas used the information from this work in his *Purchas his pilgrimes, a History of the world* (1625). Simão Vasconcellos (1596–1671), from a family that had settled in Brasil, was professor at the college of Rio de Janeiro. He wrote his *Chronica da Companhia de Jesu do Estado do Brasil* (Chronicle of the Society of Jesus in the State of Brazil, 1663), where he included descriptions of the nature and climate of the region. He claimed that in his descriptions of Brazil "one can see the quality of the land, the disposition of the climate, freshness of trees, variety of plants, abundance of fruits and medicinal herbs, diversity of living beings in water and land, birds so singular and many other prodigies of nature". Vasconcellos, deeply impressed by the natural richness of Brazil, had no doubt in locating the biblical Garden of Eden there. This caused him problems with the Jesuit censors.

José Gumilla (1686–1750) made a study of the region of the Orinoco River (Colombia and Venezuela). He had arrived in Nueva Granada (Colombia) in 1705, where he studied philosophy and theology in Bogotá. The contents of his work *El Orinoco ilustrado* (The illustrated Orinoco, 1745), is given in its complete title: "Natural, civil and geographic history of this great river… the government and customs of the Indians, .. information about the animals, trees, fruits, oils, resins, medicinal herbs and roots".[9] Although entitled as "illustrated", the book does not include any maps or drawings of plants and animals. Gumilla was aware how new everything was which he is describing. He writes: "This is not only called New World because of its new discovery, but because, compared with the Old World, everything in it is new and different". The first four chapters of the first part are dedicated to the geography of the course and basin of the Orinoco River, its tributaries and the climate, which is temperate and the same during the whole year. As he says, there is no sense in talking about winter when one is so close to the equator. He claims that after careful investigation he can affirm that there is no communication between the Orinoco and the Amazon rivers. He deals with the animals of those regions, both terrestrial and aquatic, and with the great variety of plants and trees. The second part examines the customs, organization and wars of the Indians, adding a whole chapter about the manufacture of the curare poison. Gumilla was the first European to describe with great detail the process of making this powerful poison and its mortal effects. This gave him the opportunity to write about the different types of poisonous animals, especially, snakes.

Alonso de Ovalle (1603–1615) wrote about the natural history of Chile in the first two books of his work, *Histórica relación del reyno de Chile* (Historical narrative of the kingdom of Chile, 1646), as did Diego Rosales (1601–1677) in *Historia general del reyno de Chile, Flandes indiano* (General history of the kingdom of Chile, the Indian Flanders) written in 1656 and revised later, but not published until 1877. In both books there is a defense of the nature and people of Chile against opinions spread in Europe. Ovalle wrote his book in Rome and was often diverted by his interest for the unusual and marvelous. Rosales adopted a more naturalist approach and stayed close to what he was writing, giving a very complete description of the plants and animals of Chile. He paid special attention to medicinal herbs and their effects, relying on the knowledge of the Indians of the region. The natural history of the New World can also be found in the writings of Jesuits living in Europe, like Kircher, who wrote about America basing himself on information received from the missionaries. Juan Eusebio Nieremberg (1595–1658), a professor of the *Colegio Imperial* in Madrid, published a general book about natural history, *Historia Naturae* (History of nature, 1635) with abundant information about flora and fauna of the New World, based mostly on Jesuit sources.

Some Jesuits published their studies of natural history while in exile in Italy after their expulsion from Portugal and Spain and their American territories in 1759 and 1767. José Sánchez Labrador (1717–1798), had arrived to Argentina in 1734 and

---

[9] José Gumilla, *El Orinoco ilustrado*. (Introducción, notas y arreglo de Constantino Bayle). Madrid: Aguilar (1946).

Fig. 5.1 American plant in José Sánchez Labrador (1717–1798), *Paraguay Natural* (Arch. Romanum S.I.)

worked in the famous reductions of Paraguay until the expulsion.[10] He succeeded in bringing with him to Italy his manuscripts, overcoming many difficulties and prohibitions. In Ravenna he continued his work and finished it in 1772 divided in two books with the titles, *Paraguay natural* (The Nature of Paraguay) and *El Paraguay Católico* (Catholic Paraguay) where he describes the region of Río de la Plata, and the plants, animals and customs of its inhabitants (Fig. 5.1). It is a strictly naturalist work, with more than a 100 illustrations, which have not yet been published in their totality. *Paraguay natural* is divided into four parts, the first on the geography of Paraguay: rivers, lakes, climate, winds, earthquakes, volcanoes and other natural phenomena and the minerals found in the region. The second part deals with plants and trees, the third part with animals – quadrupeds, birds and fish, and the fourth with amphibians, reptiles and insects. Sánchez Labrador knew the works of Buffon (Georges-Louis Leclerc) and Carl Linnaeus (von Linné), although he did not use the Linnaean classification in his presentation of American flora.

Thomas Falkner (1702–1784) had studied medicine in England and traveled to South America in 1738 where he joined the Jesuits in 1740 in Buenos Aires. After the suppression of the Society he returned to England where he published the

[10] Héctor Sainz Ollero, Helios Sainz Olleros, Francisco Suárez Cardona y Miguel Vázquez de Castro, *José Sánchez Labrador y los naturalistas jesuitas del Río de la Plata*. Madrid: Ministerio de Obras Públicas (1989).

observations he had made during his travels throughout Patagonia, *A description of Patagonia and the adjoining parts of South America*, (1784). José Jolis (1728–1790), a native of Catalonia, arrived in Montevideo in 1755 and worked in the reductions of Paraguay. In Italy he published *Saggio sulla storia naturale della provincia del Gran Chaco* (Essay about the natural history of the province of Gran Chaco, 1779), where he gives very full information about the fauna and flora of this region, with special attention to medicinal herbs and the animals proper to it, such as, vicunas, llamas, alpacas, and guanacos. Ramón María Termeyer (1737–1814), a native of Cadiz, arrived in Buenos Aires in 1762. He became interested in the study of insects and especially of spiders, of which he gathered a collection of 3,438 specimens. He introduced into Argentina the cultivation of the silkworm. In exile in Italy he published *Oposcoli scientifici d'entomología* (Brief scientific treatises of entomology, 1807–1809) in five volumes. He also published his studies about the electric eel of Río de la Plata. Juan Ignacio Molina (1740–1829), born in Talca, Chile, a professor of mathematics and cosmography in Santiago, published in Italy after the expulsion, *Saggio sulla storia naturale del Chili* (Essay about the natural history of Chile, 1782), translated into Spanish (1788), French (1789) and English (1808) and edited again in Italian in 1810. In this work Molina tries to refute the idea suggested by Buffon (and spread by Cornelius de Paw) about the negative influence of the American climate on the fauna and on the population which make them indolent and lazy. Molina praised Chile for its climate, fauna, and flora, as well as its people, as brave and industrious. In his work *Analogia de los tres reinos de la naturaleza* (Analogy of the three kingdoms of nature, 1820) Molina proposed continuity between minerals, plants and animals that can be considered as a certain foretaste of evolution. Francisco Javier Clavijero (1731–1787), a Mexican, also wrote in Italy; in the first book of his historical work, *Storia antica del Messico* (History of Ancient Mexico, 1780; the original Spanish version was published in 1826) he writes about the natural history of Mexico. Clavijero also wrote a *Storia della California* (History of California, 1789), which he did not know personally, basing himself on the letters of Jesuit missionaries of that region. Molina and Clavijero exemplify the nationalist sentiments of the Creole society with its praise and defense of the American nature and people. These are some of the most important works written after their expulsion by the Jesuit naturalists in their Italian exile.

Jesuit naturalists also wrote about some regions of Asia. Outstanding among them is the Portuguese João de Loureiro (1717–1791) who arrived in the Kingdom of Cochinchina (South Vietnam and Cambodia) as a royal mathematician, after spending 6 years in Goa and Macao.[11] His practice of medicine, exercised in his missionary work, aroused his interest for plants. He extended his work to plants of other regions like Sumatra and Malaysia and published his study in Lisbon after the suppression of the Society with the title *Flora Cochinchinensis* (Flora of the Cochinchina, 1790). He described and classified according to the system of

[11] G. de Fonseca, Rerum naturalium in Lusitania cultores. P. Johannes de Loureiro. *Broteria*, 5 (1906), 98–114.

Linnaeus about 2,000 species of plants, 1,400 from Vietnam. These corresponded to 647 genera many of them unknown in Europe. He also left a large number of manuscripts and herbariums dispersed today in several museums. Loreiro was in contact (through correspondence) with the Royal Society of London in the person of the British naturalist Joseph Banks, president between 1778 and 1812, who admired his scientific work. He corresponded also with the Swedish naturalist Daniel Solander, a disciple of Linnaeus, who in 1768 accompanied James Cook on his journey around the world.

Georg Josef Kamel (1665–1706), a Czech naturalist, had been a pharmacist before joining the Jesuits. In 1688 he arrived in the Philippines and was pharmacist and botanist at the college in Manila, where he had a botanical garden, mostly of medicinal herbs. Kamel kept contact with the English botanist John Ray, considered to be the father of English natural history, and James Petiver, both members of the Royal Society, to whom he sent plants and animals of the Philippines. His work *Descriptiones fructicum et arborum Luzoni*s (Description of the fruits and trees of Luzon), was published as an appendix of Ray's book *Historiae Plantarum* (Histories of plants, 1704). Other works by Kamel about the plants and animals of the Philippines were published from 1702 in the *Philosophical Transactions* of the Royal Society. Kamel was the first to describe the camellia, so Linnaeus gave it this name in his honor.

5.3 The Interest for Geography

At the beginning of the sixteenth century, geography as a science was a part of what was called "cosmography" or description of the universe, and its scope was to describe and, as they said, "to put under our eyes", the terrestrial globe. From Greek antiquity, two traditions were transmitted, one descriptive, indebted to Strabo, and the other mathematical, deriving from Ptolemy. Ptolemy, basing himself on the works of Hipparchus and Eratosthenes of Cyrene , was among the first to locate points on the surface of the Earth by their latitude and longitude, and to explain how to project the spherical Earth onto a plane. These two traditions were present in the teaching given in Jesuit schools, where descriptive geography was part of natural philosophy together with the commentaries of Aristotle and treatises on heaven and meteors. Mathematical geography came under mathematics with the application of spherical geometry to the Earth, its projection onto a plane and the determination of geographical latitude and longitude.[12] Next to the two great classical works of geography by Strabo and Ptolemy the Renaissance produced works, for example those by Francesco Maurolicus, *Cosmographia* (Cosmography, 1578) and the new maps by cartographers like Gerhard Mercator in 1569, with his new cylindrical projection. In 1570 Abraham Ortelius published the first universal atlas, *Theatrum orbis terrarum*

[12] François de Dainville, *La géographie des humanistes*. París: Beauchesnes, (1940).

(Theater of the terrestrial orb), with the first world maps showing the American continent, and in 1650 Bernhard Varenius produced his monumental geography *Geographia univeralis* (Universal geography). Jesuits not only dedicated their energy to geography as an academic discipline, but were also active as cartographers in drawing up maps, above all of the lands discovered in their missionary work in America, Asia and Africa, which were unknown in Europe. This work combines that of explorers not moved exclusively by geographical interests. There is also certain continuity between the field work of Jesuit geographers and explorers and the academic geography of the colleges. In both the two approaches appear, descriptive and mathematical, that we have mentioned. For Jesuits, geography was also an effective instrument in making known the apostolic work of the Society with maps which showed their presence throughout the world.

Already in the work of Clavius (*Commentarius in sphaeram*), which was so influential in teaching mathematics and astronomy in Jesuit schools, geography is presented in chapters II and III. These chapters dealt with the mathematical aspects of geography, such as the division of the Earth by parallels and meridians and the methods to determine geographical longitude and latitude. He also mentions that the tropics are habitable in accordance with the evidence presented by Portuguese and Spanish sailors. Riccioli, besides his work in astronomy, published an extensive geography, *Geographiae et hydrographiae reformatae libri XII* (The 12 books of revised geography and hydrography, 1661). In this book, following the tradition of mathematical geography, he focuses on the problem of measuring the Earth and determining latitude and longitude. He gives a long list of 2,800 locations with their geographical coordinates, referring longitudes to the zero Meridian at the island of Hierro (one of the Canary Islands), as was customary at that time. The section on hydrography contains an exhaustive description of oceans, rivers, and lakes. A shortcoming of this work is the lack of maps, which he excuses as due to the difficulty of finding good engravers. Both Riccioli and Grimaldi were among the first to try to measure the length of a degree of the meridian arc, using observations of terrestrial zenith distances. In 1645 they measured the distance between the cities of Bologna and Modena, obtaining 62,650 Bologna paces (approximately 122,104 m). They admitted that the rather simple method was subject to inevitable errors due to the refraction of the atmosphere. The maps lacking in Riccioli's work are abundant in those of Kircher.[13] Among these the detailed maps of Africa, Asia and South America are of special interest, with details of the courses and sources of the great rivers such as the Ganges, Nile, and Amazon. The world map, with a presentation of the ocean currents, may be the first of its kind. The maps of India, China and America were based on information Kircher received from the missionaries who explored those lands. Among his maps are those which show the origin of rivers, lakes and ocean currents in line with his theories of the *hydrophilacios* (cf. *Mundus subterraneus*).

[13] Godwin, *Athanasius Kircher's*, (Chap. 12, Maps and plans), 215–236.

Geography as an academic discipline had an important development in the Jesuit colleges of France, where humanism, natural philosophy, and mathematics were combined. Three professors merit special mention as they represent elements of what would be modern geography.[14] Philippe Briet (1601–1668), professor at the colleges of Paris and La Flèche, while still in Strabo's tradition of descriptive geography, is best considered a modern geographer. His work, *Parallela geographiae veteris et novae* (Old and new geography in parallel, 1648), compares the geography of classical and modern authors. It is an extensive work consisting of three volumes about Europe and another three about Africa and Asia, although he only published the first three. For each European country he presents the nature or physical characteristics, flora and fauna and the customs of the people. Briet describes with wit the character and customs of the people of each European country. Jean François (1586–1668), professor at the same colleges as Briet, had Descartes as a student at La Flèche. His work, *La science de la geographie* (The science of geography, 1652), is the result of long teaching experience. This is one of the first books published in France where emphasis is placed on the scientific character of geography and the need to overcome the traditional descriptive approach. Thus François states that geography is the science of places and it is based on physical and mathematical principles with quantitative and qualitative elements. Three things call his attention, namely, islands, mountains and rivers, which he describes in detail. He was especially interested in physical geography of which he was a real pioneer. For example, writing about the hydrological cycle, he stresses the importance of mountain topography and dominant winds.

George Fournier (1595–1652), professor at La Flèche and chaplain to the French Royal Navy, dedicated all his life to problems of the sea. He spent most of his time on board ships of the French Navy so that his knowledge was based on experience. His work, *Hydrographie* (Hydrography, 1643), is a comprehensive encyclopedia of all types of things that have to do with the sea and was a textbook for the students at the naval academies of Dieppe and Hesdin. A true marine geography, this work can be considered a precursor of what is called today "oceanography". Fournier divided his work into three parts: relations between men and the sea; the nature and characteristics of the sea; and geographical aspects of the art of navigation. In the first part Fournier deals with the importance of the development of navigation for commerce and politics. In the second he tries to solve the problem of the tides, which he considers the greatest mystery of the sea because nobody really knows their true causes. As a fact neither Galileo nor Descartes could explain them, and it was only Newton who gave the true solution for the first time in his *Principia Mathematica* (1687). In the last part Fournier discusses the problem of determining geographical coordinates at sea, especially, longitude. This difficult problem was not solved until much later. The book was very popular among French sailors. It is said that one captain of a French ship, when attacked defended himself using as a weapon Fournier's large book.

[14] Dainville, *La géographie,* 188–197 (Briet), 276–303 (François), 257–276 (Fournier).

Jesuits in Europe also used applied geography or cartography in carrying out geographical measurements and drawing up maps. We saw (Chap. 3) how Boscovich and Maire measured 2° of the meridian arc between Rome and Rimini and revised the map of the Papal States. In Austria, Empress María Teresa entrusted Liesganig, an assistant to Maximilian Hell at the Vienna observatory, with the measurement (almost 3°) of the meridian arc that passes through Vienna between Sobietchitz (Czech Republic) and Warasdin (Croatia). Scherffer, a professor of mathematics in the University of Vienna, participated in this project. Measurements were made between 1762 and 1766. The base line was measured along 12 km to the north of Vienna. Later measurements in 1806 showed that errors in the base line were less than 7 mm per kilometer. In 1769 Liesganig measured 2° of the meridian arc in Hungary between Czurog and Tis Telek. These measurements were published in *Dimensio graduum meridiani Viennensis et Hungarici* (Measurement of degrees of the meridian of Vienna and Hungary, 1770). In 1762 Christian Mayer also measured 1° of the meridian in Germany, in the region of Rhineland-Palatinate. In 1719, Laval carried out cartographic work on the coast of Provence, France, with observations made at land and sea. King Peter II of Portugal called Johannes König (1639–1691), a Swiss professor of mathematics at Freiburg, to help in drawing up a new map of Portugal which he did not complete. King Philip V of Spain through Zenón de Somodevilla, Marquis de la Ensenada, entrusted the drawing up of a map of Spain with modern methods and criteria to Carlos Martínez (1710–1764) and Claudio de la Vega y Terán (1680–1748), professors at the Imperial College of Madrid. The map made between 1739 and 1743 was not fully completed and was never published, although it is preserved. Tomás López, the first modern Spanish cartographer, used the originals drawings of Martínez and de la Vega for his *Atlas geográfico de España* (Geographical Atlas of Spain, 1802).

5.4 The Maps of Far-Off Lands

The most important cartographic work of the Jesuits was in drawing up the first maps of mission territories, especially, in Asia and America. This work was linked with that of exploring many unknown lands and sometimes it is difficult to tell apart geographers and explorers. Maps were an important element, in the first place, in the Jesuits' work in China. Ricci, as mentioned earlier, was the first to become aware of the shortcomings of Chinese maps. Thus with his world map (*Kunyu wanguo quantu*) he attracted the interest of Chinese scholars and literati. Ricci published the map in large format for the first time in 1584, including the translation into Chinese of all geographic terms, description of the different regions of the world, two small polar projections of the earth and a representation of the geocentric cosmology. The map was based on the world map of Ortelius. The world maps of Ortelius and Mercator, like most present-day maps, have the Atlantic Ocean at the center so that China appears on one side. Ricci put the Pacific Ocean at the center of the map, so that China was also at the center. This pleased the Chinese, who had always considered China to be the center of the world and called it the "Middle Kingdom" (*zhongquo*). Ricci's world map became very popular and was reedited at least eight times during his life. With this map he introduced into China such things as the correct spherical form of the Earth, the location of the five continents and the division by meridians and parallels. He also translated into Chinese geographic terms unknown there. His geographic work was continued by Ricci's successors. In 1623 and 1637 Giulio Alieni (1582–1649) published two books in Chinese on the geography of the world (*Zhifang waiji*) and another on the geography of Europe, both based on previous works by Pantoja and de Ursis and both with a great circulation. In 1672 Verbiest published a book on geography (*Kunyu Tushuo*), using all previous Jesuit publications, and in 1674 he produced a new world map divided into two hemispheres of one meter and a half in diameter, one with Europe, Asia, and Africa and the other with both Americas. With this map he was trying to satisfy the curiosity shown by Emperor Kangxi and the Manchu dignitaries about the rest of the world. These works also had an apostolic aim and Europe was presented in an idealized image, as a great Christian region which enjoyed harmony and stability and where all Christian kings lived in peace with each other. Jesuits also tried to introduce a different perspective to counter the Chinese centered myth, which impeded interest for anything outside China itself.[15]

Jesuits were also the first to make known Chinese geography in Europe. Ricci was the first to realize that the mysterious Cathay of Marco Polo was actually China and made clear that there was not an empire between India and China. Martino Martini (1614–1661), a student of Kircher, arrived to China in 1642. His interest for Chinese geography moved him to publish in Europe *Novus atlas sinensis* (New atlas of China, 1655). Martini estimated the population of China to be some 200 million, the first such estimate published in Europe. Michael Pyotr Boym (1612–1659), who was born in the Ukraine and became one of the few Polish missionaries in China, drew up maps of China that were published in Paris, *Abrégé de la carte de la Chine* (Compendium of the Map of China, 1670). Martini and Boym were in contact with Kircher, who used their information for the maps of China in his book *China illustrata* (1667). The renowned Dutch cartographer Joan Blaeu also used Martini's maps of China in his monumental *Atlas Novus* (New atlas, 1655). Nicholas Sanson Abbeville, a French cartographer, used maps and geographical information from Jesuits, especially Ricci, Martini and Boym, in his *Carte generale de l'Asie* (General map of Asia, 1658).

In 1702 Antoine Thomas, as already mentioned, made the first geodesic measurements of the length of a degree of the meridian arc near Peking. He used his measurements to adjust the value of the Chinese unit of length the *li*, so that there was a correspondence of 200 *li* equal to a degree, and thus a *li* was 556.5 m (today one *li* equals 500 m). Emperor Kangxi entrusted the Jesuits of the French Mission with the preparation of a new set of maps of China. Work was undertaken between

[15] Chen Minsun, Ferdinand Verbiest and the geographical works by Jesuits, 1584–1674. In Witek (ed.) *Ferdinand Verbiest,* 123–133. Theodore N. Foss, A western interpretation of China: Jesuit cartography. In C.E. Ronan and B. Oh, *East meets West: the Jesuits in China, 1582–1773*. Chicago: University of Loyola Press, (1988), 209–255.

1690 and 1718 and published as the *Atlas of Kangxi*. Jean François Gerbillon worked in the preparation of an atlas of the northern part of China, Manchuria and Mongolia which was published in Holland, *Description géographique...de l'Empire de la Chine et de la Tartarie* (Geographical description of the Empire of China and Tartary, 1736). Michel Benoist, under orders from Emperor Quianlong, made a new atlas of China where for the first time in China copper engravings were used instead of the traditional wood ones. With this atlas modern European cartography was fully accepted into China. Benoist also published a new map of the world which replaced those of Ricci and Verbiest, presenting the new discoveries and using new cartographic techniques.

Jesuits also participated in the first geographical maps of India.[16] Bouchet from his base at Pondicherry made the first maps of the south in the region of Madurai as far as latitude 14°N. He sent sketches of his maps to France where the cartographer Jean Baptiste Bourguignon used them in his map of India published in 1752. Among his other geographical works, Bouchet determined with great precision the geographical coordinates of Madras (Chennai), which d'Anville used in his cartography of India. Boudier and Pons in their long journey from Chandernagore to Jaipur in 1734 (cf. Sect. 4.7), besides making astronomical observations determined the geographical coordinates of the important places they visited and measured distances along the Ganges River and some of its tributaries. They took more detailed measurements along the way from Allahabad to Agra. The English cartographer James Rennell used their reports in his map of the north of India published in 1776. Tiffenthaler was tireless in his geographical work after the failure of his journey to the observatory of Jaipur. His interest for geography becomes clear in his words:

> Next to the salvation of souls and their conquest for God, nothing has afforded me greater pleasure than the study of the geographical position of the places, the variations of the winds, the nature of the soil, and the character and manners of the regions through which I am travelling.[17]

From his base in the College of Agra he journeyed throughout most of India. After the expulsion of the Jesuits from Portugal and the suppression of the Society in 1773, Tiffenthaler remained in India and continued his geographical studies until his death. In India he knew the French orientalist, Abraham Hyacinthe Anquetil-Duperron, who later published most of his maps. With the help of Anquetil-Duperron, Joseph Bernoulli, a professor in Berlin, published Tiffenthaler's work, *Descriptio Indiae* (Description of India), in Germany in three volumes, *Historisch-Geographische Beschreibung von Hindustan* (Historical-geographical description of India, 1785–1787). Rennell in his map of India also used Tiffenthaler's maps. The Bavarian, Francis Xavier Wendel (1726–1803), Tiffenthaler's companion and collaborator from 1758, continued his geographic work in several regions of India. He provided geographical information and maps of the region of Rajputs and the provinces to the south and southwest of Agra to Rennell who used it in his maps

---

[16] Kochar, "Secondary Tools", 177–181.

[17] Kochar, "Secondary Tools", 180.

of India. Wendel remained in India after the suppression of the Society and was one of the last members of the Jesuit Mogul mission. Pedro Murillo Velarde (1696–1753), entrusted by the General Governor of Philippines Fernandez Valdez, drew one of the first maps of the archipelago (*Carta hydrographica y chorographica de las yslas Filipinas*, Hydrographical and geographical map of the Philippine Islands, 1743). In 1752 he published in Madrid a general book on geography (*Geographia historica,* Historical geography) in ten volumes.

The first maps of the Spanish territories of America were rather inexact, drawn up from information provided by those who had traveled through their interiors, but without proper cartographic field work. The Jesuits were among the first to produce maps based on direct observation with geographic coordinates that were fairly exact for latitudes, and the measurements of distances. The maps of Jesuits covered the regions from California and Arizona to Patagonia, and were, in many instances, the first to describe the regions with different scales and some detail (Fig. 5.2).[18]

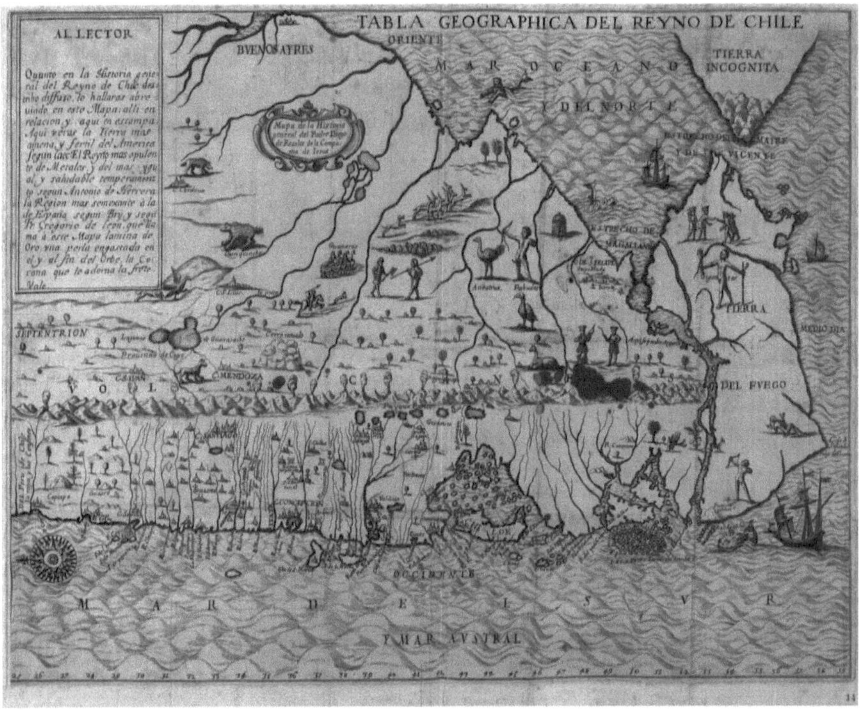

Fig. 5.2 Map of the south of Chile made by Jesuits (north is to the *left*) (Archivum Romanum S.I.)

[18] David Buisseret, Jesuit cartography in Central and South America. In J.A. Galiano and C. E. Ronan. *Jesuit encounters in the World. Jesuit chroniclers, geographers, educators and missionaries in the Americas, 1549–1767.* Rome: Institutum Historicum S.I. (1997), 113–162. Ernest J. Burrus. *La obra cartográfica de la Provincia Mexicana de la Compañía de Jesús.* Madrid: José Porrua Turanzas (1967).

Juan María Ratkay (1647–1683), a Croat who arrived to Mexico in 1680, carried out important cartographic work in the mountainous region of Sierra Madre. The Czech Adam Gilg (1653–1710) extended this work to the north to the Sonora River. Gilg collaborated with Eusebio Kino, the great cartographer and explorer of northern Mexico, California and Arizona, about whom more will be said later. In 1720 Laval, a professor at the French naval school of Toulon, was entrusted with carrying out cartographic work on the coasts of the French territory of Louisiana, published in 1728 in his book, *Voyage de la Louisiane* (Journey to Louisiana).

Sánchez Labrador not only worked as a naturalist but was also a good cartographer; he prepared many maps for his (partly unpublished) works. Among them are maps with details (along with measurements of distances) of the latitude and longitude of positions on the Paraguay River between latitudes 14°S and 24.5°S, and also of the Moxos mission, the Chiquitos region, the territory between the Paraguay and Paraná Rivers, the land to the south of the city of Asunción as far as the Munday River, and of the Uruguay River.[19] Samuel Fritz (1651–1725), a Czech who had arrived to Quito in 1685, carried out extensive cartographic work in the region of the Amazon River with astronomical observations and hydrographic descriptions which he published in 1707 with a very detailed map, *El gran río Marañón o Amazonas con la misión de la Compañía de Jesús (*The great Marañon or Amazon River with the mission of the Society of Jesus). In the maps Fritz marked the position of the frontier between the Spanish and Portuguese territories, a source of controversy at that time. On the Portuguese side, the Italian Jesuit, Domenico Capassi (1694–1736), was assigned in 1729 to draw up the maps of this frontier region. He also completed cartographic work in the regions of Río de Janeiro, Colonia de Sacramento and Río Grande do Sul. Diego Soares (1648–1748), a professor in Lisbon and Evora, collaborated with Capassi in Brazil between 1729 and 1748. After Capassi's death, Soares continued to study the cartography of the region from Rio de Janeiro to Río de la Plata and of São Paulo and Minas. José Quiroga (1707–1790), who had been in the Spanish navy before joining the Jesuits, arrived to Buenos Aires in 1728, bringing instruments for cartographic work, which he carried on for many years. He was appointed professor of mathematics in Tucumán (Argentina). From there Quiroga explored and drew up maps of the surrounded regions, though many of them have been lost. In 1753 he published in Rome a detailed map of the regions of Paraguay and Río de la Plata. In 1745 Quiroga was appointed by King Phillip V, together with José Cardiel (1704–1781) and the Austrian Matthias Strobel (1696–1769) to explore the coast of Patagonia. The expedition boarded naval ships in Buenos Aires and sailed southwards to San Julian Bay at latitude 49.5°S, making observations and carrying out cartographic work. Quiroga and Strobel took part in drawing up the treaty of limits of 1750 between the Spanish and Portuguese territories. As this treaty affected the Jesuit mission of Paraguay, with the information from their maps they challenged the Portuguese position in the treaty. Cardiel, who spent a long time in the missions of Paraná, later published in Italy his maps of these regions. Strobel also carried out cartographic

[19] Sainz Olleros *et al. José Sanchez Labrador* (1989), 305–318, 299–304.

work in the region of the Gran Chaco. Joaquín Camaño Bazán (1737–1820) spent a great part of his life in the reductions of the Indian Chiquitos (Bolivia). There besides studying their language, he made maps of the region. At their expulsion from the Spanish territories of America the Jesuits left a large collection of maps covering areas from Mexico to Patagonia.[20]

5.5 Entering Unknown Lands

In their journeys Jesuit missionaries were obliged to enter unknown territories, where they were often the first Europeans to arrive. Though their main aim was always apostolic – to bring the Christian message – they became true explorers. Thus, as we have seen, they became interested in the nature of the lands they crossed, their geographical conditions, fauna, flora and the customs of the inhabitants. Often they also made pioneer cartographic work and drew up an abundant collection of maps. As has been seen, it is difficult to separate the naturalists and geographers from those whom one might consider primarily as explorers. The latter were seeking above all to open up new paths of evangelization and therefore they entered areas in Asia, Africa and America that were unknown to Europeans. Some of these Jesuit explorers left us astonishing narratives of their journeys, often made with very precarious means. The ease we have today to travel to any part of the world was completely unknown in the sixteenth to eighteenth centuries. Journeys were made on foot, on horseback or in fragile boats, at times led by untrustworthy guides and facing all kinds of dangers.

The Jesuit mission in India began with the arrival in 1545 of Francis Xavier in Goa. From this Portuguese post, Jesuits began to enter the large territories of India. In 1578, the great Mogul Emperor Akbar, who in 1560 had united India as a vast empire and was interested in religious dialogue, asked for two Jesuits to come to his court in Fatehpur Sikri, near Agra, in the north of India, to explain Christian religion. Antonio de Montserrat (1536–1600), who had arrived to India in 1574, and Rodolfo Acquaviva (1550–1583) were chosen. They left Goa in November 1579, accompanied by a Persian convert named Henriquez as translator and by an ambassador from Akbar's Court. Montserrat wrote a narrative of the journey and the stay in Akbar's Court, *Mongolicae legationis commentarius* (Commentary about the Mongolian embassy), which was not published until 1914.[21] The expedition from Goa skirted the coast to Daman and from there took the path to Fatehpur Sikri, where they arrived three months later. Montserrat and Acquaviva were welcomed with kindness by Akbar and remained at the court for two years. During their stay they dialogued

[20] Guillermo Furlong, *Cartografía jesuítica del Río de la Plata*. Buenos Aires: Jacobo Peuser (1936).

[21] Josep Lluis Alay (edition and introduction) *Antoni de Montserrat, Embajador en la corte del Gran Mogol. Viajes de un jesuita catalán del siglo XVI por la India, Paquistán, Afganistán y el Himalaya*. Lerida: Milenio (2006).

with Muslim Imams, Hindi and Persian priests, and Buddhist monks. However, their hopes of a conversion of the Emperor to Christianity soon faded. In 1581 Montserrat accompanied the Emperor on a military campaign to the regions of Punjab, Pradesh and Kashmir as far as Kabul, and he was able to survey these remote lands. From his observations Montserrat drew up the first map which covered the whole northwest of India, Kashmir, Afghanistan and the west of Tibet. To the map he added a list of 180 geographical coordinates (latitudes and longitudes, the latter referred to the Azores as the zero meridian). After his return to Goa, Montserrat set out on a new journey this time to Ethiopia, accompanied by Pedro Páez (1564–1622), about whom more will be said later. The mission was a total failure. In February 1589 they sailed from Goa to the port of Dhofar (Oman) where they were betrayed and sent as prisoners to the Sultan of Hadhramauth. They were taken to Yemen, after crossing the desert – first to San'a and then to the port of Mocca (Al Mukha). During this journey, they were the first Europeans to taste coffee. "An infusion in water of a fruit drunk very hot," as Paez described it in his account of their journeys and misfortunes. After seven years in hard captivity, Paez and Montserrat were ransomed on payment of thousand ducats, and in 1596 were able to return to Goa. Montserrat never recovered from the illness during his captivity, but he had time to write the final version of the narratives of his journeys.

An extraordinary saga was that of the Portuguese Jesuit brother Bento de Goes (1562–1607). Bento had arrived as a soldier in India and in 1584 joined the Jesuits in Goa. His first journey was in 1595 to Lahore and Agra, as a companion of Jerónimo de Ezpeleta y Goñi (1549–1617), known as Jerome Xavier, to the Court of the emperor Akbar. There he remained until 1601 and learned the Persian language and Muslim customs. Ricci had sent letters to Goa asking for a Jesuit to try to travel from Goa to China in order to investigate the existence between these two countries of the kingdom of Cathay, about which Marco Polo had written. Bento volunteered for this mission and in 1602 left Goa for Agra and Lahore. From Lahore, he dressed as an Armenian merchant and took the name of Abdulla Isai, travelling with a caravan to Yarkand (Soch'e), in the region of Uyghur (north of Pakistan) and from there to Kabul (Afghanistan). He took as companions two Greeks and an Armenian, named Isaac, the only one who completed the journey with him. From Kabul Bento crossed the Hindu-Kush toward the north through the Pamir Mountains reaching Yarkand, where he remained for a year, in 1603. In November 1604, he joined another caravan which was going to China. In Korla (Mongolia) he met some travelers from China who informed him about the Jesuits in Beijing. With great difficulty Bento reached the northeastern part of the Great Wall of China and finally the city of Suzhou in the Sinkiang region. There he failed to obtain permission to continue to Beijing. From Suzhou Bento sent letters to Ricci, who finally received one of them and sent a Chinese Jesuit brother with an Italian-Portuguese name, Giovanni Fernandes, to meet him. In Suzhou Fernandes met Bento, now seriously ill, who died 11 days later. It seems that he had been poisoned by some of his fellow travelers who owed him money and who also destroyed part of his diary. Only the Armenian Isaac had remained with him. Fernandes was able to listen to the story of his journey and collect as much as he could of Bento's diary. Then after his death he returned to Beijing with Isaac. Ricci reconstructed the story of the journey from the narratives of Isaac and Fernandes and from the parts of Bento's diary that had been preserved. Thus, Ricci was convinced that "Cathay" did not exist, but was really the name given by Marco Polo to China, and that its capital city, Cambaluc, was Beijing.[22]

The long sea route from Lisbon to Macao was full of dangers: shipwrecks, sickness and attacks by corsairs and pirates. Only about half of the missionaries who left Lisbon arrived in China. Schall thought of the possibility of establishing a route by land and he sent Gruber and Albert Dorville (1621–1662) with this mission. In 1661 they left Beijing toward the southwest crossing the province of Qinghai and arrived in Lhasa in Tibet. From Lhasa they went westwards to Katmandu, Nepal where they followed the course of the Ganges River to Patna and then to Agra, where the Jesuits had a mission at the Court of the Great Mogul. It had taken them 17 months. Dorville exhausted from the journey remained in Agra, where he died a short time later. Heinrich Roth (1620–1668) took over from him and accompanied Gruber for the second part of the journey. Gruber and Roth continued to the west crossing Persia and Turkey, and finally arrived in Rome in 1664, three years after Gruber's depart from Beijing. In Rome Gruber told the story of his journey to Kircher, and gave him notes and maps that the latter used in his work *China illustrata*.

The Jesuits in the north of India had heard that further north, behind the high mountains there was a region where some Christians lived. Montserrat in his map of India had noted this. The search for these supposed Christians moved the Jesuits to set out for Tibet. They were the first Europeans to visit this secluded kingdom, after the visit in the fourteenth century of the Italian, Franciscan Oderico de Pordenone, sent as ambassador of the Pope to the Great Khan during the Mongol rule of China. In 1624 the Portuguese Antonio de Andrade (1580–1634), who had arrived in India in 1600, departed from Delhi with brother Manuel Marqués (1596–1642) on the long journey to Tibet. Dressed in Indian clothes they joined a group of Hindu pilgrims going to the sanctuary of Bradinah near the Himalayas. From there, Andrade and Marqués traveled to Srinagar (Kashmir), at that time the capital of the Garval kingdom, then followed the course of the Ganges River to the north. They crossed the Mana pass, 5,600 m high, and after a very difficult journey through the highest mountains of the Earth, overcoming snow and high altitude sickness, they arrived in Tsaparang, the capital of Guge, in the west of Tibet (today only ruins remain). In Tsaparang, the King welcomed Andrade and Marqués and showed his admiration for the long journey they had made. Andrade returned to India, after promising the King that he would return next year. He returned in 1625 and with the King's permission he established a stable mission in Tsaparang, with a small chapel dedicated to Our Lady of Hope. In 1629 Andrade returned to Goa, but he died of poison in 1634 while he was preparing another journey to Tibet. Marqués, who had returned to India, made another journey to Tibet in 1640, where he was made prisoner

[22] C. Wessels, *Early Jesuit travellers in Central Asia, 1603–1721*. Delhi: Asian Educational Service (1997), 1–42.

and was never heard of again. Andrade wrote three accounts of his journeys to Tibet, the first published in 1626 in Portugal, *Novo descobrimento do Gran Cathayo ou Reinos de Tibet* (New discovery of the Great Cathay or the Kingdom of Tibet).[23] The first news about Tibet arrived in Europe through these narratives. Andrade described the difficulties of the journeys, the nature of the land, the customs of the Tibetan, and his contacts and discussions with the Lamas. With Andrade's stories, the mysterious world of Tibet had reached Europe.

In 1631 the Portuguese Francisco Acevedo (1578–1660) set out from Agra for Tibet following the same route as that taken by Andrade through Srinagar to Tsaparang. Before his arrival the King, who had been so favorable to the Jesuits, had been defeated by the King of Ladakh, who incorporated those lands in his kingdom. Acevedo decided to go to Leh, the capital city of Ladakh (Indian Kashmir), to meet and bring gifts to the King. In Leh the King gave Acevedo permission to remain in the kingdom and to preach the Christian religion. Acevedo returned to Agra taking a new route going directly to the south crossing the passes of Bara Lacha (5,600 m high) and Rothang (3,900 m high). He arrived in Nagar (north of Pakistan), the capital city of the kingdom of Culu. He remained there for some time, then crossing the kingdom of Madi arrived in Bainara, a small kingdom which belong to the Great Mogol. From there he set forth to the south by what was called the "royal road" to Delhi and Agra, where he arrived in January 1632. Acevedo described his journeys in a long letter entitled "From Agra to Tibet".[24] In 1626 João Cabral (1599–1669), Estevão Cacela (1585–1630) and the Jesuit brother Bartolomeu Fonteboa traveled to Tibet from Dacca (Bangladesh), searching for the legendary kingdom of Cathay, which they still thought existed. They followed the courses of the Ganges and Brahmaputra Rivers and crossed the Himalayas at Butan, arriving in the city of Shigatse in central Tibet, capital of the kingdom of Utsang. There the King and the great Lamas welcomed them. They returned through Nepal, crossing by Katmandu, to arrive 2 years later in Dacca. In 1629 Cabral, Cacela and Fonteboa returned to Tibet and Cacela died in Shitgatse the following year. In 1631 Cabral returned to Tibet for a third time. Both Cacela and Cabral left narratives of their journeys.[25]

The existence of a Christian kingdom somewhere in Africa or Asia, ruled by the legendary priest-king, known as "Prester John", was already known in Europe during the Middle Ages. The Portuguese identified him with the Negus or Emperor of Ethiopia and made the first contacts in 1520, at the time of King Manuel I and the Emperor of Ethiopia, Lebna Degal. In 1546 Pope Julius III and the King of Portugal John III asked Ignatius of Loyola for the collaboration of the Jesuits to bring about the union of the Coptic Church with Rome. In 1554 on their first visit to Ethiopia a group of Jesuits founded a mission in Fremona (Maiguagua) in the region of Tigray, in the north of Ethiopia near Lake Tana. For the project three Jesuits were

---

[23] Hugues Didier (translation and presentation) *Les portugais au Tibet. Les premières relations jésuites (1624–1635)*. Paris: Chandeigne (2002), 29–139; Wessels, *Early Jesuit travellers*, 43–93.

[24] Didier, *Les portugais au Tibet*, 141–209.

[25] Didier, *Les portugais au Tibet*, 213–257.

ordained bishops, but only Andrés de Oviedo (1518–1577) arrived in Ethiopia from Goa (1557) and settled down in Fremona, where he died. This first phase ended in 1596. In 1603 Páez, who was recovered from the hardships of his first attempt and 7 years of captivity (1589–1596), tried again to reach Ethiopia.[26] He sailed from Goa and reached Massawa (Eritrera) through the strait of Aden; from there he went to Fremona. Páez had a great facility for languages and soon gained the favor of the Emperor Susinios (Melek Segued), who retained him at Court and took him on his military campaigns. Páez was also a good architect and built some churches. The Emperor entrusted him with the building of a new palace, the ruins of which can still be seen today.

The source of the Nile River had been a mystery for a long time. Few people know that Páez was the first European to find it. In the company of a group of soldiers of the Emperor going south near Mount Gish, he discovered (21 April 1613) the source of the Blue Nile and explored its course (the Nile has in its southern part two branches, the Blue and White Nile). He wrote: "I discovered two sources each of four palms and I saw with pleasure what so many had desired to see, what neither Cyrus the King of Persia, nor Cambyses, nor Alexander the Great, nor the famous Julius Caesar could discover". Páez wrote a story of his journeys and discovery, *Historia de Etiopía* (History of Ethiopia), which was not published until 1905, but in his work *Oedipus Aegyptiacus* Kircher did mention the discovery of the source of the Nile by Páez. On two occasions Páez obtained from the Emperor the union with Rome, but it did not last. Convinced of the uselessness of his work Páez left the Court and retired to Fremona where he died.

The Portuguese Jerónimo Lobo (1595–1678) traveled through much of Ethiopia. In 1624 on his first journey he tried a new southern route, sailing from Goa to Malindi (Kenya), but he could not find guides to take him to Ethiopia by land and had to return. The next year he succeeded using the northern route crossing the Danakil desert. After Páez, Lobo was the second European to explore the source of the River Nile. He wrote about his discovery: "to the east of the kingdom of Goiania, in the slope of a mountain, the sources of the Nile are to be found, which had been searched for with so much labor. Much has been written about them, with all kinds of speculations but without success". Lobo described his journeys in Ethiopia in his *Breve relação do Rio Nilo de sua fonte e corrente* (Brief story about the Nile River its source and course). The original Portuguese text was only published in 1971, but an English translation, *A short relation of the Nile,* was published in 1669, and later Italian, French, German and Flemish translations appeared. In Europe Lobo's book was the main source of knowledge about Ethiopia. Lobo also corresponded with the Royal Society of London about Ethiopia. In 1634 the Emperor Fasilidas expelled all the Jesuits from Ethiopia and Lobo returned to Goa. He was ordered to return to Portugal to report about the mission of Ethiopia, but Lobo's adventures

[26] George D. Bishop, *A Lion to Judah. The travels and adventures of Pedro Páez S. J,*. Anand (Gujarat): Gujarat Sahitya Prakash (1998); Charles Libois, Pedro Páez, descubridor de las fuentes del Nilo. In *Exploradores españoles olvidados de África*. Madrid: Pedro Páramo (2001), 89–112; Javier Reverte, *Dios, el Diablo y la aventura*. Barcelona: Plaza Janes (2001).

were not finished. His ship was captured by Dutch corsairs off the Brazilian coast and he was left on a desert island. He finally escaped and arrived in Lisbon 2 years later. In 1640 Lobo returned to India where he worked until 1656.

5.6 Exploring the New Lands of America

The work of Jesuit naturalists and geographers in the newly discovered lands of America has already been discussed, and as noted earlier, it is difficult to separate this from the work of Jesuit explorers, since it was on their journeys that they collected information about the geography and natural conditions of the various territories. To give an overall view of Jesuit exploration in America, it is necessary to return to some of the persons already mentioned. In 1573–1574, on his first journey, Acosta traveled from Lima crossing a great part of southern Peru to Cuzco and from there to Arequipa. Then he crossed the high plateau to Lake Titicaca and arrived in La Paz, from where he continued south to the region of the Pilcomayo River in the south of Bolivia. Heading north he returned to Lima passing through Potosí. On this journey he was one of the first Europeans to cross the Andes and to experience high altitude sickness, which he described in his writings. In 1639, Acuña, as we saw, traveled from Quito down the Amazon River to its mouth in Brazil, and in 1730 Gumilla explored the regions along the Orinoco River in what is today Venezuela.

Many Jesuits, who worked at the famous Indian "reductions", explored the regions surrounding the Paraguay River. Cardiel, for example, traveled several times through all the missions. In 1762–1765, Quiroga, Cardiel, and Strobel explored the southern regions of Patagonia. About the same period Jolis explored the region of the Chaco. The need for a land route from Buenos Aires (Argentina) and Asunción (Paraguay) to Potosí (Bolivia) and Lima (Perú) was felt to be important. Already in 1560 a long and tiring route was established around the Chaco at its southern border. Several Jesuits, like Bartolomé Jiménez in 1703, tried to find a direct way through the Chaco, following the basins of the Paraguay and Pilcomayo Rivers, but without success. In 1715 José Francisco de Arce (1651–1715) and Bartolomé de Blende (1675–1715) with a group of guarani Indians tried to reach the Chiquitos Mission in Bolivia from Asunción, but a group of payagua Indians murdered them when they were half way there. In 1720 further attempts were made with no success. Finally in 1766, a year before the expulsion of the Jesuits, Sánchez Labrador succeeded in finding a direct route from Asunción to Santa Cruz de la Sierra in Bolivia.

The Italian Eusebio Francisco Kino (1645–1711) stands out for his explorations in North America.[27] In 1681 he had arrived in Veracruz (Mexico), from where he

[27] Ernest J. Burrus, *Kino and Monje. Explorers of Sonora and Arizona. Their vision of the future. A study of their expeditions and Plans*. Saint Louis: Jesuit Historical Institute (1971); Charles Polzer, *Kino, a legacy: his life, his works, his missions, his monuments (A Kino Guide II)*. Tucson: Jesuit Fathers of Southern Arizona (1998).

began his journeys toward the north. The first in 1683–1685 to California was part of an official mission which failed to give the hoped-for results. He then decided to act by himself beginning a program of more than 40 expeditions, between 1687 and 1711, to the regions of Sonora, Arizona and California. In his expeditions, Kino, established missions, founded schools and carried with him cattle and plants. At the same time he drew up maps of the new regions. Maps were for him an important part of his work as a missionary and explorer. Kino was the first to discover that Baja California is a peninsula and not an island, as was believed at the time. He explored and drew maps of the region following the courses of the Gila, Colorado, Santa Cruz and San Pedro Rivers. In 1965 in recognition of Kino's work in the United States a statue was dedicated to him in Washington in the Capitol and two others in the State of Arizona at Tucson and Phoenix. The Croat Fernando Consag (1703–1759), who had arrived in Mexico in 1731, continued Kino's work. He settled in Baja California and from there in 1746 explored the region to the north to the mouth of the Colorado River; then in 1751 and 1753 he went on to the coast of the Pacific Ocean. Consag's maps, covering the northern part of Baja California, were copied and reproduced many times.

French Jesuits arrived to Canada in 1625 and they were the first missionaries in what was called the New France. They dedicated themselves to work with the Huron and Algonquin Indians. The wars between the Huron and Iroquois between 1630 and 1660 ended with the practical annihilation of the former, and this also affected the Jesuit missionaries with the martyrdom of Jean de Brébeuf (1594–1649) and his companions. In 1665 the Jesuits extended their missionary work to the west and southwest with the explorations of those regions.[28] Jacques Marquette (1637–1675), who had arrived in Quebec in 1666, began to explore and draw up maps of the region of Lake Superior. His curiosity was aroused by tales from the Illinois Indians of a great river which runs to the south. In 1673 he, together with the French explorer Louis Jolliet and some Indians, boarded two canoes and began their journey at Green Bay on Lake Michigan and went down the Fox River to Lake Winnebago. From there they descended by the Fox and Wisconsin Rivers until they reached the great Mississippi River at what is today the town, Praire du Chien (Wisconsin), where the Jesuits now have a boarding school. They continued down the river to the point near which it is joined by the Arkansas River. They were sure that the Mississippi went on southwards to the Gulf of Mexico. Thus Marquette and Jolliet were the first European to explore the Mississippi. They returned upstream and spent the winter of 1674 at a place where the city of Chicago is located today. Marquette died during his return journey on the shore of Lake Michigan, near today's city of Ludington, where he was buried. The story of his journeys with the first map of the Mississippi River was published later in 1855, *Récit des voyages et découvertes de R. P. Jacques Marquette en l'année 1673* (Story of the journeys and discoveries of the Rev. Jacques Marquette in the year 1673).

[28] Elisabeth Jones, *Gentlemen and Jesuits: Quest for glory and adventure in the early days of New France*. Toronto: University of Toronto Press (1992).

Claude Allouez (1622–1689), arrived to Canada in 1658 and was a companion of Marquette. In 1676 he drew a detailed map of Lake Superior and explored the region of the lakes; from Green Bay he entered the region of Illinois. Charles Albanel (1614–1696) arrived in Quebec in 1645 and explored the regions to the northwest. In 1671 he left Quebec by the Saint Lawrence River and traveled to Tadoussac. From there, following the Saguenay River, he arrived at Lake Saint Jean, went west to Lake Mistassini and Lake Nemiscau, and finally going down the Rupert River he arrived in James Bay, part of the great Hudson Bay, from where he returned to Quebec. In 1674 on one of his journeys he was taken prisoner by the British troops and sent to England. Once released, he crossed to France and in 1688 returned to Canada. Lake Albanel near Lake Montessini is named after him. In the 1635 Jesuits founded a college in Quebec, where a chair of hydrography was established in 1708. Joseph Pierre de Bonnécamps (1707–1790) occupied the chair between 1741 and 1759 and carried out cartographic work of the region surrounding Quebec with very exact particulars of its geographic longitude. The French astronomer Jean Nicolas Delisle presented this information at the *Académie des Sciences* of Paris. These are just a few examples of the many Jesuits who entered unknown lands, rivers and lakes, from Canada to Patagonia, well before any other European, moved by their desire to bring the Christian message.

Chapter 6
The New Observatories

6.1 A New Beginning

The expulsion of the Jesuits (from Portugal in 1759, from France in 1764, and from Spain and its American colonies in 1767), and the suppression of the Society of Jesus by Pope Clement XIV in 1773, put an end to their scientific work. Suddenly the activity of so many Jesuit scientists, in colleges, universities and observatories in Europe, America and Asia, was cut short. Ironically, the suppression of the order coincided with a moment of special development in their scientific work, with the creation of observatories in many colleges and universities and the introduction into their teaching of Newtonian physics and astronomy, along with modern mathematics and geography. But now the end has come to what, as we have seen, had been the dedicated efforts over many years of many Jesuit scientists. Some of them, it is true, continued their work after the suppression in universities and observatories that were now state institutions. However, these were only occasional personal survivals from a tragic shipwreck marking the end of an age.

In 1814 Pope Pius VII restored the Society of Jesus and a new chapter in the history of the Jesuit contribution to the natural sciences began. During the almost 50 years that the Society was suppressed, important steps had been taken in the development of science and in society in general. This new situation conditioned the Jesuits' scientific work in this period, so that it has a different character from that of the previous one. Science and technology had rapidly established the foundations for the industrial revolution, which transformed modern society. In almost all fields of science (e.g. in physics and chemistry), both fundamental theoretical developments and new practical applications were to be found. In the past the Society of Jesus had been present at the birth and first steps of modern science. Now Jesuit scientists found a developed level of science which was more difficult for them to integrate. This may explain why Jesuits were more to be found in sciences like meteorology and seismology, which were then beginning. Universities in Europe, which had suffered a general decline during the eighteenth century, were in a process of reform,

A. Udías, *Jesuit Contribution to Science: A History*,
DOI 10.1007/978-3-319-08365-0_6

beginning in Germany and France. In this process, most universities put the emphasis on scientific research and became in great part state institutions. Theology had lost its importance at universities and in some cases was completely suppressed, being replaced by the natural sciences. The Society had lost more than 600 schools and universities in Europe, and had to begin its educational work in a very different situation. Education had become a state function, with imposed norms and study programs. Most new Jesuit schools were involved with what is called secondary education and, except for North America, only in a few cases reached university rank. However, in spite of all these difficulties, the Jesuits did not forget their scientific tradition and, taking into account the new situation, tried to extend their glorious past.

Giovanni Antonio Grassi (1775–1849) presents a curious example which shows the difficulty of establishing a link between the scientific work of the old and that of the restored Society and how impossible it was for Jesuits to restart what had been lost with the suppression. Grassi had joined the group of Jesuits in Belarus, where Catherine II had refused to promulgate the decree of suppression. Former Jesuits in China wrote to the Jesuits in Belarus asking for some members to come to China to continue the astronomical work in the imperial observatory of Beijing. They thought it was possible to continue the previous work. Grassi with two companions tried to reach Beijing from Russia. They traveled first to Portugal and then to England, but they had to abandon the project. Grassi eventually traveled to North America, where he became professor of mathematics and later rector of Georgetown University in Washington. He realized that it was not possible to take up the old work and that something new had to be started. Georgetown University had been founded in 1789 by John Carroll (1738–1815), first bishop of Baltimore and a former Jesuit, and was managed by former Jesuits. After the restoration of the Society it became the first Jesuit University in North America and it was here that one of the first new Jesuit astronomical observatories was established.

In 1814, after the restoration of the Society, the Jesuits faced a difficult situation, especially in Europe, with continuous economic, social, political and religious revolutions taking place. For some years Jesuits were expelled from countries like Germany, France and Spain. But soon they took up again their work in two main fields, namely, schools and foreign missions. This would condition their future scientific work, centered on the colleges and the missions. The Superior General Jan Roothaan (1785–1853), who ruled the Society from 1829 to 1853, began the revision of the *Ratio Studiorum*, which needed to be adapted to the new times.[1] An important point was the need to increase the role of mathematics and natural science. However, the revision soon proved to be an impossible task. Educational programs at universities and secondary schools were established by governments and varied greatly from one country to another, making it impossible to impose a unique model. The need to adapt Jesuit schools to legal educational norms, with different types of programs in different countries, resulted in the old *Ratio* losing its practical relevance. The teaching program established in the *Ratio* was only followed in a modified form in the

[1] See "Educación", *DHCJ*, 1202–1214.

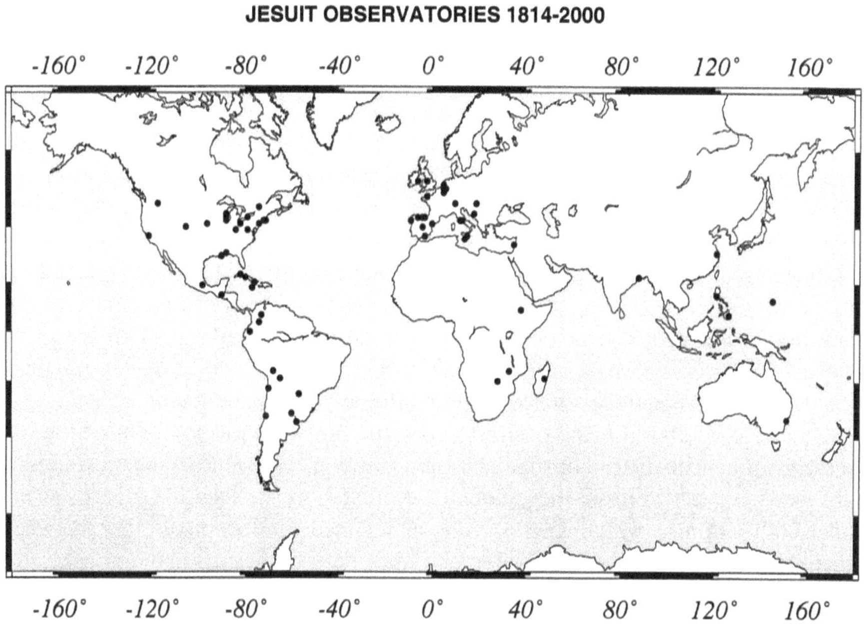

Fig. 6.1 Jesuit observatories (1814–2000) (Udías 2003)

training of Jesuits, completely separated now from lay students. This program established a first level of humanist studies based on the Latin and Greek classics, a second of scholastic philosophy, where some scientific subjects were included, and a third of theology. With regard to the teaching of science, a distinction was made between schools for lay students, generally in secondary schools and in some cases at university level, and the centers for the philosophical training of Jesuit students. In the latter the level of science teaching varied from one country to another. It included lectures on mathematics, astronomy, physics, chemistry, biology and geology, but their importance in the programs diminished with time.

The most remarkable feature showing Jesuit dedication to science in this new period was, without doubt, the establishment of observatories world-wide, a total of 74 between 1814 and 1970 (Fig. 6.1).[2] This is an impressive enterprise that has failed to receive adequate recognition by the historians of science. The first observatories were astronomical and later diversified into meteorological, geomagnetic and seismological. Their geographical distribution was as follows: 26 in Europe, 21 in North America, 18 in Central and South America and 9 in Asia, Africa and Australia. They were of different types, some being for a time internationally recognized research

[2] A general information on the observatories can be found in: Pierre de Vregille, Les observatoires de la Compagnie de Jésus au debut du XXe siècle. *Rev. des Questions Scientifiques*, 59 (1906), 10–72, 493–579; Agustín Udías, *Searching the Heavens and the Earth: The History of Jesuit Observatories*. Dordrecht: Kluwer (2003).

centers in astronomy, meteorology or seismology, while others were small meteorological or seismological stations. The establishment of observatories, in spite of their relatively large number, did not result from an explicit program from the central Jesuit government in Rome, but was the result of local and individual initiatives. The attitude of the Superior Generals in Rome varied – from that of Roothaan, who was opposed to the establishment of the observatory at Georgetown University, to that of Luís Martín (1846–1906), who encouraged the work of Jesuits in observatories. There is only one official document, that of 1934, which recommended establishing observatories and laboratories in the faculties of philosophy, as a help for teaching science to Jesuit students. The first observatories were astronomical, but they also made geophysical observations. As costs increased for more advanced telescopes, observatories focused on other observations, such as meteorological, seismological and geomagnetic, with more affordable instrumentation.

Other factors, besides the desire to continue the scientific tradition of the old Society, influenced the establishment of new observatories. One of the most important was the wish to resist the rationalist current, spread since the middle of the nineteenth century, which aggressively maintained an incompatibility between science and the Christian faith. The scientific work of the Jesuits in their observatories was presented as a practical refutation of such ideas. The new situation gave the new Jesuit observatories a certain apologetic character in defense of the Christian religion, which had not been present in the previous period. They were presented as a clear proof that Christian religion and the Catholic Church, in particular, were not against scientific progress. The scientific standing of some Jesuit observatories showed that one could be at the same time a religious and a scientist and that science does not lead necessarily to atheism. The tendency to attack religion using science was especially widespread at the middle of the nineteenth and beginning of the twentieth centuries. It coincided, as we will see, with the greatest development of Jesuit observatories. This situation began to change after the Second World War. More critical positions regarding science replaced its standing as the only valid vision of the world and attacks on religion lost much of their force. New attitudes in the Church, which took their origin from the Vatican II Council, have changed the apologetic stance. Instead there is recognition of the mutual autonomy of science and religion and the need for a dialogue between them. The modern practice of science has also changed with its concentration in great institutions with large economic resources. This situation made ownership by the Jesuits themselves unfeasible. This affected in the first place the astronomical sections of such observatories, and later also those concerned with geomagnetism, meteorology and seismology. In the last two fields the development of national services of meteorology and seismology made redundant and unnecessary the work of Jesuit observatories. The process of independence of the new nations of Africa and Asia has affected the role of observatories in mission countries, often linked formerly with the colonial administration and now transferred to state institutions. Changes in the priorities of the Society itself, after the General Congregation 32 (1974–75), which emphasized social work and the pursuit of faith and justice, questioned the continuity of the observatories. In some cases Jesuit scientists looked for a different contribution to science outside their own institutions. The decrease in Jesuit vocations in Europe and North America also had an influence and reduced the role of Jesuits in the natural sciences. Thus, due to a number of causes, external and internal, Jesuit observatories entered a crisis between 1970 and 1980 leading to the closure of most of them. At present only four or five continue active, but with very limited Jesuit presence, except for the special case of the Vatican Observatory. The golden years of Jesuit observatories can be placed between 1860 and 1960, as we will see in the rest of this chapter. During those years Jesuits without an explicit general program established by the central authorities ran observatories throughout the world, moved by a variety of motives linked to their apostolic work. A complete study of this interesting Jesuit scientific work has not yet been done.

6.2 Returning to Astronomy

As we have seen in the previous chapters, since the first years of the Society astronomy had been one of the preferred fields for Jesuit scientists. One need only recall the names of Clavius, Scheiner and Riccioli in Europe, and those of Ricci, Schall and Verbiest in China. It is not surprising that, maintaining the tradition of the old Society, Jesuits interested in science turned once again to astronomy. Thus, only a few years after the restoration of the Society, Jesuits began to plan the establishment of observatories in their colleges and faculties of philosophy. The Jesuit interest in astronomy and geophysics in this new period is manifested by the names of 31 Jesuits who became members of the Royal Astronomical Society of London between 1841 and 1977; three were "associate members" and 19 were recorded in obituaries in the publications of this society.[3]

The first of the new observatories was installed at the Roman College (Fig. 6.2). An observatory had been established there in 1787 when the college was managed by the secular clergy and its first director was Giuseppe Calandrelli. In 1824 the college was returned to the Jesuits and the French Jesuit Dominique Dumouchel (1773–1839), who had studied science at the *École Politechnique* of Paris, took charge as director of the observatory. It was then one of ten observatories in Italy. Dumouchel installed a new Cauchoix equatorial telescope, an Ertel meridian circle and other instruments. He was one of the first to forecast in 1835 the return of the comet Halley. In 1840 Francesco de Vico (1805–1848), who had collaborated with Dumouchel in the determination of the orbit of the comet Halley, succeeded him. Vico gave a great impulse to the observatory and began the compilation of a new catalog of stars, observations of the satellites of Saturn, and of eight comets between 1844 and 1847. With Vico, between 1843 and 1847 Benedetto Sestini (1816–1890) made a catalog of the color of 2,450 stars, separating them into white, blue, yellow, orange and red stars. This was a pioneer work of this type, together with those of John Herschel and Wilhem Struve. In 1848, the Jesuits were expelled from Rome

[3] Agustín Udías, Serving God and Science, *Astronomy and Geophysics*, 42 (2001), 23–24.

Fig. 6.2 Interior of the dome and Merz telescope of the Observatory of the Collegio Romano (1854) (Udías 2003)

after the revolution which installed the short-lived *Republica Romana*. Vico and Sestini moved to Washington, where Vico took charge of the recently installed observatory of Georgetown University. However, Vico died that same year in London, where he had gone to buy some instruments. A year later the Jesuits returned to Rome. Pietro Angelo Secchi, who had collaborated with Vico at Georgetown Observatory, was appointed director of the observatory of the Roman College. Secchi is one of the great Jesuit scientists of this new period, as we will see in Chap. 9. In 1854 the observatory was moved to a new site on the roof of the Church of Saint Ignatius according to the old project of Boscovich.[4] It rested on top of the four massive columns which supported the false dome of the church. Three telescopic domes were installed on three columns: the main one for the Merz equatorial telescope of 24 cm diameter and the other two domes for the Cauchoix telescope and the Ertel meridian circle. On the forth column were the magnetic instruments. The new observatory counted with the support of the Jesuit general superior Pieter Beckx (1745–1887) and the generosity of Pope Pius IX who awarded

[4]Angelo Secchi, *Descrizione del nuovo Osservatorio del Collegio Romano.* Rome: Tipografia delle Belle Arti (1856).

the observatory the title of "Pontifical" and entrusted it with the keeping of official time for the Papal States. Each day at noon a ball would fall from a mast on the roof of the church of Saint Ignatius marking the exact time. Work at the observatory did not last long; in 1870 the troops of King Vittorio Emanuelle entered Rome, the Jesuits were expelled, and the Roman College and its observatory confiscated by the state. However, Secchi, thanks to his international prestige, was allowed to continue as director until his death. He was succeeded by Pietro Tacchini, an astronomer who had worked at the observatories of Modena and Palermo and collaborated with Secchi. Such a compulsory expropriation (in this case of this first modern Jesuit observatory) by a government moved by anti-religious and anti-Jesuit prejudices, belongs to a pattern that will be repeated in other cases, as we will see.

The second Jesuit observatory of this period was established in 1838 at Stonyhurst College, Lancashire (England).[5] This college had a long history; it was founded in 1593 with the support of King Phillip II of Spain in St. Omer, France, for the education of young English Catholics, and was moved to England in 1794, at the time of the French Revolution. In 1843 faculties of philosophy and theology for the training of Jesuits were established nearby at St. Mary's Hall. The observatory began with meteorological observations and in 1845 and 1848 the first Jones equatorial telescope was installed. In 1846 Alfred Weld (1823–1890) was appointed director; he began the real scientific work with a program of astronomical, meteorological and geomagnetic observations. In 1860 Stephen J. Perry was appointed director and he, together with Secchi, is another of the great Jesuit scientists of this period, as we will see also in Chap. 9. With Perry as Director, the observatory had its most productive years and served as the place for the training of many Jesuits who became directors of other observatories in different parts of the world. Perry's main work was research on the relation between solar activity and the magnetic field of the Earth. After his death this work was continued by his collaborator and successor Walter Sidgreaves (1837–1919), who also introduced research on solar and stellar spectroscopy. These lines of research were continued by Aloysius Cortie (1859–1925) and Edward O'Connor (1874–1954) until 1932. Cortie carried out studies of solar spectroscopy during his observations of the solar eclipses of 1905 (Vinaroz, Spain), 1911 (Tonga Island) and 1914 (Sweden). He also studied the relation between sunspots and variations of the terrestrial magnetic field. In 1947, after the Second World War, the observatory was closed due to financial problems. In 1957, for the International Geophysical Year, there was an attempt to reopen the observatory for magnetic observations, but the project did not last long and in 1974 it was finally closed. At present, students of the college and amateur astronomers use the telescope of the observatory.

The most important Jesuit astronomical observatory in North America was that of Georgetown University.[6] The observatory was inaugurated in 1844 in a building

[5] Stephen J. Perry. L'Observatoire de Stonyhurst. *Annales de la Soc. Scient. de Bruxelles*, 4 (1880), 147, 281–302.

[6] Francis J. Heyden, *The beginning and end of a Jesuit observatory (1841–1972)*. Manila: Publ. of Manila Observatory (1975).

expressly built for this purpose with a large dome where in 1849 an equatorial telescope of 12 cm diameter from Troughton and Simms was installed. The observatory was the fourth to be set up in the United States only 6 years after the first, the Hopkins Observatory in Williamstown. The first director was James Curley (1796–1889). Vico was also for a short while the director, and Secchi collaborated in the work of the observatory, when both were expelled from Rome. Johann Georg Hagen (1847–1930) was director between 1888 and 1906, during which time he continued the research on variable stars that he had begun in a small observatory installed by him in the Jesuit school at Praire du Chien (Wisconsin). In 1899 Hagen began the publication of *Atlas stellarum variabilium* (Atlas of variable stars), and he continued this after moving in 1906 as director to the Vatican Observatory. He also began the publication in Berlin of the four volumes of the mathematical work *Sypnosis der hoeheren Mathematik,* that he completed in 1930. Paul A. McNally (1890–1955) was director between 1927 and 1948. He took part in several expeditions to observe solar eclipses, for example to Siberia in 1936 and to the South Pacific in 1937. Francis J. Heyden (1907–1991) was the last director. He tried to give new life to the observatory with a graduate program in astronomy. With the collaboration of his students he began a series of research projects on solar spectroscopy, radio astronomy and the study of planetary atmospheres. Heyden participated in expeditions to observe solar eclipses, in 1947 in Brazil, and in 1948 in China. In spite of these promising projects, the university decided in 1972 to close both the department of astronomy and the observatory. Heyden moved to the Philippines, where he continued his scientific work in Manila Observatory for 20 more years.

In 1879 the Haynald Observatory was inaugurated in the Jesuit college of Kalocsa, Hungary.[7] Cardinal Lajos Haynald, Archbishop of Kalocsa, who had a great interest in astronomy, was behind its founding. Carl Braun (1831–1907), who had studied astronomy in Paris, was its first director. He installed a Merz equatorial telescope of 18 cm diameter and a Hilger spectroscope for the study of the Sun following in Secchi's footsteps. His successor Gyula Fenyi (1845–1927), director from 1885 to 1913, focused his investigation on the study of the Sun. His role in Kalocsa was the same as that of Perry in Stonyhurst. Fenyi, a tireless observer, made about 40,000 solar observations of which 6,000 were published. He centered his studies on the nature of sunspots and solar prominences and on fulgurations and their relation with magnetic storms, about which he published (between 1886 and 1925) 145 scientific articles in Hungarian, German, French and Italian. At that time, the nature of these solar phenomena was still a debated question. Fenyi proposed that their origin was due to movement of the material in the solar chromosphere with currents of incandescent gasses passing in different directions. The observatory continued working after the Second World War until in 1950 the communist government confiscated it together with the college.

The observatory of Valkenburg (Holland) was established as part of the science program taught to the Jesuit students in the faculty of philosophy.[8] German Jesuits

[7] Julius Fényi, Die Sternwarte in Kalocsa, das Haynald observatorium. *Nachr. der öster-ungarische Provinze*, 3, 14–23 (1906).

[8] Udías, *Searching*, 94–96.

had founded this faculty in 1872 after being expelled from Germany by the Bismarck government. The observatory, inaugurated in 1896, was used for both teaching and research. It was situated on a 30 m tower with a dome of 5 m diameter housing a Saegmüller and Clasey equatorial telescope of 23 cm diameter, along with some other minor telescopes and astronomical instruments. The main research was in the study of variable stars in collaboration with Hagen. Its most active directors were Joseph Hisgen (1868–1910), Alfred Bauer (1872–1912) and Michael Esch (1869–1938). Observations were published in the seven volumes of *Veröffentlichungen der Sternwarte des Ignatiuskollegs Valkenburg* (Publications of the Observatory of Saint Ignatius College, Valkenburg). In the last years, until 1938, Esch worked alone on the observation of variable stars. In 1941 German troops invading Holland occupied the observatory and the college. Nothing is known of what happened to the astronomical instruments. After the war, the Jesuit faculty moved to Frankfurt, but astronomy was no longer taught.

The Observatory of Cartuja was founded in 1902 in the Faculty of Philosophy and Theology of Granada, Spain, with three sections for astronomy, meteorology and geodynamics.[9] It housed a large dome of 8 m diameter resting on a circular tower which occupied the center of the building. A Mailhat equatorial telescope of 32 cm diameter was installed together with other smaller telescopes and a photoheliograph for solar observations. Though its founder was Juan de la Cruz Granero (1849–1917), rector and professor of cosmology, the first director was Ramón Martínez (1847–1911), who from 1904 shared the direction with José Mier y Terán (1878–1942). The work on astronomy centered on the study of the Sun, but this was soon abandoned and the observatory focused its work on seismology, as we will see in Chap. 8. In 1967 the new director, Teodoro Vives, tried to revitalize the astronomical work with the installation of an observatory in Sierra Nevada at a height of 2,506 m with a Cassegrain telescope of 32 cm diameter given by the Georgetown Observatory which had closed. Vives established relations with several European observatories interested in the altitude and good climatic conditions of the new observatory. However, hopes for a renewal of Jesuit work in astronomy soon diminished when in 1971 Vives left both the observatory and the Jesuit order. In that year the observatory was transferred to the University of Granada.

Interest for the scientific training of Jesuit students inspired the foundation of three laboratories for physics, chemistry and biology in the Faculty of Philosophy in Roquetas (Tarragona, Spain). The laboratories of chemistry and biology began to function in 1904, but in 1914 moved to Barcelona and became schools for lay students. They joined with one another and continue today under the name *Instituto Químico de Sarriá*. The physics laboratory was changed into an observatory dedicated to the study of the relations between solar activity and terrestrial magnetism, on the initiative of Ricardo Cirera (1864–1932), who had worked in the Manila Observatory. The new observatory was inaugurated in 1904 as the *Observatorio del*

[9]Antonio Due Rojo, El cincuentenario del Observatorio de Cartuja. *Urania* 234, 67–81, (1953). *Observatorio de Cartuja, 1902–2002*, Granada: Caja de Granada (2002).

Ebro de Física Cósmica.[10] It began with the observation of the solar eclipse in Spain, 30 August 1905. We will see later the contribution of this observatory to the study of the Earth's magnetic field. Luís Rodés (1881–1939), director between 1920 and 1939, had studied astronomy and astrophysics in the United States at the universities of Harvard and Chicago, and had worked at the observatories of Yerkes and Mount Hamilton. Rodés began a line of research on solar and stellar spectroscopy, solar physics, and the relation of solar activity to variations in the Earth's magnetic field. He published a popular book of astronomy *El Firmamento* (The firmament, 1927), written in a scientific but accessible language, in which he presented some of the results of his research. At the end of the Spanish Civil War the observatory was badly plundered. Antonio Romañá (1900–1981), director between 1939 and 1970, reconstructed the observatory and began a new program of astronomical and geophysical observations. He was an important figure in the recovery of Spanish science in the post-war period. Jesuits continued to direct the observatory until 2000 when it passed into lay hands.

There were other European Jesuit faculties of philosophy which installed small astronomical observatories, equipped with a dome and a telescope, to provide demonstrations in lectures. In 1838 the *Collège Notre-Dame de la Paix* in Namur (Belgium) and in 1884 the University of Louvain installed such observatories. The same happened in the faculty of philosophy in Oña (Burgos, Spain) with a small dome on the roof of the main building. This observatory was abandoned when the Jesuits were expelled from Spain in 1931. The *Stella Matutina* College for lay students had a small observatory installed in 1889; it was used by astronomy professors till 1936. Similarly the diocesan seminary in Oudensbosch (Holland) had a small observatory installed in 1891. The Jesuits left the seminary in 1929 and the observatory was closed.

In the United States the Jesuits founded small observatories in some of the colleges and universities. Between 1880 and 1888 Hagen used one in the college of Prairie du Chien (Wisconsin) where he began to observe variable stars. In 1885 Joseph F. Rigge (1842–1913) and his brother William F. Rigge (1857–1927) installed an observatory in Creighton College (Omaha, Nebraska) with several telescopes and they collaborated with Hagen. In 1895 Jerome S. Ricard (1850–1930) installed a small observatory in the University of Santa Clara (Santa Clara, California) with sections devoted to meteorology and astronomy. In 1900 he installed two telescopes and in 1928 a new building was added with a large dome 15 m in diameter. Ricard thought of installing a large reflector telescope with a 1.5 m diameter mirror; it would had been one of the largest at that time. However, he was cheated by the manufacturer, who took the money and never delivered the mirror. In 1899 the Faculty of Philosophy of St Mary's, Kansas, installed a small observatory, but more important was the one installed at Marquette University, (Milwaukee, Wisconsin) with two telescopes, 15.2 cm and 13.3 cm in diameter.

---

[10] Ignacio Puig, *El observatorio del Ebro. Idea general sobre el mismo*. Tortosa: Imprenta Moderna del Ebro de Algueró y Baiges (1927); Manuel García Doncel and Antonio Roca, *El Observatorio del Ebro. Un siglo de historia (1904–2004)*. Roquetas: Observatorio del Ebro (2007).

James L. McGeary (1869–1945) was the first director; he made observations of sunspots, comets and planets. A third telescope was added in 1937, but interest for astronomy diminished, and the observatory, having been used mainly for teaching purposes, was closed in 1983.

6.3 Observatories in Mission Lands

Towards the end of the nineteenth and the beginning of the twentieth century, Jesuits also established observatories in mission lands. These had different characteristics from those in Europe and North America. Instead of having an apologetic aim, they tried to show the activity of the Church in scientific areas and to help with the scientific development of countries which lacked proper institutions. Observatories were, thus, an important help in missionary work. Quite often and for long periods, Jesuit observatories were the only scientific institutions that existed in many countries in Africa, Asia and Central and South America. They had the support of the colonial administrations and they provided the international scientific community with valuable information from places from which otherwise nothing was available. They also had a practical function, for example, providing the basis for time systems and also carrying out the first geodetic measurements necessary for drawing the first maps. Their meteorology and seismology departments, in countries prone to frequent hurricanes and earthquakes, provided invaluable service. Four such observatories, in Manila (Philippines), Zikawei (China), Tananarive (Madagascar) and Ksara (Lebanon), are considered to have been the great mission observatories.[11]

Manila Observatory was installed in the Jesuit school known as the *Ateneo de* Manila. It began in 1865 from very modest beginnings, making only meteorological observations.[12] This was one of the first observatories established in the Far East, before those of Jakarta and Tokyo. Federico Faura (1840–1897), who took charge of the observatory in 1867, was a key person in its beginnings and can be considered as its founder. He renewed the instrumentation and began astronomical observations with those of the solar eclipse of 1868. Faura traveled to Europe and visited the observatories of the Roman College and Stonyhurst. In 1879 he began his studies of tropical hurricanes (to be seen in Chap. 7). In 1897 José Algué (1856–1930) succeeded Faura and gave a new impulse to the observatory. He had studied astronomy at Georgetown University and worked in the observatory with Hagen. That year the astronomical section was enlarged with a new building that

[11] Pierre Lejay, Trois observatoires de Mission. *Études* 182 (1925), 594–602.

[12] Manuel Saderra-Masó, *Historia del Observatorio de Manila*. Manila: E. C. McCullough (1915); William C. Repetti, *The Manila Observatory*, Washington (1948); James J. Hennesey, The Manila Observatory. *Philippine Studies* 8 (1960), 99–120. Aitor Anduaga, Spanish Jesuits in the Philippines: Geophysical research and synergies between science, education and trade, 1865–1898. *Annals of Science* (2013) doi: 10.1080/00033790.2013.841289. 1–25.

had a rotating dome of 10 m diameter housing a Merz telescope of 48 cm diameter. The section of astronomy took charge of the time service used for a large part of the Far East, keeping time with an accuracy of one tenth of a second. Algué had to solve the problems that followed the occupation of the Philippines by the United States. In 1901 the North American administration recognized the scientific work of Jesuits in Manila and entrusted to the Observatory the direction of the new Philippines Weather Bureau. The observatory became the center of a network of meteorological stations. On this occasion the Observatory prepared two long publications about the natural conditions of the Philippines: *El archipiélago filipino* (The Philippine Archipelago) and *Atlas de Filipinas* (Atlas of Philippines), published by the government of the United States.

In 1906 Juan Comellas (1863–1942), who had studied with Hagen in Georgetown, took charge of the astronomy section of the observatory. Miguel Selga (1879–1956), who had studied astronomy at Harvard and worked in the Lick and Lowell observatories, succeeded Algué in 1926 as director, and was the last Spanish director. Beginning in 1920, American Jesuits began to arrive at the Observatory. In 1926 Charles Deppermann (1889–1957), arrived and he became director from 1945 to 1957. He was the most important figure in meteorology, as we will see in the next chapter. In 1928 William Repetti (1884–1966) arrived in Manila. He had studied at Saint Louis University with Macelwane (see Chap. 9) and specialized in seismology (see Chap. 8). In 1942 the Japanese army occupied the Philippines and work ceased at the Observatory, which was totally destroyed in 1945, when American troops were already near to Manila. The fire was deliberately provoked and it destroyed the buildings, instruments and the library with the whole collection of astronomical, meteorological and seismologic observations carefully made during 80 years. After the war and the independence of the Philippines, the orientation of the observatory changed. The meteorology section, which had been its most important part, was closed due to the founding of the new official Philippines Weather Bureau. The observatory was organized along research lines in seismology, astrophysics, ionospheric and solar physics. In 1951 new buildings were built at Mirador, Baguio and in 1962 at Quezon City. James J. Hennesey (1909–1987) became director in 1957 and gave a new impulse to the Observatory with new instruments for solar physics, such as a large spectroheliograph and a Razdow solar telescope. As happened with other Jesuit observatories, work in Manila began to decline in the 1970s and today it is devoted to environmental studies. In the next chapter we will discuss the contributions to meteorology made by Faura, Algué and Deppermann.

When French Jesuits returned to China in 1843, the memory of the great astronomers Schall, Verbiest, Kögler and Hallerstein, who had been directors of the Imperial Observatory of Beijing during the seventeenth and eighteenth centuries, went with them. The effort to recover this tradition began in 1871 with the establishment of the Observatory of Zikawei (Xujiahui) near Shanghai, devoted mainly to meteorology.[13] The heirs of the family of Xu Guangqi, who had helped Ricci in the translation of scientific works into Chinese and had been an important figure in the

[13] *L'Observatoire de Zi-ka-wei. Cinquante ans de travail scientifique.* Paris: Imp. d'Art G. Boüan (1930).

first Chinese Christian community, donated the land on which the observatory was built. In 1876 Marc Dechevrens (1845–1923) became director and the true founder of the observatory. He gave it a solid scientific character with its two sections of meteorology and geomagnetism. Stanislas Chevalier (1852–1930), who succeeded Dechevrens in 1887, established in 1901 an astronomical observatory on the Zose (Sheshan) hill about 25 km to the southeast of Zikawei. The observatory had a large dome of 10 m diameter which housed a Gautier double equatorial telescope of 40 cm diameter. For 28 years Chevalier carried out an intense program of astronomical observations, especially of the orbits of asteroids and their perturbations, which he published in 17 large volumes. We will see in the next chapter the work related to cyclones begun by Dechevrens and carried on by Louis Froc (1859–1932), director between 1896 and 1931. The observatory also had a time service which was much appreciated in the Far East. Between 1926 and 1933, the observatory participated, as one of the important stations, in the international project for the revision of longitudes. In one of the experiments Zikawei formed a large triangle with Algiers (North Africa) and San Diego (North America) to measure differences in longitude with great accuracy. One of these measurements was the difference in longitude between Zikawei and Greenwich with an accuracy of five thousandth of a second of arc. In 1934 Zikawei Observatory published the results of this project and gave the determination of the difference of longitude with respect to other observatories and possible sources of errors. One of the aims of the project was to verify Alfred Wegener's theory of continental drift. The short duration of the Project, less than 1 year, could not confirm or disprove the proposed motion of the continents.

In 1931, Pierre Lejay (1898–1958), was appointed director. His doctorate was from the University of Paris where he had presented a thesis on atmospheric electricity. In Zikawei he introduced new lines of research on atmospheric and ionospheric physics and gravity measurements, on which he published a popular textbook, *Développements modernes de la gravimétrie* (Modern developments of gravimetry, 1947). The Japanese-Chinese war and the occupation of Shanghai by Japanese troops in 1938 cut short Lejay's efforts to promote the scientific progress of the observatory. Lejay returned to France in 1939 and continued his scientific work there, but he never returned to China. The observatory continued working until December 1949 when it was occupied and confiscated; the Jesuits were expelled, by the communist government of Mao Zedong, yet another example of a Jesuit observatory to end tragically. The last director had been the Italian Ernesto Gherzi (1886–1976), whose work on tropical cyclones and microseisms will be discussed in the next two chapters.

The most important Jesuit observatory in Africa was established in Madagascar (Fig. 6.3).[14] The influence of France in that country had increased after the conflict between them in 1884. The French government wanted to establish an observatory and entrusted this project to the Jesuits. Elie Colin (1852–1923), a professor of science, was the person chosen. He prepared for the task with Perry in Stonyhurst. In 1889 Colin arrived to Madagascar and began the construction of the observatory

[14] Charles Poisson, *Un cinquantenaire, l'Observatoire d'Abohidempona, Tananarive, Madagascar, 1889–1939*. Paris: Editions Dillen (1939).

Fig. 6.3 Observatory of Tananarive (Madagascar) about 1890 (Udías 2003)

on a height, Ambohidempona, situated on a plateau 1,402 m above sea level. At that time it was the highest observatory in the world. The observatory had a large dome 8 m in diameter with an Eichen telescope 20 cm in diameter, and it also had very complete meteorological and geomagnetic instruments. The observatory would provide the basic information for the first geodetic measurements to be made in order to draw up the maps of the island and also ensure the time service for the region. In 1895 relations between the governments of Madagascar and France were broken off and the Jesuits were expelled from the observatory and from the country. In the ensuing armed conflict, which ended with the occupation of Madagascar by France, the observatory was practically destroyed.

In 1896 Colin returned to Madagascar and began the reconstruction of the observatory on a more modest scale; it was inaugurated in 1902. Between 1896 and 1906, Colin together with Désiré Roblet (1828–1914) established the geodetic base-lines for the maps of the region. They measured with great accuracy a base-line of 5,571 m and carried out the triangulation of the Imerina region, which covered a surface of 32,000 km^2. For this work they received in 1898 the Herbert-Fournet Award of the Geographical Society of Paris. In 1899 Colin was nominated correspondent member of the *Académie des Sciences* and in 1921 officer of the *Légion d'Honneur*. In 1923 Charles Poisson (1882–1965) succeeded Colin. He had been an officer in the French Navy and took part in the First World War. Although Poisson renovated some of the seismological and geomagnetic instruments, and continued the service forecasting cyclones and also the time service, the scientific work and importance of the observatory began to decline. In 1945 it was clear that the observatory had become outmoded, but funds were not available for its renewal. Lack of funds and the independence of Madagascar from France in 1960 complicated the situation for the observatory. Finally, the Jesuits left in 1967 and the observatory was transferred to the University of Madagascar.

In 1875 French Jesuits, who had been in Syria since 1843, founded the university of Saint Joseph in Beirut, Lebanon with plans for an observatory. In 1906 the observatory was established in the town of Ksara in the Bekaa Valley.[15] Bonaventure Berloty (1856–1934), a doctor in science from the University of Paris, was the first director. He had prepared for this post by working at the observatories of Stonyhurst and Ebro. The observatory had four sections: astronomy, geomagnetism, meteorology and seismology. In 1913 a new telescope was installed in a new tower with a dome. These promising beginnings were cut short by the events which followed the First World War. Lebanon was part of the Ottoman Empire which took part in the war on the side of Germany. The French Jesuits were expelled from Lebanon and the observatory plundered by Turkish troops. After the war Syria became a French protectorate and Jesuits were allowed to return. In 1918, Berloty returned to Ksara and the observatory was reconstructed with help from the French government. New instruments were installed, among them a new equatorial telescope of 20 cm diameter. In 1921 the observatory was entrusted to the direction of the new Meteorological Service of Syria with 14 stations in the territory. As in Madagascar, the Ksara Observatory took an active part in the geodesic investigation for the maps of Syria. The Jesuits were responsible for measuring the two base-lines and some of the triangulations of the region.

In 1925 Charles Combier (1880–1950) succeeded Berloty as director of the Observatory. A group of new Jesuits joined the observatory and took charge of the sections of geomagnetism, meteorology and seismology. The observatory kept up its scientific work, except for astronomy which practically ceased after Berloty's death. The independence of Lebanon in 1944 provoked a change in the observatory's financial and scientific reliance on France. In 1950 Combier died in a car accident and Jacques Plassard (1911–2005) took his place and was the last Jesuit Director. He planned to update the work of the observatory focusing on geophysical observations. Together with the French geophysicist Pierre Stahl, he carried out extensive field work in gravity measurements over several areas of Lebanon. Between 1967 and 1979 the observatory worked on the synthesis and publication of a long series of observations, along with a climate atlas, catalogs of earthquakes, and different types of maps. No new Jesuits joined the observatory and finally in 1979 it was transferred to the Lebanese government. A few years later it was practically destroyed during the war that affected great part of the Bekaa Valley.

The first Jesuit meteorological observations in mainland Africa were those of some Hungarian Jesuits, who established stations at Borona in 1883 and at Zumbo in 1890, near the Zambezi River in what is today Mozambique. Kalocsa Observatory

[15] Bonaventure Berloty, L'Observatoire de Ksara (Liban). *Ciel et Terre*, 23 (1912), 103–108.

published in its bulletins the results of their observations, including those stations, as part of its work. The stations were closed in 1905. In 1903, Jesuits had established another observatory in Bulawayo (Zimbabwe), the third in continental Africa. Edmund Goetz (1865–1933), a native of Alsace who had joined the Jesuits in England, was its founder. He had been a professor of science and worked at the observatories of Stonyhurst and Georgetown. The observatory had sections for astronomy, geomagnetism and meteorology. In 1913 Goetz installed a new Cooke equatorial telescope of 16 cm diameter, which he used to observe variable stars in collaboration with Hagen. However, his most important scientific contribution was in climate studies of the region and in magnetic surveys. The work of the Jesuits at the observatory did not last long, and in 1926 the observatory was transferred to the colonial administration. At present it continues as the Goetz Observatory, named in honor of its founder. The telescope, however, was moved to St. George's Jesuit College in Harare where it is still used by the students.

Jesuits managed another observatory in Africa at a state institution in Addis-Ababa (Ethiopia).[16] The government of Haile Selassie had decided to establish an observatory, following the recommendation made during the International Geophysical Year in 1955 for an observatory near the magnetic equator. Although its main interest was geomagnetism, the observatory also had sections for meteorology and seismology. In 1958, Pierre Gouin (1917–2005), a Canadian Jesuit professor of physics in the University College of Addis Abeba, was appointed director. He had taken a scientific degree at Boston College and had been trained in geophysics by Daniel Linehan at Weston Observatory (see Sect. 8.3). Gouin carried out continuous magnetic observations as well as other observations, such as those of the solar eclipses of 1959 and 1961, and field gravity surveys of different regions of Ethiopia. In 1978 the new government of Ethiopia, which had overthrown the Emperor, removed the Jesuits from the Observatory and expelled Gouin from the country. Gouin spent some time in Manila Observatory (1979–1981) and afterwards as a professor of physics in the University of Nairobi (1982–1986), until finally he retired to Montreal, where he continued his work in the Jesuit Geophysical Observatory. More will be said at the end of this chapter about his contributions to geomagnetism, and in Chap. 8 we will see those he made to seismology.

In 1866 the Belgian Eugène Lafont (1837–1908), a professor of science at the Jesuit St. Xavier College in Calcutta (Kolkata), installed a meteorological station to promote the science teaching. In 1870 he improved the station with the installation of a Secchi meteorograph. In 1874 the Italian astronomer Pietro Tacchini, director of the Palermo Observatory, had come to India to observe the transit of Venus. He invited Lafont, who was rector of the College, to take part in the observations. Impressed by India's good climate conditions for astronomical observations, Tacchini convinced Lafont to establish an observatory at the College.[17] With

[16] Paul Mohr (ed.), *Geophysical Observatory of Addis Ababa University: 50th Anniversary Symposium Proceedings*. Addis Ababa: Addis Ababa University (2007).

[17] Ileana Chinnici, An "Italian observatory" in India: The history of the Calcutta Observatory. *Studies in History of Medicine and Science* 14 (1996), 1–21.

Tacchini's help, Lafont installed in the College a dome with a Merz equatorial telescope of 18 cm diameter. The observatory was used mainly for solar observations and with Alphonse de Penaranda (1834–1896) it collaborated in observing solar eclipses in various places of India. In 1889 the sections of meteorology and astronomy became separate with different directors. From 1920 the observatory began to decline and in 1939 it was closed. After 70 years the observatory reopened in 2005 devoted to the study of solar physics and to teaching in the astrophysics department.

Edward F. Pigot (1858–1929), an Irish Jesuit who had been a medical doctor, was first appointed to the Zikawei observatory but in 1907 he moved for health reasons to St. Ignatius College in Riverwiew, New South Wales (Australia).[18] There he founded a small meteorological and seismological observatory and in 1922 installed a telescope and began a program for the observations of variable stars. Interest for astronomy and meteorology soon diminished and the observatory concentrated on seismology, as we will see in Chap. 8. In 1972 the meteorological station was closed but the seismological station continues to function today, supported by the Australian Geological Survey Organization.

6.4 Observatories in Central and South America

There were not many Jesuit observatories in Central and South America and in general they were relatively small, compared with those in other countries. They concentrated mainly on meteorology and seismology, which is not surprising given the frequent occurrence in the region of tropical hurricanes and earthquakes, the cause of many personal losses and great material damage. The lack of private or state scientific institutions at the end of the nineteenth and beginning of the twentieth century also explains why the Jesuit took over this responsibility. Although astronomy was not totally neglected in these observatories, it had a secondary role, except for the Observatory of San Miguel in Argentina. Again, little social interest for this particular scientific work and the high cost of telescopes may also have had an influence.

One of the few astronomical observatories managed by Jesuits, although for only a short time, resulted from the effort of Gabriel García Moreno, President of Ecuador, to modernize the scientific institutions of his country. In 1869 he founded in Quito the *Escuela Politécnica* (Polytechnic School), taking as a model the *École Polytechnique* of Paris. To run the school he asked the help of the Jesuits, who sent six young German professors of science from Valkenburg. Two years later, García Moreno decided to establish an astronomical observatory and asked Johann B. Menten (1838–1900), one of the German Jesuits, to take charge of the project. Buildings for the Observatory were finished in 1875, but unfortunately President Moreno was murdered in that year. The new government withdrew its support as

[18] David Brannagan, Earth, sky and prayer in harmony. Aspects of the interesting life of father Edward Pigot, S.J., B.A., M.B., B.C.H. (1858–1929). A Jesuit seismologist. Part I. *Earth Science History*, 29 (2010), 69–99.

they identified the Jesuits with the previous president, whose policies they opposed. The *Escuela Politécnica* was closed and the Jesuit professors returned to Germany. Menten, however, remained in Quito, where he left the Jesuit order and until 1882 continued in charge of the astronomical observatory that had been inaugurated in 1877. The *Escuela Politécnica* was not reopened until 1935, and today it recognizes the role played by the Jesuits in its foundation.

The meteorological observatory of the College of Belén in Havana (Cuba) was remarkable for the study and forecasting of Caribbean tropical hurricanes (see Chap. 7). In 1882, Benito Viñes installed a Cooke equatorial telescope of 12 cm diameter. He used it for his collaboration with Perry in the observation of the transit of Venus, 19 May 1884. In 1897 the observatory moved to the new building of the college where two towers were added, one with a dome for the telescope.[19] In 1925 the college was moved to a new site, where the telescope was not installed and the observatory restricted to meteorology.

In Mexico, the Jesuits established two observatories in Puebla and Saltillo. The most important, founded by the Italian Pedro Spina (1839–1925), opened in 1877 in the college of Puebla with sections for astronomy, meteorology, geomagnetism, and seismology. In 1880 the observatory installed a dome with a Dollond telescope which was replaced 2 years later by a more powerful one. This was the second astronomical observatory to be installed in Mexico, only 2 years after the National Observatory of Chapultepec. Gustavo Heredia (1869–1926), director from 1905 to 1910, continued the meteorological, geomagnetic and astronomical observations. He was a member of the Royal Astronomical Society of London. His studies focused on the observation of sunspots and features of the solar chromosphere. The observatory was closed down together with the college in 1914 during the Mexican revolution.

The Jesuits began to keep meteorological observations in a small observatory in the *Colegio de San Calixto*, in La Paz (Bolivia). Pierre M. Descotes (1877–1964) was its first director. He had spent some time at Cartuja (Granada) preparing for his work. The observatory's main contribution was in seismology, as we will see in Chap. 8. Descotes installed a small telescope and took charge of the time service for Bolivia and in 1926 participated in the International Project of Longitudes. As a contribution to this project he determined with great accuracy the geographical coordinates of the observatory by astronomical observations. In the course of these observations he discovered that the longitude of the observatory of Santiago de Chile had an error of 1.18 s. The existence of this error was confirmed by new observations made in the observatories of Santiago de Chile and Buenos Aires. As a consequence of this earlier error, the accepted geographical longitudes of the South American continent had been misplaced by 400 m to the west and needed to be corrected, as was done.

The great interest for astronomy of Fortunato J. Devoto, Auxiliary Bishop of Buenos Aires, lay at the origin of the Jesuit Observatory of Cosmic Physics of San Miguel, near Buenos Aires (Argentina).[20] The bishop thought that the work of the

[19] Manuel Gutiérrez Lanza, *Apuntes históricos acerca del Observatorio del Colegio de Belén, Habana*. Habana: Avisador Comercial (1904).

[20] Ignacio Puig, *El Observatorio de San Miguel, República Argentina*. San Miguel (Argentina): Observatorio de San Miguel (1935).

two astronomical observatories already existing in Argentina needed to be complemented by that of a new observatory devoted to astrophysics and entrusted the management to the Jesuits. They enthusiastically accepted and offered a site near the Faculty of Philosophy and Theology in San Miguel. Ignacio Puig (1887–1961), who was assistant director of the Ebro Observatory, was appointed director of the new observatory which was inaugurated in 1935. It occupied a three-story building and was devoted to astrophysics, cosmic rays, solar physics, and terrestrial electromagnetism. Up-to-date instruments were acquired and soon observations began.

In 1943 Puig returned to Spain and Juan Antonio Bussolini (1905–1966), an Argentinian, was appointed director. He greatly encouraged the scientific work and in 1960 installed new instruments, such as a solar telescope with Lyot filter, a spectroheliograph and two radio-telescopes of 5 m diameter. A new building was added and solar physics became the main research project. Bussolini's early death seriously affected the future of the observatory. In 1968 Mariano Castex, son of a famous Argentinian doctor and who had also studied medicine before joining the Jesuits, became director. He began a rapid expansion of the observatory which in the long run proved to be fatal. New buildings and instruments were added, and as Castex had good political relations he was able to secure funds from the state. Then the observatory lost its Jesuit character and was renamed as the National Observatory of Cosmic Physics. In 1970 political changes led to Castex being removed as director and he left the Jesuits. In 1977 the observatory was officially transferred to the National Commission of Space Research of Argentina.

6.5 The Vatican Observatory

The Vatican Observatory has a special place among the observatories managed by Jesuits.[21] First of all, it is not a Jesuit observatory, but an observatory entrusted to the Jesuits by the Popes. Clement XIV considered the Observatory of the Roman College, established in 1774 during the suppression of the Society, a Papal observatory. In 1854 Pius IX gave it officially the title "Pontifical Observatory". Its expropriation by the Italian state in 1870 left the Holy See without its own observatory. At a time when the Church was accused of being opposed to science, the Popes considered it important to show that the contrary was true with the work of their own astronomical observatory. For this reason, in 1890 Pope Leo XIII decided to found a new Pontifical Observatory named the *Specola Vaticana* (Vatican Observatory). At its inauguration in March 1891, Leo XIII formulated the reasons for its foundation with these words: "those borne of darkness are accustomed to calumniate her (the Church) … and they call her the friend of obscurantism, one who nurtures ignorance, an enemy of science and of progress.... This project is simply that everyone might see clearly that the Church and her Pastors are not

[21] Sabino Maffeo, *In the service of nine popes. 100 years of the Vatican Observatory*. Vatican City: Vatican Observatory and Pontifical Academy of Sciences (1991) and *La Specola Vaticana. Nove Pappi, una missione*. Vatican City: Specola Vaticana (2001).

Fig. 6.4 Johann Hagen at the Vatican Observatory (Maffeo, Vatican Observatory)

opposed to true and solid science, whether human or divine, but that they embrace it, encourage it, and promote it with the fullest possible dedication".[22] The observatory with a dome was built at the Vatican on top of the so called Tower of the Winds. The Barnabite Francesco Denza was appointed the first director. From its founding, the observatory collaborated with the international astronomical project *Carte du Ciel* (Map of the sky), involving 19 countries, to make a photographic atlas of the sky and a new catalogue of stars down to magnitude 11 (*Astrographic Catalogue*). For the project a double telescope of 33 cm diameter was installed in a new dome built on top of the Tower of S. Giovanni on the Leonine Wall.

Jesuit participation in the observatory began in 1906 when Pope Pius X appointed as director Johann Hagen, the director of Georgetown Observatory (Fig. 6.4).[23] Anticlerical circles in Italy commented that the Jesuits were invading Rome via the astronomical observatory. Hagen focused the work of the observatory on astronomy

[22] León XIII, Motu propio *Ut Mysticam* (1891), English translation in Sabino Maffeo, *In the service,* 207–210.

[23] H. Dopp, Johann Georg Hagen, S. J., 1847–1930. *Revue des Questions Scientifiques* 99 (1931), 5–37; Maffeo, *La Specola* (2001) 69–98; Sabino Maffeo, *J.G. Hagen S.I. Astronomo e direttore spirituale della beata E. Hesselblad.* Rome: Edizioni AdP (2006); Arndt Latusseck, *Via Nubila am Grund des Himmels. Johann Georg Hagen und die kosmichen Wolken.* Doctoral Thesis, Hamburg University (2009) 26–55.

and closed down the meteorological and magnetic sections. In 1909 a Merz equatorial telescope of 40 cm diameter was installed in a new dome on another tower of the Leonine Wall and the Tower of the Winds premises were vacated. In 1928 Hagen finished the part assigned to the Vatican Observatory for the *Astrographic Catalogue* with entries for 481215 stars. In 1930 at Hagen's death the observatory had completed 107 maps of the *Carte du Ciel* and part of the assigned section of the *Astrographic Catalogue*. At the Vatican Observatory Hagen continued his observations and studies on variable stars and the classification of star colors. He strongly defended his observations and interpretation of "obscure nebulae", he called "cosmic clouds" (first observed by Herschel), which were sill a debated question. Hagen also published a detailed presentation of the proofs of the rotation of the Earth, *La rotation de la Terre. Ses preuves méchaniques ancienne et nouvelles* (The Earth's rotation. Old and new mechanical proofs, 1911). Johan Stein (1871–1951), a professor of mathematics and physics in Amsterdam and during some years Hagen's assistant between 1906 and 1910, succeeded him.[24] The increased night lights in Rome forced him to look for a new site for the observatory outside the city. In 1935 the observatory was moved to Castel Gandolfo, at the palace of Villa Barberini, with two domes on its roof where new telescopes were installed. Alois Gatterer (1886–1953) began working on stellar spectroscopy and in 1933 installed an astrophysical laboratory. On the occasion of the inauguration of the new premises and instruments, Pius XI formally and definitively entrusted the direction and managements of the observatory to the Jesuits. Between 1935 and 1940 some ten Jesuits worked there.

During this time, the main work of the observatory was to finish the catalogue of variable stars which Hagen had begun. The *Atlas stellarum variabilium* (Atlas of variable stars) was finished in 1941 and published in nine volumes with 579 maps of 384 variable stars and 24,000 reference stars. In 1955 the part of the *Carte du Ciel* assigned to the Vatican Observatory was finished. At that time the observatory installed a collection of 1022 meteorites, a gift of the widow of Adrien Charles, Marquis de Mauroy, and began a research project on the structure of our galaxy. Daniel O'Connell (1896–1982), who had been Director of the Riverview Observatory in Australia, succeeded Stein. He renewed the instrumentation with a Schmidt telescope of 65 cm diameter, installed in a new dome in the gardens of the Villa Barberini. O'Connell focused the work of the observatory on the study of binary stars, stellar spectroscopy and polarimetry. In 1970 Patrick Treanor (1920–1978) was appointed Director. He had obtained his doctoral degree at Oxford University and had worked at the observatories of Lick and Yerkes in the United States. Treanor had begun working at the observatory in 1961 and focused his research on stellar spectroscopy and the interstellar medium. After his early death in 1978, George V. Coyne (1933-), a former professor at the University of Arizona, Tucson (United States) became director. In 1980 Coyne, aware of the worsening conditions for astronomical observations in Castel Gandolfo, established an agreement with the

[24] Jules de Kort and Jacques Marie Adrian, Obituary: In memoriam Father Johan Stein, S. J. (1871–1951) *Richerche Astronomiche* 2 (1952), 372–374; Maffeo, *La Specola* (2001) 117–124.

University of Arizona, for the joint use of the facilities of the Steward Observatory. Later an observatory was installed at Mount Graham, Arizona, where conditions for observations are very good, with a new telescope of advanced technology, which was inaugurated in 1993. Coyne initiated a new activity, the dialogue between science and religion, with the organization of a series of international seminars and publications, together with the Center for Theology and the Natural Sciences (CTNS), Berkeley (California). The part taken by the Vatican Observatory in the dialogue between science and religion shows the increasing interest of Jesuit scientists in this subject. Already in 1986, a congress organized by the observatory to commemorate the third centenary of Newton's *Principia Mathematica* was the occasion for a letter by Pope John Paul II on the relation between science and religion, one of the most important papal documents on this subject. In 2006 the Argentinian José Gabriel Funes (1963-) became Director and at present 15 Jesuit astronomers are actively engaged in the work at Castel Gandolfo and Arizona.

6.6 The Earth's Magnetism

As we have already seen, Jesuit scientists of the old Society, among them Garzoni, Cabeo, Kircher and Grandami, showed a special interest in magnetism. Kircher devoted a bulky treatise to this phenomenon and conceived of magnetism as a cosmic force which permeates the world and which he identified with love. He thought of God as the great magnet that attracts everything to him. Jesuit scientists of the restored Society also showed a special interest in these phenomena, but now from a purely scientific point of view with special emphasis on terrestrial magnetism.[25] In 1858 Secchi installed a complete set of magnetic instruments, magnetometers, inclinometers, and declinometers in the observatory of the Roman College. A special room built with non-magnetic material was prepared on the roof of the Church of Saint Ignatius. Secchi studied the periodic variations of the geomagnetic field and tried to relate them to changes in solar activity, assuming the Sun to be a great magnet. Also in 1858, Edward Sabine, one of the pioneers in the study of terrestrial magnetism (elected president of the Royal Society in 1871), installed a temporary magnetic station in the Stonyhurst Observatory. At that time he was encouraging the installation of a network of magnetic stations in both the British Isles and worldwide. He convinced Weld to begin these observations with a permanent station. Weld accepted the suggestion and in 1858 installed a complete magnetic station at Stonyhurst and began observing the absolute values of the components of the geomagnetic field and the temporal variations. Thus Stonyhurst and the Roman College were among the first geomagnetic observatories. Sabine's prestige helped Stonyhurst to be recognized in scientific circles for its work. In 1863 Sidgreaves began to carry

[25] Pierre de Vregille, Le jésuites et l'étude du magnétism terrestre. *Études* 104 (1905), 492–511; Agustín Udías, Jesuits, Role in Geomagnetism. In D. Gubbins y E. Herrero-Berrera (eds.) *Encyclopaedia of Geomagnetism and Paleomagnetism*. Dordrecht: Springer (2007), 460–462.

out monthly systematic observations of the components of the geomagnetic field, which in 1865 Sabine presented to the Royal Society as an example of accurate observations. These observations continued until 1974, when the observatory was closed, and they constitute today one of the longest series of magnetic observations at the same site. As we will see in Chap. 9, between 1868 and 1871 Perry, together with Sidgreaves, carried out magnetic field surveys in France and Belgium. In 1879 Braun began magnetic observations in Kalocsa, which Fenyi continued until 1913. Braun and Fenyi were more interested in solar activity and tried to find its relation to the variations of the geomagnetic field.

From its foundation in 1904 the Ebro observatory took as one of its main tasks the study of the relation between solar activity and the Earth's magnetic field. In 1920 Rodés, as a result of his study of magnetic storms, proposed that the Sun influences the geomagnetic field through the emission of clouds of electrically charged particles. This theory may be considered a precursor of the modern concept of the solar wind. In 1950, after the gap in scientific activity caused by the Spanish civil war, the observatory renovated its instrumentation and in 1954 installed the first ionospheric sounder in Spain. Lejay came to the observatory on this occasion. He had begun his studies of the ionosphere in Zikawei and designed a new type of ionospheric sounder. He continued his studies after his return to France as Director of the *Bureau Ionosphérique Français*. Romañá and José Oriol Cardús (1914–2012) continued Rodes' research into terrestrial magnetism with special attention to magnetic storms and periodic variations. The Ebro Observatory continued to renew its instruments between 1960 and 1990 with new ionospheric sounders, magnetometers, and solar radio telescopes.

Terrestrial magnetism was also a research subject in observatories in mission lands. In 1887 the Manila Observatory established a magnetic section and put in charge Martín Juan (1850–1888), who had been trained by Perry in Stonyhurst. The following year Martín Juan carried out a field survey of magnetic observation covering several islands of the archipelago. He died during this work, but it was completed by Cirera, who continued it on some coastal areas of China and Japan. Cirera published his observations and those of Martín Juan in *El magnetismo terrestre en Filipinas* (Terrestrial magnetism in Philippines, 1893). Because of magnetic interferences, in 1945 Manila Observatory moved its magnetic station to Antipolo outside the city. After the war magnetic stations with modern instruments were installed in Baguio and Davao. In 1952 the Observatory began a program of ionospheric studies with an ionospheric sounder in Baguio and installed two more in Balara in 1962.

In 1877 the Observatory of Zikawei acquired magnetic equipment which in 1908 moved to Lukiapang and in 1933 to Zose. Observations were published in a series of 39 volumes, *Études sur le magnetisme terrestre* (Studies of terrestrial magnetism, vol. 1–39, 1908–1932). In 1932 Joseph de Moidrey (1858–1936) completed one of the first determinations of the secular variation of the Earth's magnetic field for the Far East region. The Observatory continued its magnetic observations until it was closed in 1949. Maurice Burgaud (1884–1977), who began working at the observatory in 1924, was the last director of the magnetic station. After the occupation of the observatory he was imprisoned by the communist government until

1955. Belen Observatory in Havana began magnetic observations in 1877, and these continued until 1925. In 1877 the observatory of Puebla, México, also began magnetic observations, which lasted until 1914, when the observatory was closed.

Jesuit magnetic observations in Africa were among the first ever made in that continent and are of special interest. In 1889 the first were made in the Observatory of Tananarive, where Mascart magnetometers were installed. After the destruction of the observatory in 1896, observations began again in 1902 and continued under Jesuit supervision until 1967. In 1955 Burgaud, expelled from China, took charge of the magnetic section and renovated the equipment, with a proton-magnetometer among other instruments. The magnetic station continues to function today, having been upgraded in 1983 and 1993. Ever since its foundation in 1903 the Bulawayo Observatory has had a magnetic station installed in a separate building. In addition, in 1909 and 1914 Goetz carried out two long field magnetic surveys along two lines. The first along a line of 330 km distance from Broken Hill (Kabwe, Zambia) to the mine site called the *Star of the Congo* (near Lubumbashi in the Democratic Republic of the Congo), and the second in the region of Barotseland (Zambia) between Kazungula and Lealui (400 km distance). These magnetic profiles were among the first to be made in Africa. As we saw, the Observatory of Addis-Ababa took for its main work magnetic observations. During the 20 years under Gouin's direction continuous and accurate observations were made of the Earth's magnetic field. Gouin focused his research on the study of magnetic storms and temporal variations of the magnetic field. Between 1958 and 1962 Gouin completed the magnetic map of Ethiopia. His work was interrupted in 1978 when he was removed as director and expelled from Ethiopia.

Pierre Noël Mayaud (1923–2006) collaborated with Gouin in magnetic research. He had obtained his doctoral degree in Paris in 1955 with a thesis on the magnetic activity at the polar regions, using the magnetic observations made during the French Polar expeditions of 1954–1955. Mayaud was an expert on the time variations of the Earth's magnetic field, especially on the magnetic indexes K_n, K_s and K_m which describe these variations, and they were the subject of his publication *Derivation, meaning and use of geomagnetic indices* (1980).

Chapter 7
Meteorology. Tropical Hurricanes

7.1 Pioneers and First Observations

The term "meteorology" and the science linked with it have their roots in ancient Greece. The Greeks spoke of "meteors" to refer to any phenomena which took place above the surface of the Earth and beneath the orbit of the Moon. To this category belonged rain, winds, thunder and lightning, but also other phenomena like earthquakes, volcanoes and comets. Aristotle formalized the study of these phenomena in the four books of his *Meteorologica*, one of the treatises he wrote on natural philosophy. Until the seventeenth century this treatise laid the foundations in the West for knowledge of these phenomena. The *Meteorologica* of Aristotle was being taught at universities as part of the programs of philosophy, and in the Middle Ages such authors as Albert the Great and Thomas Aquinas wrote long commentaries on it. It was then a required subject in Jesuit colleges which had complete programs of philosophy, and Jesuit professors also wrote commentaries. Among them those by the professors of the college of Coimbra are out-standing. They were part of a collective work on natural philosophy called *Cursus Collegii Conimbricenses*, written between 1592 and 1606 and used as the textbook in many Jesuit colleges. It followed Aristotelian doctrine with very few references to actual observations. Kircher in his *Mundus Subterraneus* and Schott in *Iter extaticum* both dealt with some subjects of the *Meteorologica* introducing new approaches and some recourse to observations and experiments.

In the middle of the seventeenth century, Galileo and Torricelli developed the first thermometers and barometers to measure temperature and atmospheric pressure. Their use allowed the first systematic instrumental measurements and the beginning of meteorology as an experimental science. The use of these instruments spread very rapidly and scientists as Robert Hooke, Boyle and Huygens contributed to their development. In 1654 the Jesuit Luigi Antinori (–1679) was put in charge of organizing one of the first networks of stations equipped with thermometers, barometers,

A. Udías, *Jesuit Contribution to Science: A History*,
DOI 10.1007/978-3-319-08365-0_7

and hygrometers. The network was established around Florence and Pisa by the *Accademia de Cimento* with the support of Ferdinand II, the Grand Duke of Tuscany. This is the first mention of Jesuits in relation to the establishment of meteorological observations. Several networks of meteorological stations were established in Europe, among the first being that of the *Académie des Sciences* in France in 1688 and that of the *Royal Society* in England in 1723. The state networks, usually called Meteorological Services, began later toward the end of the nineteenth century. These dates help one to situate historically the meteorological observations made by Jesuits. As we saw earlier, the first Jesuit observatories were astronomical, but they also began to use instruments for meteorological records of temperature, atmospheric pressure, air humidity, rain-fall and wind direction. This was the start of contributions by Jesuits to this new science.[1]

In 1746 Beraud began meteorological observations in the observatory of Lyon which he carried on for 20 years. He also wrote a work about the thermometer. Pezenas kept records of continuous meteorological observations in Marseille and published in 1731 *Météorologie,* one of the first books on meteorology in French and the first book written by a Jesuit about meteorology in its modern sense. Between 1763 and 1779, Lagrange carried out systematic observations in the observatory of Milan and at about the same time Belgrado was doing this in Parma. In the observatory in Prague Stepling also kept records of temperature, atmospheric pressure and air humidity, and in the observatory in Vienna, Pilgram established a program of meteorological observations. He became interested in the problem of weather forecasting and he wrote in 1788 *Untersuchungen über das Wetterkunde* (Inquiry about meteorology). Jesuit records kept in China, beginning in 1757, have a special interest. In 1774, Jean Joseph-Marie Amiot (1718–1793) published them in France in *Observations météorologique faites à Pekin* (Meteorological observations made in Beijing). These may be the first systematic meteorological observations made in China. Verbiest, as we already have seen, had introduced meteorological instruments into China and he wrote a small treatise about the thermometer and the hygrometer.

In the eighteenth century Jesuit scientists became interested in the aurora borealis or Northern Lights. In 1716 this interest increased when unusual aurorae were observed all over Europe at relative low latitudes, as noted by Edmond Halley and Jean Jacques de Mairan. The problem of the nature of such aurorae was known in antiquity and is mentioned by Aristotle in his *Meteorologica.* Grassi was probably the first Jesuit to discuss it in 1620 in his controversy with Galileo. Between 1730 and 1785 Jesuits published 17 works on this phenomenon, especially those observed in 1739 and 1750. Among these publications were those by Beraud (1750), Boscovich (1748), Stepling (1761) and Ximenes (1750). They tried to find an explanation in the refraction of sun-light on ice particles in the atmosphere. All the meteorological observations made by the early Jesuits stopped with the suppression of the Order in 1773.

[1]Agustín Udías, Jesuits' contribution to meteorology. *Bull. American Meteorological Society,* 77 (1996), 2307–2315.

7.2 Meteorology in Modern Observatories

As was shown in the previous chapter, most Jesuit observatories founded in the middle of the nineteenth and early twentieth centuries had sections devoted to meteorology, which was then starting as a modern science.[2] The Jesuit dedication to this science can be explained by the relatively low cost of the instrumentation and by the need for well attended meteorological stations at a time when state meteorological services were not fully established. In most countries such services began about 1850, and the contribution of the Jesuit stations – together with other private institutions including some run by other religious orders – was much appreciated. Jesuits also saw this as a way to contribute to modern scientific work, with its emphasis on the empirical aspects of instrumental observations, and it was a complement to the science lectures given in Jesuit colleges and faculties of philosophy for the training of Jesuit students. Thus, along with sections on meteorology in their astronomical observatories Jesuits established many meteorological stations, some lasting only a short time, which provided data for the national services. The Jesuits established 47 meteorological stations, if one includes those in observatories, which kept records during sufficiently long periods of time. Geographically there were 17 in Europe, 6 in North America, 15 in Central and South America Central and 9 in Asia, Africa and Australia. The most outstanding for their observations, and forecast and research on tropical hurricanes were the observatories of Belén (Cuba), Manila (Philippines), Zikawei (China) and Tananarive (Madagascar).

In 1824 the first meteorological station was installed at the observatory of the Roman College with the most modern instruments of that time. From 1824 Bernardino Gambara (1814–1884) was in charge of the station with six daily observations. In 1858 Secchi installed his "Universal Meteorograph" which simultaneously recorded atmospheric pressure, temperature, humidity and wind velocity and direction. Meteorological observations were continued under Secchi's direction until 1878. Between 1862 and 1877 Secchi published his works on meteorology in the 16 volumes of the *Bollettino Meteorologico* of the Observatory. At Stonyhurst meteorological observations began in 1838 in a pavilion built for this purpose in the garden of the school and instruments were installed there. In 1842 Weld began systematic observations and in 1850 he became a founder member of the British Meteorological Society. In 1866 under Sidgreaves' direction the Board of Trade of the United Kingdom supplied the observatory with modern equipment which raised the level of the observations. The same year, at Sabine's recommendation, Stonyhurst was chosen as one of the seven top-class meteorological stations of the British Meteorological Office. From 1932 James P. Rowland (1875–1948) focused the work of the observatory on meteorological observations. Rowland, under the name of "Wizard of the North", published observations and weather forecasts in the local newspapers. Although the observatory closed in 1947 it continued sending

[2] Jesuit observatories at which there were meteorological sections can be found in Udías, *Searching*, Part II, 183–292.

meteorological observations to the British Meteorological Office until 1960. Continuous observations from 1848 to 1942 were published in the bulletin of the Observatory and constitute a very valuable series.

In 1860 the first meteorological station managed by Jesuits was installed in Spain in the Faculty of Philosophy of Saint Mark in León. At that time it was one of the 28 stations that existed in Spain, but the Central Meteorological Institute was not established in Spain until 1887. In 1868 the León station closed, when the Jesuits were expelled from Spain. In 1880, on their return to Spain, they established a meteorological station at the new Faculty of Philosophy in Tortosa (Tarragona), which in 1904 moved to the Ebro Observatory. In 1880 a station was installed at the school of La Guardia (Pontevedra). Between 1888 and 1907 Baltasar Merino (1845–1917) carried out studies of the climatology and storms on the coast of Galicia. The station closed in 1916, when the school moved to Vigo. In 1870 similar stations were installed at the school of Orduña (Vizcaya) and in 1882 at the Faculty of Philosophy in Oña (Burgos). These were closed in 1931 when the Jesuits were expelled from Spain. The same happened to a station installed in 1900 at the Seminary and later Pontifical University of Comillas (Santander). This was reopened after the Spanish Civil War and continued to function between 1942 and 1967. In 1902 the Spanish Jesuits proposed an interesting project: the establishment of a network of meteorological stations taking advantage of their schools in different regions. The idea was suggested by Algué, director of the Manila Observatory, while on a visit to Spain. The central station was installed in Madrid (*Colegio N. S. del Recuerdo*). The network was formed of eight to nine stations which sent their observations to Madrid where a bulletin was published. However, the project did not last long and ended in 1906, although some of the stations continued to function until 1931.

In 1868 in Italy Jesuits established a meteorological station, called the *Osservatorio Tusculano*, at the college of Mondragone (Frascati). Secchi had suggested this idea because he thought these observations necessary for the study of the climatology of the Lazio region. The observatory maintained its observations until 1944. In 1915 another station was installed in the college of Acireale, Catania (Sicily) and it was active until the college closed in 1978. In Malta the Jesuits established two meteorological stations, one on the island of Gozo in 1882, which lasted until 1909 in the Diocesan Seminary, and another between 1877 and 1907 in Malta itself at Saint Ignatius College, founded by English Jesuits. This latter station was an initiative of Perry and its observations were published in the Stonyhurst bulletins. French Jesuits installed a station on the island of Jersey, near the coast of France where since 1880 they had a faculty of philosophy after their expulsion from France. In 1893 Dechevrens, after his return from China, moved to Jersey and installed the station in a separate building with a metallic tower 50 m high to study wind components at different heights. Dechevrens studied the motion of air masses and the distribution of temperature and atmospheric pressure in relation to height, a research he had started in Zikawei. In 1923, after Dechevrens' death the tower was dismantled, but regular meteorological observations continued even during the war when the island was invaded by German troops. In 1954 the Jesuits left the island, but Charles

Rey (1897–1981), who had succeeded Dechevrens in 1934 and managed the station during the war, remained and continued to maintain the station until 1979. Since 1945 the station has belonged to the British Meteorological Service and continues today after 120 years of uninterrupted service under the name of St. Louis Observatory, in memory of the Jesuits who managed it for 86 years.

In North America Jesuit meteorological stations belonging to colleges and universities were not very numerous. At Saint Louis University meteorological observations began in 1860 and from 1925 were part of the activities of the Department of Geophysics which offered courses in meteorology. Since 1944 the university has organized a graduate program in meteorology at master and doctorate levels. Saint Louis University is the only Jesuit university with such a program. In 1964 the name of the department was changed to Department of Earth and Atmospheric Sciences. In the observatory of the University of Santa Clara Ricard's main interest was really meteorology and in 1895 he began to keep records. He thought that there was a correlation between weather and sunspots, so that weather could be forecast from solar observations. Ricard was a stubborn man, defending his ideas in spite of opposition from scientists like George E. Hale, director of Mount Hamilton Observatory. Ricard studied the climatology of California, specially its rainfall pattern, and published a local weather bulletin so that he was known popularly as the "Padre of the Rains". In 1895, Frederick L. Odenbach (1857–1933) installed a meteorological station at Saint Ignatius College (today John Carroll University) in Cleveland (Ohio), where he especialized in the study of atmospheric electrical phenomena. In 1898 he installed a Secchi Meteorograph and developed an instrument he called a "ceraunograph" to detect lightning. Other meteorological stations were established in 1910 at Marquette University, Milwaukee (Wisconsin) and Spring-Hill College, Mobile (Alabama). In 1955 Gherzi, having been expelled from Zikawei, installed a meteorological station at Jean de Brébeuf College, Montreal (Canada). There he continued his work in meteorology and in 1958 installed an ionospheric sounder. With his empirical approach to science Gherzi was a pioneer in the study of atmospheric electricity in Canada.

In several countries of Central and South America, Jesuits laid the foundations for what would be the national meteorological services. In recognition of the Jesuit experience in this science, the governments of Ecuador (in 1870) and of Colombia (in 1922) entrusted the Jesuits with the organization of their meteorological services. In 1864 the Jesuits had begun meteorological observations in the *Colegio de San Luis* of Quito (Ecuador). Seven years later this became the central station of the National Meteorological Service of Ecuador. In 1870 Menten became director of the Service until 1875, at the death of the President García Moreno, when the station was closed. In 1920 the President of Colombia entrusted Simón Sarasola (1871–1947) with the organization of the National Meteorological Service. Sarasola had worked at Belén Observatory in Cuba from 1905 and in 1910 had established a small observatory in Cienfuegos. The central station of the Service was established at the Jesuit college of San Bartolomé in Bogotá. The project called for 20 secondary stations, but only four actually worked. Sarasola was the first to carry out studies of climatology in Colombia. In 1939 he left this post and in 1943 returned to Cuba as director of the Belén Observatory. In 1941 Sarasola, along with Jesús Emilio Ramírez, about whom more will be said in the next chapter, founded the *Instituto Geofísico de los Andes Colombianos*, which had a meteorological section. In 1949 Ramírez became director of the National Meteorological Service of Colombia.

In Cuba, the Belén Observatory played an outstanding role in the study and forecasting of tropical hurricanes, as will be seen more in detail later. In México the observatories of Puebla and Saltillo had meteorological stations. Both sent their observations to the Meteorological Service of Mexico, They closed in 1914. In 1892 the Jesuits began meteorological observations in La Paz, Bolivia, at the observatory of the *Colegio de San Calixto* and these continue today. Small meteorological stations were established by Jesuits in several other countries of Central and South America. In 1851 one was set up in Guatemala, but it lasted only until 1870. That of Santiago de Chile also only lasted only from 1867 until 1872. The duration of the station of the college of Santa Fe (Argentina) was longer from 1904 to 1967. In 1914 the Jesuits installed a meteorological station in Sucre (Bolivia), with observations continuing until 1966. In Asunción, Paraguay, the Jesuits kept meteorological records for only a short time, from 1935 to 1940. Many of these observatories were due to personal initiative and lasted only during the life-time of their founders. However, they indicate the interest for scientific research in the Jesuit colleges in these countries which lacked a strong scientific tradition.

The most important Jesuit meteorological observatories in Asia and Africa were in Manila, Zikawei and Tananarive and these will be seen later in relation to the study of tropical hurricanes. In 1866, Lafont had set up the observatory of Xavier College (Calcutta) as a meteorological station. In the following year he forecast, before the official Meteorological Service did so, the arrival of a cyclone, and this called public attention to the observatory. In 1872 the observatory installed one of the few Secchi meteorographs then in existence. From 1869 Edward Francotte (1843–1923) was in charge of the meteorological observations. He studied the climatology of the region, using the long series of observations kept from 1868 to 1923. Two years after the death of Francotte, the meteorological station was closed. From its founding in 1903 the observatory of Bulawayo had a complete meteorological instrumentation. Goetz used the observations made there to study in detail the climatology of the region, especially its rainfall pattern (*The rainfall of Rhodesia*, 1909). In 1909 Ksara Observatory (Lebanon) began meteorological observations. In 1921 after the First World War, under the French Protectorate, the observatory took charge of the Meteorological Service of Syria with a network of 14 stations in Syria and Lebanon. Berloty was its first director until his death in 1934; he was succeeded by Combier and Jean Rey (1896–1979). Combier published several studies about the climatology of Syria and Lebanon. After the independence of Syria and Lebanon, the observatory limited its meteorological observations to Lebanon. Rey continued in charge of meteorological observations till 1975 and in 1951 he published the first pluviometric map of Lebanon. In 1907 Pigot began meteorological observations at Riverview Observatory (Australia), which were continued on a regular basis until 1972. From its foundation in 1958 the observatory of Addis Ababa (Ethiopia) had a meteorological station, which was under Gouin's supervision until 1978. We saw earlier that in 1883 the Hungarian Jesuits made the first meteorological observations in central Africa in the Borona mission near the Zambezi River (Mozambique). In 1893 Laszo Menyhart (1849–1897), who had worked with Fenyi in the Kalocsa Observatory, took charge of the observations. He installed new instruments and a new station in Zumbo some kilometers to the north. Observations from both stations were sent to Kalocsa Observatory and published in its bulletins. After Menyhart's death, interest for meteorology decreased, but the stations continued until 1905, when both were closed down in spite of Fenyi's interest for their continuation.

7.3 Caribbean Hurricanes

Each year a number of hurricanes or tropical cyclones cross the Caribbean region causing extensive damage and human casualties. In 1854 the Jesuits had established a school the *Real Colegio de Belén* in Havana (Cuba) and 3 years later they installed a small meteorological observatory, called the Belén Observatory, which would become famous for its studies of hurricanes. In 1870 Benito Viñes (1837–1893), born in Poboleda (Tarragona), arrived to Cuba and was appointed director (Fig. 7.1).[3] His study and forecasting of hurricanes during 24 years constitute one of the most outstanding Jesuit contributions to meteorology. In 1882 he made a study trip to Europe, when he visited several observatories and spent some time in Stonyhurst with Perry. In his first years as director Viñes gathered all the information obtained at the observatory, from the start of regular observations in 1858 until 1871, and he published this in 1872. In 1873 to upgrade the instrumentation he installed a Secchi meteorograph. On the basis of these records he examined the characteristics of the hurricanes that had affected the island of Cuba. With great thoroughness he sifted the accumulated data about the organization, circulation and translatory movement of Caribbean hurricanes. At that time not much was known about these phenomena except for the early studies of the American meteorologists, William Redfield, William Reid and James Espy. Redfield had studied a hurricane which crossed Cuba in 1844. In his studies Viñes took an empirical approach, examining with care the variations in atmospheric pressure, wind direction, and the aspect of the clouds – their form, motion and height. He published the results of his studies in two works: *Apuntes relativos a los huracanes de las Antillas en Septiembre y Octubre de 1875 y 1876* (Notes concerning the hurricanes of the Antilles in September and October of 1875 and 1876, 1877). and *Investigaciones relativas a la circulación y traslación ciclónica de los huracanes de las Antillas* (Research on the cyclonic circulation and translation of the hurricanes of the Antilles, 1895) a work written in 1893 and published posthumously. In 1898 the Weather Bureau of Washington published an English translation of the first work, *Investigation of the cyclonic circulation and*

[3]Antonio López de Santa Anna, *Contribución a una biografía completa del P. Benito Viñes Martorell, S.I., célebre meteorólogo de las Antillas*. Santander: Hermanos Bedia (1957); Luis E. Ramos Guadalupe, *Benito Viñes, s.j. Estudio biográfico*. Havana: Editorial Academia (1996).

Fig. 7.1 Benito Viñes (1837–1893) with Secchi's meteorograph at Belen Observatory (Biblioteca Comunidad ICAI, Madrid)

translatory movement of the West Indian hurricanes. A summary of this work had been published in English in the United States in 1885 and reedited in 1887 and 1889. In 1888 Everett Hayden, chief of the Meteorological Division of the U.S. Navy and editor of *Pilot Charts,* a publication of the Hydrographic Office of Washington, reproduced parts of Viñes' work in a book about hurricanes.

Viñes proposed two general laws for cyclonic circulation. The first states that winds rotate counterclockwise in the northern hemisphere and clockwise in the southern one. According to the second law, cyclonic currents are convergent at a lower height and divergent at higher heights. As a consequence of this law, there are different forms and motions of types of clouds: low clouds, high cumulus, dense cirrostratus and fine cirrocumulus. Viñes considered this second law to be fundamental for cyclonic circulation. To these general laws he added six particular laws concerning the movement and translation of cyclones in the Caribbean region. The first states that the general trajectory of cyclones in the Caribbean is parabolic in an east-to-west direction, from the Atlantic Ocean to the Antilles, and then curves to the north and finally to the northeast. The second law concerns what he called the "bending points" (*recurvas*), and determines the latitude and longitude of the point where the path of the cyclone changes from a westward to an eastwards direction.

For cyclones this point moves to the north in the period from June to September and to the south for those in October. Viñes pointed out the practical use of this law in order to forecast the approximate paths of hurricanes depending on the date of their occurrence. The other four laws establish particular characteristics of the trajectories depending on the location and the dates.

Viñes never claimed that the path of cyclones follows these laws exactly, but he thought that they were general patterns deduced from observations. He was the first to describe the anomalies produced in the trajectories when two cyclones approach each other, making clear that they never join. He suggested that changes in the motion of cyclones are due to air currents in the upper layers of the atmosphere and extended their structure in a vertical dimension. He was one of the first to study the three dimensional structure of cyclones, which he suggested extend to great heights. He also suggested that the energy of cyclones was related to the latent heat of condensation liberated by their cloudy external walls and the intensification of precipitation. His continuous and patient observations put Viñes well ahead of his time with regard to the nature of hurricanes.[4] In May 1889, the *Pilot Chart* published the following: "These important laws, established by the study and long experience of Fr. Viñes, should be thoroughly understood by every navigator and utilized by shaping his course so as to avoid a hurricane". And in a lecture to the Naval Academy the editor, Hayden, spoke of "Padre Viñes, the eminent Havana meteorologist, whose scientific ability, indefatigable energy, and unrivalled opportunities have linked his name so inseparably with all recent advances in our knowledge of these storms".[5]

Viñes applied his knowledge about the nature of cyclones to the practical use of accurate forecasting, thus preventing many casualties and material damage. The basis of his work was the constant hourly observation each day from four in the morning to ten at night of temperature, atmospheric pressure and humidity, the direction and speed of wind and variations of the magnetic field. He based his forecasts on these observations, on the state of the sea and changes of pressure, and on the aspect and movements of high altitude clouds, especially cirrus, which diverge from the vortices of a cyclone. Viñes made his first forecast in September 1875, when a hurricane struck Cuba after 4 years of relative calm. He was able to predict the arrival of the hurricane to the coasts of Cuba when its vortex was still 550 km away, and he said that it would pass through Havana 2 days later, as it actually did. He based his forecast on the sudden drop of atmospheric pressure and the observation of diverging movement of high altitude clouds. This was the first accurate modern prediction of the passing of a cyclone through a city. From then onwards, Viñes published his predictions, which were highly appreciated for their accuracy by the local authorities and maritime institutions. No doubt, many lives were saved and material damage avoided thanks to them. His study of the hurricane of October

[4] José Fernández-Partagás, *El papel del padre Viñes, S. J. en el conocimiento del huracán visto a cien años de su muerte*. (Manuscript) Miami, 1993.

[5] Walter M. Drum, *The pioneer forecasters of hurricanes*. Washington: Styormont and Jackson (1905), 13–14.

1876 also included a survey, made after four journeys to assess the damage caused, of the effects along its path across the island. In that same year 1876, Viñes became aware that he could not observe the atmospheric weather and forecast the path of hurricanes in the whole Caribbean region with only one observatory. So he determined to establish a network of stations which would send their observations by telegraph to Havana. The network was formed by 20 stations and counted with the support of maritime and telegraph companies. Seven of them in Trinidad, Barbados, Martinique, Antigua, Puerto Rico, Jamaica and Santiago de Cuba sent observations on a regular basis. The other stations only sent observations during the cyclone season. Viñes also designed two instruments, which he called *ciclonoscopio* and *ciclononefoscopio*, to detect an approaching cyclone by observing the direction of winds and the distribution and orientation of clouds. He never finished their construction, but Faura and Algué in Manila used Viñes' ideas for similar instruments, as will be shown below. In their book about Atlantic cyclones, Gordon Dunn and Banner Miller affirm that Viñes was a true scientist, who made important contributions to the theory and the art of forecasting hurricanes. He was among the first meteorologists of the Western hemisphere in the nineteenth century to contribute accurate knowledge about tropical hurricanes.[6] Ivan R. Tannehill, in his popular book about hurricanes, quotes Viñes several time and considers him preeminent among the pioneer observers of hurricanes in the West Indies.[7] Viñes himself saw his work as a service. As he wrote in reply to a query from the Havana Chamber of Commerce, "For my part I am desirous only of serving all so far as service is rendered possible by my poor health and the limited means at my disposal; nor do I wish other recompense, after that which I hope from God, than to be of use to my brethren and to do my little share for the advancement of science and the good of humanity".[8]

In 1893 after Viñes' death, Lorenzo Gangoiti (1845–1922) was appointed director and continued the observation, study and forecasting of cyclones. At the end of the Spanish administration of Cuba in 1898 the United States Weather Bureau installed an office in Havana which ignored completely the work of the Belén Observatory. Those in this agency claimed that, prior to the American arrival in Cuba, nothing really scientific had been done regarding the observation of hurricanes. In 1905 Walter M. Drum (1870–1921), a Jesuit from Georgetown University, published a book on the work done by the Jesuits in Havana which changed this view.[9] The U. S. Weather Bureau recognized the work of the Belén Observatory, closed its office in Havana, and asked the observatory to send daily telegrams to Washington about the conditions of the weather. Manuel Gutiérrez Lanza (1865–1943) and Simón Sarasola were directors of the observatory and continued the study of

[6] Gordon E. Dunn and Banner I. Miller, *Atlantic Hurricanes*. Baton Rouge: Louisiana State University Press (1960).

[7] Ivan R. Tannehill, *Hurricanes, their nature and history*. Princeton: Princeton University Press (6 ed.) (1945), 88.

[8] Richard M. De Angelis, The hurricane priest, *ESSA World*, October-1968, 16–17.

[9] Drum, *The pioneer forecasters*, 23–29.

hurricanes. Sarasola published *Los huracanes de las Antillas* (The hurricanes of the Antilles, 1925), in which he summarized the work done by the observatory during 70 years. Gutiérrez Lanza wrote a similar work *Génesis y evolución del huracán del 20 de Octubre de 1926 y catálogo de ciclones en la isla de Cuba 1865–1926* (Genesis and evolution of the hurricane of 20 October 1926 and catalogue of cyclones in Cuba from 1865 to 1926, 1928). In 1961 the observatory was closed when the government of Fidel Castro confiscated the college and the observatory. The last director was Rafael Goberna (1903–1985), who had studied at Saint Louis University. This is yet another example of a Jesuit observatory being closed by a communist government, like those of Kalocsa and Zikawei. The Jesuits established two other observatories in Cuba, one in Cienfuegos in 1910 and another in Santiago de Cuba in 1930. Both were basically meteorological stations, which sent their observations to Belén Observatory. The one in Cienfuegos also had an astronomical section with several small telescopes.

7.4 Typhoons and Cyclones in the Philippines

In Chap. 6, we saw that the observatory of Manila had been founded in 1865, mainly for meteorological observations, equipped with the proper instruments and, from 1869, with a Secchi Meteorograph.[10] In 1884 a royal decree from the Spanish government named the observatory as the official center of meteorology for the Philippines. It had the title, *Servicio Meteorológico de la Isla de Luzón* (Meteorological Service of the Island of Luzón), and a network of 13 meteorological stations, all supported with the necessary financial funding. The stations sent three daily observations – six during the typhoon season – and the observatory was in contact with others in Hong Kong, Macao, Shanghai and Saigon. The official character of the observatory is of interest because at that time there was no such official meteorological service in Spain itself; the first was founded 3 years later in 1887.

The study of "typhoons" (the name given to Pacific cyclones, also known in the Philippines as *baguios*) by Faura began 1879.[11] In July of that year he made the first forecast of a typhoon's arrival in Manila. This was probably the first official warning of a cyclone in the Far East. In November of the same year Faura gave warning of another typhoon threatening Manila 2 days ahead of its arrival, so that precautions could be taken such as the closing of the harbor. The accuracy of these predictions enhanced the prestige of the observatory, and made clear the need to pay attention to its weather reports, especially regarding the arrival of typhoons. In 1882 Faura published his work about typhoons in the Philippines, highlighting the phenomena

[10] John N. Schumacher, One hundred years of Jesuit scientists: The Manila Observatory. *Philippine Studies* 13 (1965), 258–286.

[11] Necrological Session in honor of Rev. Fr. Federico Faura, S. J. Articles by Luis Rodes, Eduardo Fontseré and Eduardo Alcobé. *Memorias de la Real Academia de Ciencias y Artes de Barcelona*, Barcelona, 22, (1930), 69–92.

Fig. 7.2 José Algué (1856–1930) and Miguel Saderra Mata (1887–1938) at the Manila Observatory (Philippines) (Manila Observatory)

which allow one to predict their path, *Señales precursoras de temporal en el Archipiélago Filipino*, (Precursor signs of storms in the Philippine Archipelago, 1882). Faura, who was in contact with Viñes, soon became aware that typhoons in the Philippines were of the same type of meteorological phenomena as hurricanes in the Caribbean. Thus, one could apply to typhoons the same laws that Viñes had proposed for hurricanes. Faura's approach, like that of Viñes, was basically empirical, based on the observations made at the observatory. His careful analysis allowed him to put forward some pioneer notions on the importance of the motion of air masses, and on the concept of fronts, their formation and development. Basing himself on ideas suggested by Viñes, Faura also developed an aneroid barometer adapted to detect cyclones.

Algué, who succeeded Faura in 1897, continued to study and forecast typhoons (Fig. 7.2).[12] In 1898 he published, *Baguios o ciclones filipinos. Estudio teórico práctico* (Baguios or Philippine cyclones. A theoretical and practical study), a comprehensive work based on the observations made at the observatory from its founding. The Spanish edition was followed by one in French and in 1904 by an

[12] Ignacio Puig, El P. José Algué, S. J., *Ibérica*, 833 (1930), 392–398.

English translation: *The cyclones of the Far East.* In this work Algué describes the main characteristics of typhoons, the areas where they originate and the trajectories they follow. He identifies two types of typhoons: those with parabolic trajectories, which move clockwise around a center located in the north Pacific; and those with a straight trajectory, which move westward from the Philippines to the southern coast of China. For many years, this was the most important work on typhoons in the Philippines. Paul Bergholz, a German meteorologist, copied sections of Algué's book in one that he published about cyclones in the Far East. His book appeared first in German in 1900 and afterwards in English, but its debt to Algué's book was not acknowledged. Algué improved the instrument designed by Faura, adding an adjustable scale and an indicator of the wind direction to the barometer. This instrument, which Algué called a *barociclonómetro*, was commercialized in 1905 and for many years was very popular among navigators in the Pacific to warn them of the proximity of a storm. Bergholz also reproduced almost exactly the instrument of Algué, and sold it under his own name. Bergholz' book and instrument were extensively used in German ships in ignorance of the fact that they were copies of Algué's works.

In 1899 the Philippines became dependent on the United States and this had an effect on the situation of the observatory. After several visits by Algué to Washington and hesitation on the part of the American administration, the latter acknowledged the work done by the Manila Observatory and in 1901 handed over to it the direction of the newly created Philippines Weather Bureau, appointing Algué as director. Thus the observatory retained its official character. William Doberck, Director of the Observatory in Hong Kong, who had already shown his animosity against the Jesuits in his relations with Faura and Dechevrens, took a negative view over this process. He denounced the Manila Observatory to the U. S. Weather Bureau for its lack of professionalism, and for a time it was prohibited from giving cyclone warnings outside the Philippines. However, when full support from the United States Administration and official approval were regained, the observatory enlarged its network of meteorological stations, establishing 51 new stations, 9 of first order, 25 of second and 17 of third.

Selga's contribution to the study of typhoons was his extensive catalogue from 1848 to 1934, which included important historical research. His work, *Charts of remarkable typhoons in the Philippines 1902–1934. Catalogue of Typhoons 1848–1934* (1935), contains an atlas with maps of the trajectories of the most important typhoons. Selga had to handle the difficult situation that arose during the Japanese occupation of the Philippines, when the American Jesuits were imprisoned, and he witnessed the destruction of the observatory. Toward 1920, American Jesuits had begun to replace Spaniards at the observatory. Exceptional among them was Deppermann, who had gained his doctoral degree from Johns Hopkins University and worked at Lick Observatory and the U. S. Weather Bureau.[13] In 1932 he spent some time at the Geophysical Institute of Bergen (Norway), where he came into

[13] James J, Hennessey, Charles Deppermann S. J., Philippine scientist. *Philippine Studies*, 5 (1957), 311–335.

contact with the Bjerknes' school of meteorology. Vilhem Bjerknes and his son Jakob had developed between 1917 and 1940 the fundamental ideas of modern meteorology, introducing theories of circulation and atmospheric fronts. Deppermann, after his return to Manila, began to apply the ideas of the Bjerknes' school to the formation and evolution of typhoons. This was his most important contribution to meteorology.

Deppermann continued Algué's classic work on typhoons in the Philippines. For the first time he applied modern ideas of frontology to tropical storms and the analysis of air mass with reference to the genesis and paths of typhoons. This was the first such quantitative analysis applied to tropical cyclones. Two of his most important works are *Outline of Philippine frontology* (1936) and *Some characteristics of Philippine typhoons* (1939) published in Manila by the Bureau of Printing. In his works Deppermann describes the physical characteristics of typhoons, such as barometric oscillations and waves and the relation between barometric gradient and wind strength. He discussed the differences between tropical and extra-tropical storms and the use of the motion of cirrus clouds to prognosticate the movement of cyclones. He also proposed a theoretical model of typhoons in terms of a Rankine vortex with two convection rinks. Deppermann based his studies on a broad collection of observations. Considered the leading authority on tropical cyclones, his studies cast light on the characteristics of typhoons, their nature and dynamics. In the *Compendium of meteorology* (1951) edited by T. F. Malone, Deppermann is cited more than 40 times. Gordon E. Dunn, in his chapter on tropical cyclones, asserts that tropical meteorology is largely indebted to Deppermann for his detailed description and analysis of typhoons in the Philippines. In the chapter on aerology and tropical storms, Herbert Riehl comments, "Deppermann is one of the few writers who had made a detailed effort to calculate radial and tangential velocity components and made a theoretical model…Apart from Deppermann, writers have contented themselves with application of simple hydrodynamics." Other authors in the same book praise Deppermann's contributions to the study of tropical storms.[14] The Japanese meteorologist, Kazuo Ogasahara, acknowledges in his handbook of meteorology (*Kishogaku Tsuron*) that Deppermann had shown a deep comprehension of the contributions introduced by the Bergen school to meteorology.[15] After the war, the independence in 1946 of the Philippines from the United States had an effect on plans for the renewal of the observatory in the field of meteorology. The new Philippine government established a meteorological service which did not include Manila Observatory. Thus the observatory closed the section of meteorology and focused on seismology, ionospheric studies, solar physics and astrophysics.

[14] T. F. Malone, *Compendium of meteorology*. Boston: American Meteorological Society, (1951) (G.E. Dunn, Tropical cyclones, 887–901; H. Riehl, Aerology of tropical storms, 902–916).

[15] Hennessey, "Deppermann", 317–319.

7.5 Cyclones in the China Sea and the Indian Ocean

As already mentioned, meteorology was the first purpose of the Observatory at Zikawei. From its foundation, it had a complete set of meteorological instruments, including a Secchi meteorograph, a second being added in 1881. From 1876 Dechevrens gave the observatory a solid scientific standing in meteorology and established an active network of international relations.[16] In 1879, he published a detailed study of the nature and effects of a typhoon which had caused heavy damage in Shanghai.[17] This helped to bring the work of the observatory to public notice. Thus, the study and forecasting of typhoons, which devastate the coasts of China each year between July and September, became the main work of the observatory. Although considered a private institution, the observatory was associated from 1876 with the Chinese Maritime Customs Administration. This collaboration and that of the telegraph and maritime companies turned the observatory into a true meteorological service. In fact it was the only one which actually worked at that time in China with connections with the other meteorological stations in the Far East. In 1882, Zikawei began publishing a daily bulletin in the newspapers of Shanghai, giving the weather conditions and a forecast for the following days. From 1890 the observatory began to issue synoptic weather maps for the whole coast of China and to send by telegraph warnings of storms. Dechevrens also proposed to the British authorities in Hong Kong a detailed plan for collaboration between all the meteorological stations in China centralized in Zikawei, but unfortunately the plan was not accepted. William Doberck, Director of the Observatory in Hong Kong, was especially opposed to this plan, because he wanted his observatory to coordinate the work of all the other observatories in the region. Doberck tried to put all sorts of obstacles to the diffusion in Hong Kong of the information sent from Zikawei and Manila, and to discredit the importance of their work.

In 1885, 6 years after the work of Faura, Dechevrens began his study of typhoons using the long series of observations available in Zikawei. He put special emphasis on the vertical distribution of temperature and its effect in the formation of cyclones. Basing himself on these records he developed his own theory about the structure of cyclones. He suggested that they are formed on the boundary between high and low pressure centers by a vortex of turbulence in the high regions of the atmosphere, between 500 and 7,000 m. This attracts to its center the air masses from below. From this level, air is dispersed horizontally and at a certain distance it descends to converge to form the storm center.[18] Dechevrens called attention to the importance of winds in the high atmosphere and the formation of clouds at those levels. He related both phenomena with changes in the path of cyclones, as Viñes had done. Together with Faura and Algué, Dechevrens was one of the first to study the nature and

[16] Henri Gauthier, Un maître en physique du globe. *Études* 178 (1924), 272–291.

[17] Marc Dechevrens, Le typhon du 31 juillet 1879. *Observatoire de Zi-Ka-Wei, Mémoir 4* (1879).

[18] Marc Dechevrens, *Sur la variations de temperature observé dans les cyclones.* Shanghai: Observatoire de Zikawei (1887).

characteristics of typhoons in the Far East. Dechevrens gave special importance to the study of the vertical component of winds. He studied this phenomenon over many years and designed an instrument for this purpose which he called a "clino-anemometer". In 1889, during the World Fair in Paris, this instrument was installed on the top of the Eiffel Tower. Dechevrens returned to France in 1887, and installed an observatory in the Island of Jersey with a 50 m high tower which he used to continue his study of the vertical component of winds. In 1883 Dechevrens also contributed to the practical aspects of meteorology by developing a code of visual signs to warn mariners of weather conditions, especially during the typhoon season. For this purpose a mast was installed in the harbor of Shanghai, where the signs were displayed, linked to the observatory by a telegraph line.

Froc, Dechevrens' successor, perfected the system of storm signs. From 1896 the new system was put in place in all the more important ports of China. In 1915 Froc added to the visual system a cannon shot in the harbor of Shanghai to mark the imminent arrival of a typhoon as indicated by the observatory. In 1918 the system was enlarged to include more information and lighted signs were added that were visible during the night. The system was adopted in many ports of the Far East coast. The occurrence of more than 30 cyclones each year on the coasts of China indicates the great need for such a system. It helped to prevent many casualties, the loss of ships, and great material damage. The Observatory of Zikawei was in continuous contact with other observatories, with shipping companies, and later, by radio, with ships sailing through nearby waters. Froc became famous for his accurate forecasts and warnings of typhoons, sometimes two days ahead of their arrival on the coasts of China. So much so, that he was given the name, *Père des Typhons.* In Hong Kong, Doberck's animosity toward the Jesuits motivated his prohibiting the publication of the warnings from Zikawei. This brought grave consequences with the typhoon of 1906; Froc had warned of this but Doberck refused to acknowledge the warning, with resultant serious damage and casualties. The Governor of the British colony investigated this case, and thus brought to light the bad relations between Doberck and the Jesuits of Manila and Zikawei. Froc collected into an atlas the observations made at the observatories with the trajectories of 620 typhoons for the period 1893–1918.[19]

Gherzi had arrived in Zikawei in 1920 and from 1940 took charge of the sections of meteorology and seismology.[20] In 1927 he began to give meteorological information by radio and kept contact with ships passing through nearby waters, warning them about storms. He had to cope with the difficulties caused by the Japanese occupation and by the war that followed between nationalists and communists. In spite of all this, Gherzi kept the observatory in working order and had time to continue the study of typhoons. He was especially interested in their surface characteristics and elaborated some new hypotheses about their formation. Instead of applying the theory of atmospheric fronts to tropical cyclones, as Deppermann and

[19] Louis Froc, *Atlas of the tracks of 620 typhons*. Shanghai: Observatoire de Zikawei (1920).

[20] Jean-François Gauvin, La physique du globe tous azimuts: La carrière scientifique d'Ernesto Gherzi, S. J., 1910–1973. *Archivum Historicum Societatis Iesu*, 72 (2003), 45–80.

other meteorologists had done, Gherzi maintained that the theory of fronts was applicable to storms only outside the tropics.[21] He kept records of the observations made by the ionospheric sounder of solar radiation and of the amount of ozone in the atmosphere. He thought he could find a correlation between variations of the ionospheric layers and the motion of air masses which might allow better forecasts of the weather. Gherzi tried also to find a correlation between storms in the atmosphere and the generation of microseisms in the Earth's crust, as will be discussed in the next chapter. After the occupation of Zikawei by the Communist government, Gherzi moved to Macao, where, between 1950 and 1953, he reorganized the meteorological service. There he published an extensive work in two volumes, *The meteorology of China* (1951). In 1954, Gherzi moved again, first to Saint Louis University (Missouri) and later to Montreal (Canada) where he continued his work in meteorology.

Tropical cyclones similar to those in the Caribbean and the Pacific are also frequent in the Indian Ocean, where they cause damage in Madagascar and on the east coast of Africa. In general, they are produced to the east of Madagascar between October and March and propagate from east to west in a nearly straight path. From its foundation in 1889, the observatory of Tananarive, had paid special attention to the study of cyclones and the prediction of their arrival on the island. The observatory installed a network of secondary meteorological stations throughout the island, increasing them from 11 at first to 25. These stations depended on the Observatory until 1920, when they passed to depend directly on the colonial administration. Colin published his first study of cyclones in 1902 (*Les cyclons du 9 au 12 décembre 1902 à Ambohidempona,* The cyclones of 9–12 December 1902 in Ambohidempona). In 1905 he published a general work on the cyclones of this region, *Les cyclons à Madagascar,* (The cyclones in Madagascar). He continued to publish reports on the most important cyclones affecting the coasts of Madagascar. Colin observed that the path of these cyclones generally followed three courses: at a tangent to the Eastern coast; describing a parabola; and penetrating into the interior of the island by descending through the Mozambique Channel and entering via the south-west coast of Madagascar. He found that when cyclones enter the high plateau of the island at altitudes of 1,000 and 2,000 m, they rapidly lose strength and the wind velocity diminishes radically.[22]

Poisson continued the study of cyclones, observing and forecasting them and taking note of their path. In 1927, after a cyclone had caused severe damage, Poisson was officially appointed by the French colonial administration the technical Director of meteorology in Madagascar. In 1931 he published, *Les cyclones tropicaux* (Tropical cyclones), which received an award from the Colonial Academy of Science. From 1903 to 1943 the Observatory of Tananarive was officially put in charge by the colonial government of forecasting the arrival of cyclones in Madagascar. This service and the time service were the only ones supported

[21] Ernesto Gherzi, On the constitution of typhoons. *Bulletin American Meteorological Society*, 19 (1939), 59–66.

[22] Elie Colin, *Les cyclones à Madagascar*. Tananarive: Imprimerie Officielle (1905).

economically by the colonial authorities. The observatory never had sufficient funds to update and renew its instruments and the work was reduced to making observations. The Jesuits found it difficult to carry on scientific research in Madagascar owing to lack of personnel; Colin and Poisson worked practically alone, and they were subject to the routine demands of observations with great difficulty in establishing contact with other scientists. Thus scientific work in Tananarive did not have the importance of that in Belén, Manila, and Zikawei, but it deserves to be mentioned as another example of Jesuit dedication to the study of tropical cyclones.

Chapter 8
Earthquakes and Seismology

8.1 Early Jesuit Studies of Earthquakes

Jesuit scientists showed a special interest in seismology – the science that studies earthquakes, their causes, and the propagation of seismic waves in the interior of the earth – so much so that it has occasionally been called "the Jesuit science".[1] The work of the Jesuits on earthquakes during the seventeenth and eighteenth centuries will be presented as the first part of this story.

When the first Jesuit colleges were established, the study of earthquakes and volcanoes was contained in the second book of Aristotle's *Meteorologica,* a required subject in the programs of natural philosophy. As mentioned in a former chapter, the most used commentary on this work in Jesuit colleges was known as the *Conimbricensis,* published in 1593 by the professors of the college of Coimbra. In this work the explanation of earthquakes is contained in the third volume, treatise 11, and follows Aristotle's ideas with minor modifications. Earthquakes were thought to be produced by winds trapped in cavities inside the earth which shake it when trying to escape. This could be considered official Jesuit doctrine about earthquakes at that time. In his book *Mundus subterraneus* (1664) Kircher developed this doctrine in a different form, linking the occurrence of earthquakes with the air and fire conduits (*aerophilacios* and *pyrophilacios*), which he proposed existed inside the earth. He added to the Aristotelian doctrine the modern idea of the explosion of inflammable materials and compared earthquakes with the explosion of mines. Kircher presented, as an example, the earthquake of 6 April 1667, which partly destroyed

[1] Manuel M. Sánchez Navarro-Neumann, Os Jesuitas e a sismologia. *Broteria*, 24 (1937), 145–151; Daniel Linehan, Jesuits in seismology. *Earthquake Information Bulletin*, 16 (1984), 156–165; Agustín Udías and William Stauder, The Jesuit contribution to seismology. *Seismological Research Letters*, 67 (1996), 10–19; John Ebel, The Jesuit science. *Company* (Summer) (2009) 28–31; Agustín Udías, Jesuits' studies of earthquakes and seismological stations. In M. Kölbl-Ebert (ed.) *Geology and Religion: A history of harmony and hostility*. London: The Geological Society S.P. 310, 135–143 (2009).

A. Udías, *Jesuit Contribution to Science: A History*,
DOI 10.1007/978-3-319-08365-0_8

the city of Ragusa (Dubrovnik), causing about 6,000 deaths. In 1648 Martin Lister in England had suggested the explosive origin of earthquakes, an idea that Nicolas Lemery spread in France. Newton and Buffon supported this idea and it took the place of the Aristotelian explanation among the authors of modern science. Some Jesuits incorporated these new ideas in their works about the origin of earthquakes. Italy is one of the countries in Europe most prone to earthquakes. It is hardly surprising that several Italian Jesuits wrote on this subject. Michele de Bono (1697–1775), professor in the college of Palermo, wrote a book about the origin of earthquakes: *Discorso sull'origine di terremuoti* (1750), using both Aristotelian and modern ideas. Belgrado, professor at Parma, described the damage to this city caused by an earthquake in 1738 and speculated about its causes. Giulio Cesare Recupito (1581–1647), Giuseppe Giulio (1747–1832) and a Spaniard exiled in Italy, Francisco Gusta (1744–1816), wrote about earthquakes in Sicily and Calabria, two of the most seismic regions of Italy. Outside Italy Johann Baptista Grossinger (1728–1803), professor at the Jesuit College in Vienna, wrote in 1738 about the earthquakes in Hungary.

The earthquake in Lisbon on November 1, 1755, followed by a large tsunami, caused the destruction of this city and damage and casualties in the western coastal towns of Portugal, Spain, and Morocco and was felt in a large part of Europe. This earthquake aroused a new interest in Europe for the study of these phenomena and marks a landmark in the history of seismology. It also aroused philosophical and theological debate, in which Immanuel Kant and Voltaire (François-Marie Arouet), among others, took part. Jesuits could not fail to provide interpretations of this event. The most tragic case was that of the Italian Gabriel Malagrida (1689–1761), who had been a missionary in Brazil and returned to Portugal. After the earthquake he insisted in his sermons and in his book, *Juizo da verdadera causa do terremoto* (Judgement about the true cause of the earthquake, 1756), that it had been a punishment by God. This provoked the wrath of the First Minister of the King of Portugal, Sebastião Jose de Carvalho, Marquis of Pombal, who accused him of terrorizing the people with his sermons and thus hindering the recuperation of the city. Malagrida was first banned to Setubal and later, in 1759, he was accused of participating in an attempt against the king. He was put in prison, judged of heresy by the Inquisition, and executed in 1759. It seems, however, that the real motives were Malagrida's strong criticism of the colonial practices in Brazil of the Captain General, a brother of Pombal, and the influence he had at court as confessor of the king's mother. Other Jesuits considered the earthquake from the scientific point of view. José de Mesa (1705–1784), a professor at the college in Cadiz, described in detail the advance of the sea into the city due to the tsunami. Far from Lisbon, Stepling wrote about the changes produced by the earthquake in the thermal springs of Tepliz near Prague and he raised questions about the origin of earthquakes in his *Fragen uber das Erdbeben* (Questions about the earthquake, 1784). On this occasion Gaspar Sagner (1720–1781), a professor of mathematics at Prague, also published a treatise on the causes of earthquakes: *Discursus de terrae motus causa* (1756). After the Lisbon earthquake, other Jesuits wrote about the destruction of Lisbon: Johann Schwab (1731–1795) a professor at the University of Heidelberg, and Antonio Pereyra (1693–1770), both in the same year, 1783.

Central and South America are prone to the occurrence of strong earthquakes. The Jesuits who wrote about the natural history of these regions, as we saw in Chap. 6, dedicated parts of their books to the description of these earthquakes. Acosta, in *Historia natural* (Libro 3, Cap. 26), begins by pointing out, against the traditional Aristotelian doctrine, that earthquakes and volcanoes are not always related, as in America there are large earthquakes in regions where there are no volcanoes. However, with regard to the origin of earthquakes he maintained the Aristotelian doctrine. His descriptions of large earthquakes are of greater interest, for example, one in Chile in 1575 and two in Peru, one in 1582 in Arequipa and another in 1586 in Lima. He also describes the tsunamis which followed the earthquakes in Chile and Peru. Describing the one in Lima, he says that after the earthquake the sea advanced with strength over the beaches and covered the land for almost two leagues, the height of the waves reaching 14 *brazas* (about 26 m). He holds that earthquakes are more frequent near the sea shore and in years of very abundant rain. He justifies this by saying that water blocks the exit of wind inside the earth and causes its violent shaking. Cobo in his *Historia*, dedicates a chapter to earthquakes in Peru (Book 3, Chapter 20), describing in detail the earthquake of Arequipa of 24 November 1604, which he considers the most powerful. This earthquake was followed by a tsunami which flooded the shore with waves up to 20 m high. He refers to earthquakes in Lima in 1606 and 1609, Trujillo 1619, Cuzco 1650 and Santiago de Chile 1647. He also concluded from his experience that earthquakes are more frequent near the sea shore. In a letter that he published Juan González Chaparro (1581–1651) describes in great detail the earthquake which destroyed Santiago de Chile, on 3 May 1647. Descriptions of other earthquakes in Chile are found in the histories by Ovalle and Rosales. Pedro Lozano (1697–1759), a professor at the college of Córdoba (Argentina) wrote a detailed description of the Lima earthquake of 28 October 1746 which caused great damage to the city and was followed by a tsunami which flooded the port of Callao. In Germany, Joseph Pfriem (1711–?) a professor at the colleges of Mainz, Bamberg and Spira, also wrote about this earthquake basing himself on the news received from the Jesuit missionaries in Peru.

8.2 Seismological Stations in Europe

Seismology is a relatively modern science dating from toward the end of the nineteenth century with the development of the first seismographs, instruments made to detect and record the ground motion produced by earthquakes. The first such instruments were developed in Italy and Japan, two seismically active countries. Between 1880 and 1920 seismological stations were established throughout the world. Thus the restoration of the Society of Jesus coincided with the beginning of seismology as an instrumental science. Jesuits began to install seismographic stations in the new observatories we have described in Chap. 6. In 1888 Giovanni Egidi (1835–1897) installed the first of these instruments in a Jesuit observatory, the *Osservatorio Tusculano* in Frascati, which became one of the earliest Italian seismological stations. Egidi himself constructed this instrument, a "seismoscope", still lacking a graphic recorder. He collaborated with the Italian seismologist Michele Stefano de Rossi in keeping seismological observations that were published in the journal *Bolletino del Vulcanismo Italiano.* In 1908 the observatory installed two new Cancani and Agammenone seismographs, provided by the *Uficio Centrale de Meteorologia e Geodinamica.*

In 1902 the Cartuja Observatory (Granada) installed two Vicentini and Stiattesi seismographs, some of the first in Spain. The Observatory was located in the most seismic area of Spain and the seismological station soon acquired great importance. In 1908 Manuel Sánchez Navarro-Neumann (1867–1941), took charge of the station and was director of the observatory between 1915 and 1931.[2] Sánchez Navarro-Neumann is thought to be one of the great early seismologists in Spain. As lack of funds prevented the acquisition of new instruments, he designed and built them himself, following the models produced in Germany by Mainka and Wiechert. Thus he built a mechanical seismograph with a mass weighing 3,000 kg and a ground amplification of 1000, a great achievement for an instrument of that type. In 1924 Sánchez Navarro-Neumann also built two electromagnetic seismographs of the type newly developed by Boris Galitzin in Russia and from that time replaced the mechanical instruments. His principal seismological work was the study of earthquakes in the south of Spain. Between 1917 and 1921 he published the first modern catalogue of earthquakes in the Iberian Peninsula. Aware of the lack of seismological literature in Spanish, he published the first modern book about earthquakes and their effects on buildings, *Terremotos, sismógrafos y edificios* (Earthquakes, seismographs and buildings, 1916). Antonio Due Rojo (1898–1975) tried to keep the station in good working order in the difficult years after the Spanish civil war. He continued the work of Navarro-Neumann and published studies of the seismicity of southern Spain. Lack of funds prevented the renewal of the instruments and the development of the State Seismological Service diminished the importance of the observatory. In 1904 a seismological station had been established at the Ebro Observatory with Vicentini seismographs. The instrumentation kept being renewed and at present the station continues in operation, but with no Jesuit participation.

Interestingly, in 1908 a seismological station was installed in Ireland where there are seldom any earthquakes. The station at Mungret College, near Limerick had two Bosch-Omori horizontal seismographs. In 1911, William J. O'Leary (1869–1939), a professor of physics, built a seismograph of his own design consisting in an inverted pendulum with a mass of 600 kg held by three wires, which recorded both horizontal components. In 1915, O'Leary moved his seismograph to the Jesuit Faculty of Philosophy in Rathfarnharm Castle, and there he built a new instrument of the same type with a larger mass of 1,500 kg. In 1929 O'Leary moved to Australia where he became director of the Riverview Observatory. In 1935 Richard E. Ingram (1916–1967) became director of the seismological station of Rathfarnharm Castle.

---

[2] Antonio Due Rojo, *Labor científica del R. P. Manuel Sánchez Navarro, S. J.* Granada: Observatorio de Cartuja, (1940) and El R. P. Manuel M. Sánchez Navarro, S. J. *Boletín Real Sociedad de Historia Natural*, 39 (1941), 117–120.

He had studied mathematics at University College in Dublin, obtained his doctoral degree in mathematics from Johns Hopkins University (Baltimore) and trained in seismology at the California Institute of Technology (Pasadena).[3] Ingram renewed the instruments of the station, but he was more interested in mathematics and the theoretical aspects of seismology. He published several articles on seismic wave propagation and the mechanism of earthquakes. Ingram was a professor of mathematics at University College Dublin, and contributed to the promotion of mathematical teaching in Ireland. He took part in the edition of the mathematical works of William R. Hamilton. In 1962 he moved to Georgetown University as professor of mathematics and collaborated in seismological problems with the U.S. Coast and Geodetic Survey and the Dominion Observatory of Canada. After his death in 1967 the seismological station at Rathfarnharm Castle was closed. In 1909 a Milne horizontal seismograph had been installed at Stonyhurst Observatory. In 1920 and 1924 new instruments were acquired: a Wiechert horizontal seismograph and two Milne-Shaw horizontal seismographs. Work in seismology was reduced to recording observations and the station was closed in 1947. In 1920 a Mainka seismograph was installed in the observatory on the Island of Jersey which operated as long as the Jesuits were there, so until 1979. Between 1924 and 1928 the college-seminary of Travnik (Bosnia) installed a Conrad horizontal seismograph.

In conclusion, although Jesuits established and managed several seismological stations in Europe, real seismological research was only carried out in the Cartuja Observatory, especially by Sánchez Navarro-Neumann, who was a true pioneer of seismology in Spain. In the Ebro Observatory, although they had up-to-date seismological instrumentation, their work never went beyond recording observations. Ingram's theoretical seismological work was not really related with the seismological station of Rathfarnharm Castle.

8.3 Seismology in North America. Jesuit Seismological Association

In 1900 Odenbach, who had established a meteorological station in Saint Ignatius College (today John Carroll University) in Cleveland, began to get interested in seismology and installed two seismographs of his own design. He became aware of the publicity generated by the news about earthquakes in the newspapers and thought that this could be used to promote Jesuit colleges and universities. At that time Catholic educational institutions were not well considered in the United States, and Odenbach believed that their prestige could be increased in this way. He conceived the idea of creating a seismological network taking advantage of the geographical distribution of Jesuit colleges and universities in the United States. Such a network did not exist at that time. In addition, he thought that a uniform instrumentation

[3] Thomas Murphy, The Rev. R. E. Ingram, S. J., *Quarterly Journal of Royal Astronomical Society* 9 (1968), 226–227.

would be a very positive aspect of the network, since seismographic stations were then using different types of instruments. In 1908 Odenbach presented his idea to the Jesuit superiors and presidents of colleges and universities. In the presentation of the project he insisted on the great benefit this would provide to the Society of Jesus and to the Catholic Church. From the apologetic point of view, the project would also show that the Church was not an enemy of science and progress, a criticism that was often made. The answer was very positive and 15 Jesuit universities and colleges joined the project, namely: Cleveland (Ohio), Georgetown (Washington, D. C.), Brooklyn (New York), Fordham (New York), Worcester (Massachusetts), Buffalo (New York), Spring Hill (Alabama), New Orleans (Louisiana), Saint Louis (Missouri), Chicago (Illinois), Milwaukee (Wisconsin), St. Mary's (Kansas), Denver (Colorado), Santa Clara (California) and Spokane (Washington). In Canada the college of Saint Boniface (Manitoba) also joined the project. Thus the stations practically covered the whole continent of North America. Odenbach selected for their common equipment the German Wiechert horizontal seismograph, a modern instrument at an affordable price.

In 1911 instruments were installed in the 16 stations and the network began working under the name "Jesuit Seismological Service" (JSS).[4] The stations of the network sent their observations to the central station established in Cleveland. The central station processed the data and sent them to the International Seismological Center (*Bureau Central International de Seismologie*) which had been created in 1906 in Strasbourg (France). In fact, the JSS was the first seismological network at continental level with uniform instrumentation. At that time the federal government of the United States did not have a similar seismological network and there were only 44 stations in the whole country, so that Jesuit stations were one third of them. However, what looked like a very promising beginning floundered very soon. Problems arose about collaboration between the stations and government institutions which promised funds. Odenbach was completely opposed, but others were in favor of this collaboration, especially Francis Tondorf (1870–1929), who was in charge of the station at Georgetown University.[5] At the end of the same year, Odenbach resigned as coordinator and proposed that the stations continue working independently and send their observations directly to the center in Strasbourg. This was the end of this first attempt.

In 1925, James B. Macelwane, about whom we will learn more in the next chapter, returned to Saint Louis University, after getting his doctorate at the University of California, Berkeley, and he became director of the Department of Geophysics. Macelwane soon began to think of the possibility of reviving Odenbach's project. Some of the stations had ceased working, but most, like those of Denver, Georgetown,

[4] N. H. Heck, The Jesuit contribution to seismology in the U.S.A. *Thought*, 19 (1944), 221–228; James B. Macelwane, The forerunner-The Jesuit Seismological Service. In J. B. Macelwane (ed.) *Jesuit Seismological Association, 1925–1950. Twenty-fifth Anniversary Commemorative volume.* St. Louis (Mo.): Saint Louis University (1950), 3–16; Udías and Stauder, "Jesuit contribution", 15–16; Carl-Henry Geschwind, Embracing science and research. Early twenty-century Jesuits and seismology in the United States. *Isis*, 89 (1998), 27–49.

[5] Geschwind, "Embracing science" 38–42.

Fig. 8.1 Blessing of a seismograph at the Seismological Station of Georgetown University (Georgetown University Library) (Blessing of seismograph)

Spring Hill and Saint Louis, continued and kept publishing their observations. The station at Georgetown under Tondorf's direction had updated its instruments with modern Galitzin electromagnetic seismographs, recently developed in Russia. Sánchez Navarro-Neumann from the Cartuja Observatory encouraged Macelwane to go ahead with the project, which he considered to be very important. Several institutions in the United States, like the Carnegie Institution and the National Research Council also supported the idea. In 1925, the project received the approval of the Jesuit provincial superiors. The network was established under the new name, "Jesuit Seismological Association" (JSA), with a central station at Saint Louis and Macelwane as president.[6] The network was formed of 14 stations: Saint Louis (Missouri), Cleveland (Ohio), Georgetown (Washington, D. C.), Brooklyn (New York), Fordham (New York), Worcester (Massachusetts), Buffalo (New York), Cincinnati (Ohio), Spring Hill (Alabama), New Orleans (Louisiana), Chicago (Illinois), Milwaukee (Wisconsin), St. Mary's (Kansas), Denver (Colorado), Santa Clara (California), San Francisco (California) and Spokane (Washington) (Fig. 8.1). The central station at Saint Louis not only collected the data from the stations, but using these observations and those from other stations determined the origin times and locations of earthquakes throughout the world. These were published in a bulletin much appreciated by seismologists worldwide. The JSA established an agreement with the United States Coast and Geodetic Survey; this body also

[6] James B. Macelwane, The Jesuit Seismological Association , En J. B. Macelwane (ed.) *Jesuit Seismological Association, 1925–1950. Twenty-fifth Anniversary Commemorative volume*. St. Louis (Mo.): Saint Louis University (1950), 17–23.

received the observations from the stations and independently determined the origin times and locations of earthquakes. With this service the JSA, under Macelwane's direction, enlarged its scope and placed itself at the vanguard of seismology. This service continued till 1963, when it was found that other institutions and agencies, like the National Earthquake Information System (NEIS) in the United States and the International Seismological Center (ISC) in the United Kingdom, were doing the same studies but with greater means. Most JSA stations updated their instrumentation and continued working until about 1980. Some even continue to the present day, but with minimal Jesuit presence. In 1962 the stations of Georgetown, Saint Louis and Spring Hill were chosen to form part of the network established by the Government of the United States around the world, known as the World Wide Standard Seismographic Network (WWSSN), which lasted until 1990. This network was formed of 125 stations using the identical latest type of seismographic instruments for short and long periods. It was established to monitor the occurrence of underground nuclear explosions during the cold war. A total of 10 Jesuit stations participated in the program. Their choice is a sign of the confidence placed by the government of the United States in the JSA and the Jesuit work in seismology.

In the context of the JSA, several Jesuits excelled for their work in seismology. Joseph Lynch (1894–1987) was Director of the Fordham seismological station, first as a young professor between 1920 and 1923 and afterwards during almost 50 years from 1928 to 1977. He updated the instruments of the stations as new models were developed. He was a very careful observer and published works of an observational character. In 1946, invited by the government, Lynch traveled to the Dominican Republic to study the powerful earthquake that had occurred that year causing severe damage in the capital Santo Domingo. No seismographic stations existed at that time and he helped to install the first. A young Dominican engineer was trained in Fordham to be in charge of the station. Lynch published an interesting account of his long life as a Jesuit seismologist.[7]

Daniel Linehan (1904–1987) was Director from 1950 to 1974 of the Weston Seismological Observatory, which was part of Boston College (Massachusetts),[8] the Jesuit university. He had studied physics first, at Boston College, and then geology at Harvard University. In 1948 Linehan returned as professor to Boston College, where he founded the Department of Geophysics, of which he was director until 1963. In the department he trained several Jesuits who afterwards worked at other observatories. Between 1955 and 1980, he enlarged the Observatory with a network, consisting first of 6 stations and finally of 30 stations, covering the northeast of the United States. In 1940 he established the first seismographic station in Jamaica which functioned until 1975. Weston Seismographic Observatory continues today as a research institution of Boston College. Linehan was among the first to study the T phase from earthquakes with their epicenter in the sea. He was interested

[7] Joseph Lynch S.J., *Watching our trembling Earth for 50 years*. New York: Dodd, Mead and Co. (1970).

[8] John F. Devane and James P. McCaffrey, Reverend Daniel Linehan (1904–1987). *Bulletin Seismological Society of America*, 78 (1988), 1618–1620.

in the applied aspects of seismology and made contacts with oil exploration companies. These companies provided him with seismic prospecting equipment that he used in many studies of the shallow structure of the Earth's crust. In 1951 Linehan with Lynch's help used this technique to carry out archaeological studies under Saint Peter's Basilica in Rome. The results of these studies helped in the discovery of a tomb identified as that of Saint Peter. Lynch from 1957 to 1970 and Linehan from 1970 to 1986, succeeded Macelwane as presidents of the JSA. Linehan was also famous as an explorer, making many scientific journeys to regions of Africa, Asia and South America, many of them on UNESCO missions. In 1954 he participated in an expedition to the Arctic, which had several scientific aims, such as the exact determination of the location of the North Magnetic Pole. After a long and hazardous journey in a small ship he located it near the Prince of Wales Island, to the north of Canada.

Between 1955 and 1958, Linehan took part in three expeditions to the Antarctic with the United States Navy. There, among other geophysical observations he determined the thickness of the ice layer using seismic exploration methods. Linehan was the first priest to celebrate Mass on both the North and South Poles and also the first to perform a baptism in the Antarctic. Linehan was an enthusiastic professor of geophysics, much loved by his students, a prolific writer of scientific and popular science articles and a frequent speaker in scientific conferences on geology and seismology. He was also frequently invited to speak on subjects related with his expeditions, especially those to the Antarctic. Two other Jesuits took part in expeditions to the Antarctic during the International Geophysical Year (1957–1958). Edward Bradley (1923–1996), director of the seismological station at Xavier University (Cincinnati) was officially the chief seismologist at the Ellworth Station (Antarctic). He determined the topography of a large area under the ice sheet using seismic exploration methods. A mountain, Bradley Nanutak, was named after him. Henry Birkenhauer (1922–2011) director of the station at John Carroll University (Cleveland), carried out seismic studies in the Antarctic for a period of 18 months.

In 1960, William Stauder (1922–2002), after finishing his doctorate in geophysics in the University of California, Berkeley, returned to Saint Louis University to take charge of the Department of Geophysics and the seismological stations which Macelwane had established.[9] Stauder's doctoral thesis on the mechanism of earthquakes was supervised by Professor Perry Byerly, whose thesis had been directed by Macelwane. Stauder gave a strong impulse to seismological studies in Saint Louis University and renewed and increased the number of seismological stations. From 1987 to 2002 he was the last president of the JSA. His studies to determine the mechanism of earthquakes from the analysis of seismic waves are recognized as an important contribution in seismology. In the 1960s the theory of plate tectonics was developed, an important landmark in earth sciences with a new understanding of the dynamics of the Earth. Stauder contributed to the establishment of this theory and with his study of the mechanism of earthquakes at subduction zones (zones of

[9] Brian J. Mitchell and Agustín Udías, William Stauder (1922–2002). *Seismological Research Letters*, 74 (2003), 373.

collision between plates) he helped to clarify the process at plate boundaries. In the 1970s he turned his attention to the seismic zone at the center of the United States. There between 1811 and 1812 four large earthquakes had shaken the region and destroyed the town of New Madrid. He installed a network of seismological stations to study in detail the small earthquakes that were still happening in that area. Together with Otto Nuttli he discovered from these studies a complex network of faults that crosses the region and that may cause a repetition of a large earthquake. Beside his scientific work, Stauder was Dean of the Graduate School of Saint Louis University and held various offices in scientific societies and commissions.

Among directors of stations of the JSA network mention should be made of Louis J. Eisele (1912–1988), director for 48 years from 1940 to 1988 of the station of Spring Hill, one of the few seismological stations in the south of the United States. He updated the station with new instruments and in 1962 supervised the installation and management of the instrument of the WWSSN station. From 1953 to 1970 he sent to the Associated Press office in Mobile information that was distributed to the press about the earthquakes recorded at the station – and was thus popularly known as "Father Earthquake".

A seismological station was established at the *Collège Jean de Brébeuf* in Montreal (Canada). In 1952 Maurice Buist (1902–1986), a professor of science at the college and its founder, had visited Weston Observatory and had been very impressed with the work Jesuits did there. Linehan encouraged him to install a seismological station in Montreal. That same year Buist installed the first seismographs and 3 years later he built an underground vault where he installed new instruments provided by the Dominion Observatory in Ottawa. Buist conceived the idea of transforming the observatory into a geophysical research center. Gherzi contributed to this idea with his studies of the ionosphere and radiometeorology. In 1960 Buist sent three young Jesuits to be trained for this purpose, but the project did not prosper. After their studies the young Jesuits preferred to teach science at state universities. The observatory continued its work in seismology only at an observational level. However, in 1975 the arrival of Gouin, after his expulsion from Ethiopia, strengthened the work of seismology at the observatory. In 1979, Gouin published the first catalogue of earthquakes in Ethiopia (*Earthquake history of Ethiopia and the Horn of Africa*) and in 2001 a catalogue of earthquakes felt in Quebec (*Historical earthquakes felt in Quebec from 1534 to March 1925*). Buist, the first and last Jesuit director of the observatory, retired in 1983. The observatory continues today as part of the college, but without a Jesuit on the staff.

8.4 Seismology in Central and South America

The west coast of Central and South America is a region where severe earthquakes are a frequent occurrence. We have seen how the naturalists of the old Society described the occurrence of important earthquakes in this region in their natural histories. Jesuits of the restored Society in Central and South America soon became involved in the study of earthquakes. In 1877 Jesuits in the observatory of the college of Puebla, Mexico, installed the first seismographs. Gustavo Heredia designed and built the instruments consisting of vertical pendulums which recorded horizontal motion on smoked glass circular plates. Heredia was able to record earthquakes in Mexico with these instruments and he published his observations for the period 1877–1906. These were in fact the first seismographs installed in the American continent, a few years before those installed in California. The seismological station was in operation until 1914 when the observatory and the college were closed. In 1906 the Jesuits installed two Bosch-Omori seismographs in Belén Observatory (Havana), the first in the Caribbean region. They were in operation until 1930. In 1880 Viñes had made a survey and published a report of the damage caused by an earthquake in the western part of Cuba. In 1920 Jesuits installed a seismological station in Sucre, Bolivia with instruments they themselves constructed using horizontal and vertical components. They were in operation for only a few years and reinstalled in 1945 only to be destroyed in the Bolivian earthquake of 1948. In 1949 German Saa (1909–1970) installed two Bosch-Omori seismographs at the *Colegio de San Luis*, Antofagasta (Chile); in 1957 he added a Wilson-Lamisson seismograph. In 1963 the station was closed because of the establishment nearby of a seismographic station of the WWSSN network. Saa intended to establish a seismographic station in Arica in the north of Chile, but this was never done.

In 1912 the Observatory of San Calixto in La Paz (Bolivia), where work in astronomy and meteorology has already been mentioned, established a seismological station that acquired a singular importance in South America.[10] The foundation of this station was due to a resolution passed at the Manchester meeting in 1911 of the International Seismological Association, at which the Jesuits were asked to install a station somewhere in the center of South America. This request shows the appreciation in the scientific community for Jesuit work in seismology. In 1912 and 1913 Descotes installed the first seismographs built by himself following Navarro-Neumann's design in the Cartuja Observatory, where he had spent some time training in seismology. In 1926 the station acquired three components Galitzin-Wilip electromagnetic seismographs, the most modern at the time. Descotes, with his very careful and exact observations, established the observatory as one of the best in South America. In their influential *Seismicity of the Earth* (1949) Charles F. Richter and Beno Gutenberg, two famous Californian seismologists, noted: "La Paz (observatory) at once became and still remains as the most important single seismological station of the world. This is a consequence of its isolated location, the sensitive instruments and the great care with which the records were interpreted and the reports issued under the direction of Father Descotes".[11] As no time signals were

[10] Ramón Cabré, 75 Años en la vanguardia de la sismología. *Revista Geofísica* 29 (1988), 47–56; Robert R. Coenraads, The San Calixto Observatory in La Paz, Bolivia. Eighty years of operation. Director Dr. Lawrence A. Drake S. J. *Journal and Proceedings, Royal Society of New South Wales,* 126 (1993), 191–198.

[11] Beno Gutenberg and Charles F. Richter, *Seismicity of the Earth and associated phenomena.* Princeton: Princeton University Press (1949).

broadcast in South America before 1926, Descotes kept exact time on the seismograms thanks to very careful astronomical observations, and over the years the station kept updating the instrumentation.

In 1959 Ramón Cabré (1922–1997) came to the observatory and became its director from 1965 to 1993.[12] With him the observatory became an important research center in seismology. In 1960 it participated in an international project to study the structure of the Earth's crust in the Bolivian high plateau (*altiplano*). In 1962 the station became one of the WWSSN stations with the best seismographs available. In 1966 Luis Fernández, after obtaining his doctorate at Saint Louis University came to the Observatory and gave a great impulse to seismological research and to the training of young researchers. He left the observatory in 1970 (and also later left the Jesuit order). Under Cabré's direction the observatory participated in several international seismological research projects. In 1972 a new seismological station was installed in an excavated tunnel in the Zongo Valley with modern equipment provided by the U. S. Geological Survey. This was 1 of 12 stations of that type installed world-wide. The following year the station was promoted and became what was called an "Adapted Seismic Research Observatory". In 1966 Cabré's prestige in South America for his work in seismology had motivated his nomination for 2 years as the first Director of the Regional Seismological Center for South America (*Centro Regional de Sismología para America del Sur*, CERESIS), a center depending on UNESCO for seismological cooperation between the South American countries. In 1993 the Government of Bolivia awarded Cabré the order of *Condor de los Andes* the highest Bolivian award.

In1992 Lawrence A. Drake (1931–2007), who from 1965 had been director of the Riverview Observatory in Australia, took over the direction of the La Paz observatory after Cabré's retirement.[13] Drake had retired as professor at Macquarie University and thought that he could do good service at La Paz. This was a very generous offer that involved changing to a new language and culture at an advanced age. In La Paz Drake applied his research on surface waves to determine the complex structure of the high plateau of the Andes. He continued to update the instruments of the station and undertook several research projects on the seismicity and structure of the Earth's crust in the Bolivian region. In 2001 for health reasons he retired as director and returned to Australia. He was the last Jesuit director of the observatory, which continues to operate but without Jesuit personnel.

In 1923 Simón Sarasola installed two seismographs at the observatory of San Bartolomé (Bogotá), the first in Colombia. Sarasola resigned from his post as director of the Colombian National Meteorological Service, to collaborate with Jesús Emilio Ramírez (1904–1981), in the establishment of a new Jesuit observatory devoted to seismology and meteorology (Fig. 8.2). In 1941 the new observatory was inaugurated with the name, *Instituto Geofísico de los Andes Colombianos*

[12] W. Richard Ott, Rev. Ramón Cabré Roigé, S. J. May 2, 1922-February 24, 1997. *Seismological Research Letters* 68 (1997), 901.

[13] Agustín Udías, Lawrence Drake, S. J. (1931–2007). *Seismological Research Letters* 78 (2007), 420.

Fig. 8.2 Jesús Emilio Ramirez (1904–1981) and the Wiechert seismograph at the *Instituto Geofísico de los Andes Colombianos* (Bogota) (Pontificia Universidad Javeriana, Bogota)

(Geophysical Institute of the Colombian Andes). In 1943 Sarasola returned to Cuba and Ramírez continued as sole director. Under the direction of Ramírez, the *Instituto* became one of the most important centers of seismological research in South America.[14] Ramírez had been trained in seismology at Saint Louis University where he received his doctorate under Macelwane's supervisión with a thesis on the study of microseisms. The seismological station of the *Instituto* was at a tunnel excavated into the rock where modern seismographs were installed. In 1957 the *Instituto* moved to a new building in the campus of the *Universidad Javeriana* (Bogotá) and in 1960 the *Instituto* became part of the University, when Ramírez was appointed its rector. In 1962 a WWSSN station was installed and in 1975 it became part of the Seismic Research Observatory (SRO) network. The SRO network was a very select worldwide network of only 13 stations established by the U.S. Geological Survey with modern seismographs installed in wells at a depth of 100 m. The choice of Bogotá as a station in these two networks indicates the prestige of Ramírez and the *Instituto* in seismological research. For a long time the *Instituto* was in charge of the Seismological Network of Colombia and had its own network of five seismographic stations, along with seven other stations with strong-motion instruments for the observation of near earthquakes.

[14] Jesús Emilio Ramírez, *Historia del Instituto Geofísico al conmemorar sus 35 años (Universidad Javeriana) I. Antecedentes y Fundación; II. Los años de la guerra 1944–1945; III. Consolidación del Instituto Geofísico*. Bogotá: Editora Guadalupe (1977).

Ramírez soon played an important role in the scientific circles of Colombia and South America. From 1947 to 1960 he was professor of geophysics at the *Universidad Nacional* of Bogotá. In 1967 with Luis G. Durán, he published *Nociones de prospección Geofísica*, a textbook of applied geophysics, the first in South America, and in 1969 the *Historia de los terremotos en Colombia* (History of earthquakes in Colombia) the most complete work on earthquakes in Colombia. Ramírez published a large number of scientific articles on the seismicity and tectonics of Colombia. From 1973 to 1979 he organized the Nariño Project, an international project for the study of the seismicity and the structure of the Earth's crust in the Colombian region. At that time this was one of the most important projects in earth science to be carried out in South America. Ramírez was very active in international seismological circles. He was president of CERESIS and from 1958 to 1966 president of the Academy of Science of Colombia. In 1971 the Colombian Government awarded him the *Orden al Mérito "Julio Garavito"*. In 1962, Goberna, expelled from Cuba by Castro's government moved to Bogotá and became assistant director of the *Instituto*. After Ramirez' death he became director and continued his work in seismology. In 1993 the last Jesuit director retired and the *Instituto* became part of the Engineering Department of the University with the new name, *Instituto Geofísico de la Universidad Javeriana.*

8.5 Seismological Stations in Asia, Africa and Oceania

In 1868 the Manila Observatory installed two home-made seismographs, the first installed by Jesuits in Asia. Faura recorded the violent Manila earthquake of 18 July 1880 with these instruments. The seismograms made then were the first of an earthquake in the Philippines. John Milne, a seismologist who worked in Japan, acknowledged the work of the Jesuits in Manila and began a correspondence with Faura. In 1881 the observatory installed new seismographs, now made in Italy, the best instruments at that time. In 1890 Miguel Saderra-Masó (1865–1939) arrived to Manila with new instruments and reorganized the seismological section of the Observatory. In 1895 he published the first work on seismology in the Philippines, *La sismología en Filipinas* and later the first catalogue of Philippine earthquakes, *Catalogue of violent and destructive earthquakes in the Philippines, 1599–1909* (1910). These two works mark the beginning of seismological studies in the Philippines. Saderra-Masó was the first to relate the earthquakes in Philippines with the tectonic accidents in the islands. He installed new instruments in Manila, among them, in 1930, the newly developed Galitzin-Wilip electromagnetic seismographs, and he established new seismographic stations in Baguio, Butuan, Ambulong and the island of Guam.

In 1928 William C. Repetti (1884–1966), who had obtained his doctorate in Saint Louis University under Macelwane's direction, arrived to Manila. His doctoral thesis, based on an analysis of the propagation of seismic waves, became one of the first studies to determine the internal structure of the Earth. Repetti took charge of the seismological section of the observatory and in 1946 published a new catalogue of earthquakes in the Philippines. After the Second World War the observatory ceased to work on meteorology and instead seismology became one of its main research areas. New seismographs were installed in Manila, Baguio and Davos. In 1962, the WWSSN program of the United States government installed two of its stations in Baguio and Davos. The seismographic stations are still in operation with updated instrumentation.

In 1904 the Observatory of Zikawei installed the first seismographs, Omori instruments provided by the Japanese government. Thus, Zikawei became the first seismographic station in China. Between 1915 and 1932 the observatory installed Galitzin-Wilip electromagnetic seismographs, the best at that time. Zikawei's observations were highly appreciated by seismologists, given the lack of other stations in the region. The same was the case for the seismological observations of the Observatory of Tananarive (Madagascar), which in 1899 had installed its first seismographs. These were the first to be installed in Africa. In 1927 the observatory updated the seismological instrumentation with Mainka seismographs. In 1910 Ksara Observatory (Lebanon) installed the first Mainka seismographs which in 1937 and 1957 were updated with modern electromagnetic instruments.

In 1907 Pigot installed the first seismographs at Riverview Observatory (New South Wales, Australia). In 1941 the Observatory acquired new Galitzin type instruments made in Australia. Thomas N. Burke-Gaffney (1893–1958), who worked at the observatory from 1946 and was director from 1952 to 1958, won the esteem of seismologists for the observatory with his careful analysis of seismological records. For this reason Riverview was also selected in 1962 as a station of WWSSN program. In 1972 Drake was appointed director after obtaining his doctorate at the University of California (Berkeley) with a thesis on the study of the propagation of seismic surface waves in heterogeneous media. Drake was also a professor of geophysics in Macquarie University and kept up contact with Berkeley with frequent visits. As was mentioned earlier, in 1992 after retiring as a professor he moved to La Paz, Bolivia and continued to work in seismology until 2001. The Riverview seismographic station continues to operate, but without Jesuits.

8.6 The Study of Microseisms

Microseisms are continuous waves observed in seismograms, with larger amplitudes for periods of between 4 and 10 s, which are generated by atmospheric perturbations on the ocean surface. Ever since the start of instrumental seismology these waves have attracted the attention of seismologists. For example, in 1903 the early seismologists, Fusakichi Omori in Japan and Emil Wiechert in Germany, studied this phenomenon and already proposed its atmospheric origin. Some Jesuits devoted particular attention to the study of microseisms. The first was Algué in Manila who in 1894 linked the occurrence of microseisms registered in the Philippines with the activity of cyclones on the sea. He dedicated a chapter to microseisms in his book about typhoons published in 1904. He suggested that seismological stations located at the coast should be used to detect the proximity of cyclones by the observation of microseisms. He also suggested that some microseisms could be caused by the effect of wind on nearby mountains, an idea that has been proved not to be correct. Gherzi in Zikawei made microseisms the main object of his studies.[15] In 1924 he announced that microseisms could be divided into four classes according to their origin: those due to cyclones or centers of low atmospheric pressure; those due to anticyclones or high pressure; those caused by cold; and those with no known cause. He considered the first type to be the most important: they are caused by sudden changes of atmospheric pressure in the vortices of cyclones that are transmitted through the water layer of the oceans to the Earth's crust. He also proposed that in stations at the sea coast, the effect of the sea waves on the coast should be taken into account. At Saint Louis University Ramírez, under Macelwane's supervision, studied the nature of short period microseisms and their generation by low atmospheric pressure over the oceans. Using the idea that microseisms are caused by storm centers, Ramirez thought it might be possible to determine the location of tropical cyclones by using a system of three seismological stations located at the vertices of a triangle.[16] The United States Navy became interested in this idea and tried its use to locate the centers of cyclones in the Pacific and Atlantic Oceans. However, in practice the system proved not to be very efficient and was abandoned.

Gherzi's interest in microseisms and his position as a member led to his suggesting to the Pontifical Academy of Sciences the organization of a study week devoted to microseisms. Francesco Vercelli, Director of the Geophysical Observatory of Trieste and also a member of the Academy supported the idea and acted as chairman. The study week took place at the offices of the Academy in the Vatican from 19 to 26 November 1951.[17] During this week 16 eminent seismologists took part; among them were Maurice Ewing of Wood Hole Oceanographic Institute, Inge Lehman of the Geodesic Institute of Copenhagen, discoverer of the inner core of the Earth, Jean Pierre Rothé of the *Institut de Physique du Globe* of Strasburg and Robert Stonneley, of Cambridge University and President of the International Association of Seismology. Three Jesuits were present: Gherzi, Macelwane and Due-Rojo. The study week tried to analyze the problem of the nature and origin of microseisms. On the first day Pope Pius XII gave an important talk on the relation between science and theology. A year later the United States Naval Research and the Geophysical Research Directorate of the U.S. Air Force organized a symposium on microseisms. Three Jesuits, Macelwane, Lynch and

[15] Ernesto Gherzi, Étude sur les microséismes. *Observatoire de Zikawei: Notes séismologiques* 5 (1924), 1–16.

[16] Jesús Emilio Ramírez, An experimental investigation on the nature and origin of microseisms at St. Louis, Missouri. *Bulletin Seismological Society of America,* 30 (1940), 34–84; 139–178.

[17] *Semaine d'étude sur le problème des microséismes.* Vatican City: Pontificae Academiae Scientiarum (Scripta Varia 12) (1952).

Ramírez were among the 31 specialists. Their participation shows that the expertise of Jesuits in this subject was generally acknowledged. The interest in microseisms shown by Jesuits was limited to those of this period. Moreover, Jesuit interest for seismology decreased drastically after 1970. From that year onwards most Jesuit seismographic stations either closed or handed over their direction to lay people.

8.7 Jesuits and Seismological Organizations

The Jesuit contribution to seismology coincided with the beginning of the twentieth century and with the establishment of the first national and international seismological organizations. In 1901 the first International Conference of Seismology took place in Strasbourg. On this occasion the establishment of the International Association of Seismology (IAS) was decided and it began in 1905. In 1903 Cirera, first director of the Ebro Observatory, had participated, as one of the two Spanish delegates, in the second International Conference of Seismology. In 1907 three Jesuits, Berloty (Lebanon), Sánchez Navarro-Neumann (Spain) and Stein (Vatican) were present at the first general assembly of the IAS in The Hague. In 1922 the IAS became part of International Union of Geodesy and Geophysics (IUGG). In 1936 the IAS created a commission which met in Edinburgh to supervise the International Seismological Summary, an agency which published seismological data. Macelwane was one of its five members. In 1939 the meeting of the IAS took place in Washington, the first in North America. Nine Jesuits, among them Macelwane, Lynch and Linehan, were among the 74 American participants. Macelwane presented a report on the problem of microseisms. In 1951 the IAS changed its name and became the International Association of Seismology and Physics of the Earth's Interior (IASPEI). The IASPEI meets every 2 years and Jesuit seismologists attended these meetings as long as seismographic stations were managed by them.

Some Jesuit directors of seismographic stations occupied important positions in national seismological commissions and other organizations. For example, Cabré and Ramírez were presidents of CERESIS and the latter was also vice-president of the Seismological Committee of the Pan-American Institute of Geography and History. American Jesuit seismologists had a special relation with the Seismological Society of America (SSA). Ricard, director of the Santa Clara Observatory, was one of the 13 founding members of this society and was elected to the first directive committee. Macelwane was a member of the committee between 1925 and 1956 and in 1928 was elected president of the Society. Between 1962 and 1967 Stauder was a member of the committee and in 1966 president. Several Jesuits were presidents of the SSA Eastern Section. In 1926 Macelwane was the first president. Other presidents were Lynch in 1930, Linehan in 1954, Birkenhauer in 1956 and Stauder in 1963. Several times the JSA held joint meetings with the SSA Eastern Section.

Fig. 8.3 Joint meeting of the Jesuit Seismological Association and the Seismological Society of América (Eastern Section) in St. Louis, 1937. Jesuits are (from *left* to *right*) *1st row*: Westland (Spring Hills), Macelwane (St. Louis), Lynch (Fordham), *2nd row*: Ramirez (Bogota), *3rd row*: Stechschulte (Cincinati), Zegers (St. Louis), *4th row*: Linehan (Weston), Delaney (Weston) (Udías 2003)

In 1937, for example, at a meeting in Saint Louis, 9 of the 22 participants were Jesuits (Fig. 8.3) and in 1948, in Cleveland, 8 of the 29. Even in 1961 at the meeting in Cincinnati 8 of the 80 participants were Jesuits. In 1991 the SSA Eastern Section established a JSA medal and award, in recognition of the Jesuits pioneer work in seismology. This is bestowed each year on seismologists distinguished for their work in observational seismology.

Chapter 9
Four Great Modern Scientists

9.1 Angelo Secchi, Pioneer of Astrophysics

Pietro Angelo Secchi (1818–1878) was born in Reggio-Emilia, in the north of Italy, between Parma and Modena (Fig. 9.1).[1] He studied at the Jesuit College and in 1833 entered the Jesuit novitiate. He then studied humanities and philosophy at the Roman College (returned to the Jesuits in 1824) from 1835 to 1840. There he had his first contact with mathematics and astronomy at the lectures of Vico and Giovanni Battista Pianciani (1784–1862). Pianciani was an important figure in the teaching of science in the Society after its restoration in 1814, as will be seen in Chap. 10. Following the expulsion of the Jesuits from Rome in 1848, Secchi traveled to England with other Jesuit students and he finished his studies of theology in Stonyhurst. Weld, director of the observatory, had just installed a new equatorial telescope, and offered Secchi an opportunity to train in observational astronomy. At the end of the year Secchi traveled to Georgetown University in Washington. There he collaborated with Vico, director of the observatory, who together with his assistant Sestini had also been expelled from Rome. Secchi's stay in Georgetown was short and the following year he returned to Rome where he was appointed director of the observatory of the Roman College.

During his return journey to Rome, Secchi visited the observatories of Greenwich and Paris in preparation for his new post. In 1854 he first installed a new observatory with new telescopes and meteorological and magnetic instruments on the roof of the Church of Saint Ignatius. Using this equipment Secchi began taking observations

[1] V. van Tricht, Le Père Secchi, *Revue des Questions Scientifiques*, 4 (1878), 353–402; François Moigno, *Le Révérend Père Secchi, sa vie*. Paris: Gauthiers-Villars (1879); Aloysius L. Cortie, Angelo Secchi, S. J. In: B. Windle (ed.) *Twelve Catholic men of science*. Londres: Catholic Truth Society (1923), 149–164; Mario Vigano, Cento anni di astrofisica. Nel centenario della morte del padre Angelo Secchi. *Civiltà Cattolica*, 129 (1978), 538–549; Aldo Altamore and Sabino Maffeo (eds.), *Angelo Secchi. L'avventura scientifica del Collegio Romano*. Foligno: Quater (2012), especially therein, Ileana Chinnici, Il profilo scientifico e umano di Angelo Secchi, 43–64.

A. Udías, *Jesuit Contribution to Science: A History*,
DOI 10.1007/978-3-319-08365-0_9

Fig. 9.1 Angelo Secchi (1818–1878), director of the Observatory of the *Collegio Romano* and pioneer of astrophysics (Maffeo, Vatican Observatory)

with the help of a few Jesuit colleagues, especially Brother Francesco Marchetti (1823–1898), his faithful companion and friend. After only 25 years, cut short by his early death, Secchi had completed some brilliant research work, which covered astronomy, meteorology, and geomagnetism. At the same time he continued to teach as professor of astronomy at the Roman College and at the University of Rome. In 1856 he was elected a member of the Royal Society of London; in 1857 of the *Académie des Sciences* of Paris, and in 1853, of the *Pontificia Accademia dei Nuovi Lincei* of which he became president.

Secchi was aware of how important it was for climatological studies to have meteorological observations for a whole region and he established stations in Ancona, Bologna and Ferrara, which sent the observations to Rome by telegraph. This was the first modern meteorological network in Italy. In 1867 Secchi designed a special recorder, called a "Universal Meteorograph", which simultaneously registered atmospheric pressure, temperature, humidity, wind direction and intensity, and rainfall. This instrument allowed a rapid visual correlation of the variations in these observations. This instrument was awarded a gold medal at the Paris World Fair of 1867. The emperor Napoleon III was so impressed by the instrument that he awarded Secchi the Cross of the *Légion d'Honneur.* Peter II of Brazil also awarded him the Cross of the Order of the Golden Rose. Only eight copies of this instrument were made; six were installed at the Jesuit observatories of Rome, Manila, Havana, Zikawei (two) and Cleveland.

In 1854 Secchi was commissioned to repeat the measurement of the geodesic base-line of 12 km along the Via Appia, used in the triangulation network of the

Papal States. Boscovich had first made these measurements in 1751, but they needed to be calculated with greater accuracy. French engineers had also made the measurements in 1810, but their work was not satisfactory. Secchi took a whole year to carry out this work, which should have been followed by a complete new triangulation of the whole area of the Papal States. However, this measurement was not made until after the unification of Italy under King Vittorio Emanuele II. The engineers in charge of this project found that the base-line measurement by Secchi had great accuracy and scientific value. Other engineering works entrusted to Secchi were the establishment of a network of lighthouses on the coasts of the Papal States and the water channeling in the Roman region.

Secchi was a prolific author with 12 books and more than 500 articles in scientific journals published between 1851 and 1878.[2] Some of his most important books are *L'unità delle forze fisiche* (The unity of physical forces, 1864), translated into French, German and Russian, about which more will be given below; *Le Soleil* (The Sun, 1870) with an Italian edition and translated into German and Spanish and *Le stelle* (The stars, 1877) translated into French and German. His writings about the observatory and his studies on astronomy, meteorology and geomagnetism were published in the *Memorie dell'Osservatorio del Collegio Romano* (vols. I–VI, 1850–1863) and in the *Bollettino Meteorologico dell'Osservatorio del Collegio Romano* (vols. I–XVI, 1862–1877). Events following the occupation of Rome by the troops of Vittorio Emanuele II in 1870, with the seizure of the Roman College and the expulsion of the Jesuits, overshadowed the last years of Secchi's life. Given his international prestige, as one of the most illustrious Italian scientists, the Italian Parliament ruled that he could continue in Rome as director of the observatory instead of being expelled with his fellow Jesuits. However, Secchi's fidelity to the Society of Jesus and to the Pope led him to renounce the honors offered by the new Italian government, although this caused him grave problems. Soon his health deteriorated with the illness that caused his early death in 1878. Pope Pius IX, who had a deep friendship with Secchi, sent him his blessing during his sickness commenting: "Father Secchi has always known how to unite science with religious virtue, but the two virtues which shone most in him were humility and obedience". After his death the Italian government finally confiscated the Observatory.

9.2 Solar and Stellar Physics

Physical astronomy, today called astrophysics, began to be developed in Secchi's time. With its experimental study of the physical nature and composition of the heavenly bodies, it marked a step forward with respect to traditional astronomy,

[2] The most important scientific journals where most of Secchi's articles were published between 1851 and 1877 are: *Comptes Rendus de l'Académie des Sciences de Paris*, (175 in vols. 36–86); *Astronomischen Nachrichten*, (121 in vols. 30–89); *Atti dell'Accademia Pontificia de Nuovi Lincei* (78 in vols. 3–30); *Memorie della Società degli Spettroscopisti Italiani* (42 in vols. 1–6); *Nuovo Cimento* (40 in vols. 1–25).

which had only been interested in their position and movement. In fact, one may trace back the birth of this science to Galileo. By observing the heavenly bodies for the first time through the telescope Galileo came close to the study of their nature. Secchi was aware that the future of astronomy lay precisely in this new orientation, in spite of the criticism of some contemporary astronomers who did not consider these studies as true astronomy. The methodology that opened the door to these studies was that of spectroscopy. At the beginning of the nineteenth century, William and John Herschel and Josef von Fraunhofer began to study the spectra of the Sun and the stars. This contributed new knowledge about their chemical composition and physical conditions. In 1860, Gustav Robert Kirchoff and Robert Wilhem Bunsen made the first chemical analyses of the solar atmosphere and introduced spectrochemical analyses into astronomy. Secchi became convinced of the potential of this new line of research and devoted great effort to the use of this methodology to investigate the nature of the Sun and stars. In 1862 he installed a Hoffmann spectroscope on a Cauchoix refractor telescope and began his studies of solar and stellar spectroscopy. He was also among the first to use the new photographic techniques for astronomical observations. In 1870 Secchi with Pietro Tacchini founded the Italian Society of Spectroscopists. In the reports of this Society he published 42 articles. His most lasting contribution to science in this field was made with the first classification of the stellar spectra.

From 1851 Secchi had followed the tradition of former Jesuit astronomers, like Scheiner, and devoted his first studies to the Sun. In 1860 he traveled to the Desierto de las Palmas, (Castellón de la Plana, Spain) to observe a total solar eclipse. On this occasion he was a pioneer in the use of photography, taking some of the first photographs of the solar corona and protuberances. Ten years later he observed another total solar eclipse in Sicily, which provided him with new data. In his work, *Le Soleil,* Secchi begins by outlining the latest scientific discoveries about the Sun and proceeds with his own observations and theories.[3] In particular, he uses the new spectroscopic techniques in his account of the nature of sunspots, protuberances and the solar corona. He describes the latter in detail and proposes the existence of an atmosphere on the Sun. He presents several original ideas about the structure of sunspots and solar protuberances and assigns values, to the order of several millions of degrees Celsius, to the distribution of temperature on the solar disk. These were considered unusually large at that time, but are accepted today for the solar corona. He discovered that the core of a sunspot is located under the surface of the Sun and radiates less heat than the surrounding photosphere. At the end of the book he considers the Sun as a star among the many millions of our galaxy, and he asks: "What should we think of these stars, which are centers of light, heat and activity and destined like our Sun to sustain the life of a multitude of creatures of every species?" He answers saying that it would be absurd to think that those vast regions are

[3] This work is based on a lecture given to students of Sainte-Geneviève School in Paris in 1867. To the first French edition in 1870 followed an enlarged second edition in 2 vols. (Paris: Gauthier-Villar, 1875–1877). It was published in Italian (*Il Sole*, Florence, 1884) and translations into German (*Die Sonne*, Braunschweig, 1872) and into Spanish (*El sol*, Madrid, 1879).

uninhabited deserts and he asserts that they must be populated by intelligent and rational beings, capable of knowing, honoring and loving their Creator – an interesting reflection given our present interest in the search for extraterrestrial life!

From 1863, Secchi began an exhaustive spectroscopic study of the stars. Relying on more than 4,000 observations, he proposed in 1867 the first classification of the spectra of stars dividing them into four classes, that can be summarized as follows[4]:

1. Spectra with strong hydrogen lines, including white and blue stars;
2. Spectra with numerous metallic lines less strong than those of hydrogen, including yellow stars;
3. Spectra with bands rather than lines, sharp toward the red and fuzzy toward the violet, including orange and red stars;
4. Spectra with bands better defined toward the violet and less toward the red, including many of the small red stars.

This classification was extended and modified between 1880 and 1901 by Edward Pickering, Williamina Fleming and Annie Cannon. It is the basis of the present standard Harvard Spectral Classification, a basic instrument in astrophysical studies. This is, without doubt, Secchi's most important and lasting scientific contribution, one by which he is internationally recognized.

9.3 Unification of Physical Forces

The search for a unity in the nature of physical phenomena has been constant in the history of physics ever since the ancient Greeks. In the nineteenth century physicists tried to reduce all forces to the mechanical. Today it is accepted that there are four physical forces: gravitational; electromagnetic; strong nuclear; and weak nuclear. The search for the unification of these forces is the core of the so-called Great Unification Theories (GUT). At present only gravity remains to be unified with the other three, since as yet there is no accepted quantum theory of gravitation. Secchi put forward his ideas on this problem in *L'unità delle forze fisiche* (The unity of physical forces), with the subtitle "An essay on natural philosophy" (translated into French, German and Russian).[5] The subtitle indicated that for him natural philosophy has to be understood in terms of modern science and in the context of nineteenth century classical physics, in accordance with the contributions of, among others,

[4] *Sugli spettri prismatiche delle stelle fisse* (Rome, 1868); *Note spettroscopiche sul sole e gli altri corpi celesti* (Rome, 1872). Richard O. Gray and Christopher J. Corbally, *Stellar spectral classification*. Princeton N.J.: Princeton University Press (2009), 2–4.

[5] *L'unità delle forze fisiche* (1st ed. Rome, 1864; 2nd ed. Milan, 1874); French edition: *L'unité des forces physiques* (1st ed. Paris, 1869; 2nd ed. Paris, 1874); German edition: *Die Einheit der Naturkräfte* (1st ed. Leipzig, 1876; 2nd ed. 1884; 3rd ed. 1892); Russian edition: *Edinstvo fiziceskih* (1st ed. St. Peterburg, 1872; 2nd ed. 1873; 3rd ed. 1880).

Lord Kelvin, James Prescott Joule, Rudolf Clausius, Faraday, André Marie Ampère and Maxwell. In the spirit of the time, unity in physics was to be sought in mechanics. Kelvin had commented that he could not understand something satisfactorily in physics until he had made a mechanical model of it. One can see how Secchi followed this line of thought in his search for the unification of physical forces.

L' Unità is divided into four books dealing with heat, light, electricity and the constitution of matter. In his prologue, Secchi insists that he is writing with a view to physics and not to metaphysics; also, that the aim of the natural sciences is to gain knowledge of the physical forces. He claims that all physical phenomena can be explained ultimately by the general laws of mechanics. He takes as his starting point the explanation of heat as motion, using the recent kinetic theory of gases, and he extends this explanation to all physical forces, including gravity and electromagnetic forces. He suggests that one must accept the existence of imponderable matter, in addition to ponderable matter. He means by this the "ether" which completely fills all space including the vacuum space. The existence of ether, accepted by the majority of physicists at that time, had its origin in the ideas of Descartes. It had reappeared in the nineteenth century with the wave theory of light and electromagnetism. Secchi asserted that matter and forces are linked together, although they are not identical and that all forces can be ultimately reduced to some form of movement. In his words "heat, light, electricity, magnetism, and molecular forces are all purely mechanical actions". He adds that even gravitation, which controls the motion of the heavenly bodies, is also of this sort, although it is difficult to explain.

A few years later, the arrival of relativistic and quantum physics would undermine this so-called "mechanicist program" of classical physics that had been followed by most physicists of the nineteenth century and enthusiastically adopted by Secchi. In 1887 the experiments by Albert A. Michelson and Edward W. Morley questioned the existence of ether and it was finally abandoned with Einstein's general theory of relativity. However, at the end of his book Secchi had envisioned this possibility: "If some day it could be proved that it is useless to accept this second condition of matter (the ether), this would only reduce the number of means that nature uses to reach its aim." However, the fundamental principles proposed by Secchi will continue to be valid, even if (as he adds) we must admit that there are many problems to be solved. Secchi's work shows the total abandon of the Aristotelian physics, doctrine current in Jesuit teaching of natural philosophy until the end of the eighteenth century, and the acceptance of the physics of his time. It will be seen how Aristotelian physics returned in the neo-Thomistic currents of natural philosophy in Secchi's time. He became the object of strong opposition from the philosophy professors who followed these currents, and was even accused of atheism. Thus, he had to defend science against those who in their apologetic fervor saw in it dangers for Christian faith. However, the fact that he could have been considered a reductionist in his physics, did not prevent him at the end of the book from turning his gaze to the Creator of this extraordinary world: the One who "as the first cause originally set the limits of their intensity and direction to all actions".

9.4 Stephen J. Perry, Astronomer and Geophysicist

Stephen Joseph Perry (1833–1889) was born in London and educated with the Benedictines at St. Edmund College (in Douai, France), where he first showed his interest for mathematics (Fig. 9.2).[6] He moved to the English College in Rome to study for the priesthood and in 1853 joined the Jesuit novitiate in England. In 1858 Perry studied mathematics and science in London and in the following year in Paris where he attended the lectures of Joseph Louiville and Joseph Bertrand. In 1860 he was appointed professor of physics and mathematics at Stonyhurst College and director of the observatory. After finishing his studies of theology, he returned there in 1868 as professor and director of the observatory where he remained until his death.

In 1859 Stonyhurst Observatory had begun systematic geomagnetic observations under Weld's direction and with the support of General Sabine. Perry took part in this program at his arrival at the Observatory. Together with Sidgreave he carried out, during the summers of 1868 and 1869, a field survey of magnetic measurements in France: 15 stations in the Western area and 21 stations in the Eastern area. They made a similar survey in Belgium in 1871 with 20 stations. In each of the station for the three surveys they measured the magnetic field declination, inclination and horizontal intensity. The results were published in *Transactions of the Royal Society* in 1870 (vol. 17) and in 1872 (vol. 20). George B. Airy, Astronomer Royal, the reviewer of this work for the Royal Society, commented that these were the best magnetic observations known to him for a limited region. The observations

Fig. 9.2 Stephen Perry (1833–1889), astronomer and geophysicist, director of Stonyhurst Observatory (Udías 2003)

[6] Aloysius L. Cortie, *Father Perry, the Jesuit astronomer*. London: The Catholic Truth Society, 2nd ed. (1890); J. Thirion, Le R. P. Perry. *Revue des Questions Scientifiques*, 27 (1890), 201–208; George D. Bishop, *Stephen Joseph Perry (1833–1889) priest, scientist, educator*. University of Manchester, Master of Science Thesis (1977) and Stephen Perry (1833–1889); forgotten Jesuit scientist and educator. *Journal British Astronomical Association*, 89 (1979), 473–484.

in Belgium clearly showed the influence of the coal mines on the magnetic field. Following his arrival at Stonyhurst, Perry made a special study of the diurnal, monthly, annual and secular variations of the magnetic field and of the magnetic storms. At that time, the origin of these phenomena was not yet clearly understood. At a conference given to the Liverpool Polytechnic Society in 1872 Perry suggested that the secular variations of the terrestrial magnetic field were due to a rotation of the magnetic pole around the geographic pole with a period of 560 years, and that the annual and semiannual variations had a solar origin. The conference, entitled *Terrestrial Magnetism*, was published in 1873 in the prestigious scientific journal *Nature.* These suggestions were fairly modern proposals at that time.

Perry also established an important program of astronomical observations, including those of the satellites of Jupiter, and occultations of stars by the Moon and asteroids. He showed that some comets disintegrate into meteorites, as happened with that expected in 1866. In November 1872 he observed a large fall of meteorites, which reached up to 40 per minute. However, the field in which he made his most important contributions was solar physics, in the tradition of Jesuit astronomers like Scheiner and Secchi. In 1881 Perry began his observations of sunspots, solar faculae and protuberances. From these observations he drew up 3,800 large-scale careful drawings of the Sun's surface (27 cm diameter) and drawings of particularly interesting sunspots to even greater scale, as well as detailed photographs. For his spectroscopic studies he installed in his telescopes two spectroscopes giving direct vision, an automatic spectroscope of six prisms (a Browning spectroscope), and other instruments. Attributed to Perry was the discovery of the so called "veiled spots" on the Sun of several kinds. His main aim was to show the influence of solar activity on the Earth's magnetic field. In 1843 Heinrich Schwabe had shown that in solar activity there is an 11 year periodicity, and 10 years later Sabine established that the maxima and minima of perturbations of the terrestrial magnetic field correspond to this periodicity. This influence was accepted by many scientists but still denied in 1892 by the influential physicist Lord Kelvin. Perry tried to demonstrate this influence experimentally by his careful magnetic and solar observations, but his early death prevented the completion of his work. It was not until 1904 that Walter Maunder proved definitively that magnetic perturbations have a solar origin.

While carrying out his program Perry collaborated with Kew Observatory in the comparison of magnetic observations and also with Lord Kelvin, Arthur Schuster and Balfour Steward. Furthermore, he was in close touch with Sabine and also kept up in the observatory a program of meteorological observations. His scientific contributions were published in 129 articles in scientific journals, memoirs and reports.[7] In addition to his scientific research, Perry continued to give classes in physics and mathematics in the college and took the students for practical classes in astronomy. He was also in great demand in the northern cities of Great Britain as a lecturer on popular science. For example, in 1884 he gave a cycle of 12 lectures in Liverpool on the solar system. Under Perry, Stonyhurst Observatory became a center for the

[7] Some of Perry's scientific articles were published at: *Monthly Notices of the Royal Astronomical Society (30)*, *Proceedings of the Royal Society of London (10)*, *Philosophical Transaction of the Royal Society of London (5)*, *The Astronomical Register (6)* and *Nature* (35).

training of Jesuits, who then directed other observatories. Among them were Dechevrens (Zikawei), Faura and Cirera (Manila), Viñes and Gangoiti (Havana), Berloty (Lebanon) and Goetz (Bulawayo). They spent a time at Stonyhurst, where Perry introduced them to the work of an observatory and helped them to acquire instruments. Perry's influence left a deep mark on modern Jesuit observatories. In 1874 in recognition of his research in the field of terrestrial magnetism Perry was elected a member of the Royal Society of London (Perry and Secchi are the only modern Jesuits to have received this honor). In 1869 he was also elected a member of the Royal Astronomical Society and he had a place on the council of these two societies. He also belonged to other scientific societies like the *Pontificia Accademia dei Nuovi Lincei* (1887) and the *Société Scientifique* of Brussels. In 1886, Perry was awarded a doctorate *honoris causa* by the Royal University of Ireland.

9.5 Perry's Scientific Expeditions

Perry's prestige due to his research on solar activity and terrestrial magnetism led to his being asked to lead four scientific expeditions financed by the British Government and the Royal Society. Two were to observe solar eclipses and the other two to observe the transit of Venus across the solar disc.[8] The areas where solar eclipses are total are sometimes of difficult access; also the duration of such an eclipse may be only 3–8 min, and observations can be impeded by clouds. For these reasons, long journeys may be needed and instruments installed in remote regions, all involving considerable hardship. The same can be said of observations of the transit of Venus. However, such observations provide very important astronomical information, so that in spite of the difficulties such expeditions are worth while.

In 1870 Perry was commissioned by the Royal Astronomical Society to lead one of four groups on the expedition to Cadiz (Spain) to observe the solar eclipse of the 22 December. On that occasion the main interest focused on observing the form and extension of the solar corona and its spectrum. Perry noticed that the form of the corona was quadrilateral and extended to seven eighths of the Sun's diameter. The spectroscopic analysis showed the presence of hydrogen lines in the whole corona. The success of this expedition assured Perry of his participation in other expeditions. The painter Paul J. Naftel, who painted several watercolors of the moment of the eclipse, accompanied Perry on this occasion.

Perry's most important expedition was in 1874. He went to the Kerguélen Islands in the south of the Indian Ocean, near the coast of the Antarctic, to observe the transit of Venus across the solar disc.[9] In Chap. 2 mention was made of the part played by Jesuit astronomers in the observation of the transits of Venus of 1761 and 1769;

[8] Corti, *Father Perry*, 51–80.

[9] Stephen J. Perry, *Notes on a voyage to Kerguelen island to observe the transit of Venus, December 8, 1874*. London: Henry S. King, (1876); Joseph Ashbrook, Father Perry's expedition to Kerguelen Islands. *Sky and Telescope*, (June 1966), 340–341. Peter D. Hingle, The priest and the stuffed penguin; Father Stephen Perry S.J. and the transit of Venus expeditions to Kerguelen Island, 1874 and Madagascar, 1882. *Journal British Astronomical Association* 115 (2005) 150–158, 168–170.

this Jesuit tradition continued in the observation of the new transits. The expedition to the Keruélen Islands was one of five expeditions organized by the British Government, the other four being to Hawaii, Egypt, New Zealand, and Rodrigues Island (Indian Ocean). Perry was the official leader of this expedition, he and Sidgreaves being the only astronomers. They took telescopes and other instruments for meteorological and magnetic observations from Stonyhurst. The expedition sailed from London in two frigates of the Royal Navy and arrived on the 8th of October at Harbour Island, one of the Keruélen Islands. The transit, which took place on the 8th of December, was observed from three sites of the island, using visual and photographic methods. These observations were later used in part to measure with great accuracy the distance from the Earth to the Sun. The expedition remained on the island until the end of February taking measurements to determine exactly the geographical coordinates of the observation points. Perry and Sidgreaves took many notes of the magnetic field and studied the meteorological conditions during their five months on the island. During the return journey they took magnetic measurements at various ports: Cape Town, Bombay, Aden, Port-Said, Malta, Palermo, Rome, Naples, Florence and Moncalieri. In Rome, Perry and Sidgreaves were received in audience by Pope Pius IX. During this long journey Perry and Sidgreaves won the sympathy and admiration of the officers and crew of the two ships. Their observations became the basis of several scientific studies. In 1875 and 1876 Perry presented the results of his work in London at the Royal Institution and in Brussels at the *Société Scientifique*.

Transits of Venus happen in pairs separated by eight years. Perry's successful observations in Keruélen led in 1882 to his being asked to head one of the four British expeditions to observe the second transit. The expeditions were to Australia, Canada, the Caribbean and Madagascar. Perry took charge of the one to Madagascar. Perry and Sidgreaves again sailed in a British Navy ship and arrived to the south of Madagascar on 22 October. The place selected for the observations was Nos Vey, a small island two miles from the southeast coast of Madagascar. The transit took place on the 6th of December and was successfully observed from two sites. Three days later the expedition left the island, but as on the Keruélen expedition, Perry and Sidgreaves had made magnetic and meteorological observations. Sidgreaves took advantage of his stay in the island to take notes on the local people and their customs.[10]

The other three expeditions were organized to observe solar eclipses. In 1886 Perry led the team to the Island of Carriacou in the Little Antilles near Grenada. This time the group of observers was larger, consisting of seven British astronomers and an Italian, Tacchini, who had succeeded Secchi at the Observatory of the Roman College. Apart from many other observations made during this eclipse, Perry took spectroscopic measurements of the solar corona to investigate the origin of the absorption lines of the spectrum. In 1887 Perry, together with Ralph Copeland, an astronomer of the Scottish Dun Echt Observatory, was invited to Russia by Professor Fedor Bredichin of the Observatory of Moscow University to observe the solar

[10] Walter Sidgreaves, The last scientific voyage of H.M.S. Fawn. *The Month*, 47 (1883), 457–482.

eclipse of the 19th of August. The place selected for the observations was near the town of Kineshma near the Volga River northeast of Moscow, but dense clouds made all observation impossible. Perry's last expedition was to observe the solar eclipse (22 December 1889) from the Salut Islands, near the coast of the French Guyana, where the French government had established an infamous prison in 1852. On 2 December the expedition reached the Salut Islands on a British Navy ship and located an observation point on the Royale Island. Once all preparations for the observations had been made, Perry took advantage of his knowledge of French to attend the sick of the hospital and the inmates of the prison and celebrated Mass for them. On the day of the eclipse, he began to feel sick but he completed his observations of the solar corona. His condition rapidly deteriorated and, attended by the Jesuit brother, John Rooney (1856–1920), who had accompanied him on this journey, Perry died of acute dysentery on board ship, while on the way to Georgetown in British Guyana where he was buried. Perry knew how to combine to the end his responsibility as a scientist, carrying out his observations, and his spiritual duties: attention to the sick and the prisoners which may have led to his death. In the words of an obituary note of the journal *Observatory*[11]:

> Father Perry's labors in solar physics and terrestrial magnetism, his ever-ready willingness to lecture in public, his cheerful acceptance of the most onerous posts in many astronomical expeditions culminating in the sacrifice of his life, are well known to all.

9.6 James B. Macelwane, a Great Seismologist

James Bernard Macelwane (1883–1956) was born in Port Clinton, Ohio, a small town on Sandusky Bay in Lake Erie, the second of nine children in a family of farmers and fishers of Irish origin (Fig. 9.3).[12] He studied at the Jesuit "Saint John School" in Toledo (Ohio) and entered the Jesuit novitiate of the Buffalo Mission run by German Jesuits. In 1910 he moved to Saint Louis University (St. Louis, Missouri) to study humanities and philosophy. There he also studied science and in 1912 obtained a Master in Science. On completing his theological studies, Macelwane began teaching science at Saint Louis University.

While still a student, he had his first contact in 1911 with seismology helping to install the Wiechert seismograph of the JSS network. On this occasion, together with Joseph Joliat (1882–1962), he wrote his first scientific article *The physics of the seismograph*. Joliat was, from 1933 to 1947, director of the seismological observatory at John Carroll University. In 1921, Macelwane went to the University of

[11] Corti, *Father Perry*, 81–91.

[12] Perry Byerly and William Stauder, James B. Macelwane, S. J., September 28, 1883 – February 15, 1956. *Memoirs of the National Academy of Science*, 31 (1958), 254–281; Victor Blum, Sketch of the life of James Bernard Macelwane, S.J., *Earthquakes Notes*, 27 (1956), 9–11; Henry Birkenhauer, Father Macelwane and the Jesuit Seismologial Association, *Earthquakes Notes*, 27 (1956), 12–13.

Fig. 9.3 James B. Macelwane (1883–1956), seismologist, founder of the Jesuit Seismological Association (Udías 2003)

California (Berkeley), to complete his doctorate studies. Ever since the 1906 San Francisco earthquake, Berkeley had developed a strong program in seismology and installed some of the first seismographic stations in the United States. There Macelwane studied the first theoretical works of seismology developed in Europe by Emil Wiechert, Karl Zöppritz, and Boris Galitzin. He defended his doctoral thesis in 1923, directed by professor Elmer Hall and with the collaboration of Andrew Lawson, on the propagation of surface waves produced by earthquakes. This was the first thesis on seismology in North America. Lawson, Head of the Department of Geology, saw in Macelwane the person to establish a program of seismological studies and asked him to stay at Berkeley for two more years as an associate professor. In Berkeley Macelwane organized a program of seismology and the installation of a network of seismographic stations in the northern part of California. A similar network had been installed in the southern part by Bailey Willis and Beno Gutenberg, at the California Institute of Technology in Pasadena. In Berkeley Macelwane began the systematic study of earthquakes in northern California, a practice that continues to the present day. Macelwane was the person who advised Perry Byerly, when he took charge of the seismographic stations in Berkeley and established a renowned center of seismological research.

After his return to St. Louis in 1925, Macelwane's first concern was to revive the Jesuit network of seismological stations, now named the Jesuit Seismological Association (JSA), as was noted in the previous chapter. As president of JSA, Macelwane was in continuous contact with these stations, and personally visited them every year. He established at Saint Louis University the Department of Geophysics, the first such department in a North American university. As Dean of

Graduate Studies between 1925 and 1933, he gave great impulse to research. In 1944 he established the university's Institute of Technology for the training of engineers and was its first dean. He devoted great efforts to the training of engineers. Moreover, on his arrival at Saint Louis he launched two main research projects: the study of the propagation of seismic waves in the Earth's interior and that of regional earthquakes in the central region of the United States, as he had done in Berkeley with those of northern California. The first produced some of the first tables and curves of travel-times, one of the basic needs for seismological research. The second continues to the present day. As was seen in Chap. 8, Macelwane also investigated the problem of microseisms and together with Ramírez, whose doctoral thesis he directed, tried to apply their observation to determine the path of tropical hurricanes. Another subject was the existence of earthquakes at great depth, a topic hardly known at that time. On this theme he directed the doctoral thesis of George J. Brunner (1882–1949), who made one of the first studies of the characteristics of these earthquakes and calculated their travel-time curves. Brunner was later director of the seismological observatory of Loyola University (Chicago) and worked on seismology at Saint Louis.

In earlier times the nature of the interior of the Earth had been a subject of pure speculation, as is seen in Kircher's proposals. In 1798 Henry Cavendish had determined the value of the mean density of the Earth (5.5 g/cm^3) larger than that of any rock found on the surface. Thus it was thought that the Earth must have a core of great density. Analysis of the propagation of seismic waves in the interior of the Earth provided the first experimental evidence that allowed the investigation of its internal structure. In 1906 Richard Oldham showed that a core really exists and in 1914 Gutenberg determined its depth at 2,900 km. In 1909 Andrija Mohorovicic discovered that the Earth's crust is separate from the material beneath, the so-called "mantle". Macelwane became interested in this problem and began to study the structure of the Earth's interior using the analysis of seismic waves. From 1928 he directed several doctoral theses concerned with this problem. The first was that of Repetti, who discovered that in addition to the discontinuity between crust and mantle there was another inside the mantle at a depth of 950 km (later better established as 650 km). In 1934 another student of Macelwane, Cornelius Dahm, proposed the existence of a transition zone above the Earth's core, and in 1952 Otto Nuttli investigated the existence of two branches in the reflected waves at the core that had been first observed by Macelwane in 1930. In 1937, Macelwane formed part of the eight-member committee established by the United States National Research Council to study the Earth's interior. Their findings were published in 1939 with a long article by Macelwane about the evidence derived from the observations of seismic waves.[13] Admittedly, Macelwane never considered sufficiently proved that the outer core of the Earth was fluid, as proposed by Harold Jeffreys, a professor at Cambridge University and also a member of the committee. But this has been clearly demonstrated by later research.

---

[13] James Macelwane, Evidence of the interior of the Earth derived from seismic sources. In Beno Gutenberg (ed.) *Internal Constitution of the Earth*. New York: McGraw Hill, 219–290 (1939).

9.7 Macelwane, Professor and Educator

An important aspect of Macelwane's role is as a professor and educator in the context of Jesuit higher education in the United States. From the nineteenth century Jesuits have established up to 28 universities in the United States. Their main aim was firstly the education of Catholic youth, and only later was research included. Macelwane played an important role in this evolution, moving the emphasis from just teaching to include research. Between 1931 and 1932 he chaired a commission established to study the quality of work being done in Jesuit universities in the United States. This commission reflected the Superior General Wlodomir Ledokowski's concern about the academic quality of Jesuit educational institutions in the United States. The long final report which recommended a greater effort in research is known as the *Macelwane report.*[14] He put his ideas into practice at Saint Louis University. On his arrival he established the Department of Geophysics and later the Institute of Technology, and he was head and professor in both. Macelwane published the first textbook of seismology in North America, *Introduction to theoretical seismology* (1936), in two volumes: the first *Geodynamics* was by himself; the second was on *Seismometry* by Frederick Sohon (1894–1972), director of the Seismological Observatory of Georgetown, and devoted to the theory of the seismograph. In the first volume Macelwane develops the theory of the origin and propagation of seismic waves from a physical-mathematical approach. He begins with the mathematical theory of elasticity and applies it to the propagation of seismic waves in the Earth's interior, the analysis of seismograms and the methods for the determination of epicenters and depths of earthquakes. At a popular level he wrote *When the earth quakes* (1947), which became a best-seller. He published more than 130 scientific articles on seismology.

Macelwane was also interested in the practical applications of geophysics to prospecting mineral resources and in the development of geophysical engineering. He gave special attention to the use of seismological methods for the detection of oil-bearing structures. In line with this, he was in contact with, and became a consultant of, several prospecting companies such as Reiber Exploration Company, General Geophysical Company and Root Petroleum Company. The establishment of the Institute of Technology in Saint Louis University was partly to train engineers for the oil prospecting companies. Macelwane also worked in the development of seismological instrumentation and collaborated with the company Sprengnether in the design of new types of seismographs.

Macelwane took an active part in many scientific commissions and congresses. He was president of various committees (seismology and physics of the Earth) of the United States National Research Council. In 1928 he was president of the Seismological Society of America and in 1953 of the American Geophysical Union that groups the most prestigious scientific societies of geophysics and seismology in

[14] Report of the commission on higher studies of the American Assistency of the Society of Jesus, Ms. 234 pp. (Saint Louis University Archives).

the United States. He was also president of the seismological section of the National Science Foundation and had good relations with the research offices of the United States Defense Department and Air Force. In 1944 he was elected a member of the National Academy of Science, the only Jesuit who has merited this honor. In 1948 he received the William Bowie Medal of the American Geophysical Union, the highest award of this society, and in 1955 the Mendel Medal of the Villanova University (Philadelphia). He received honorary doctorates from the universities of Washington, John Carroll, Marquette and Saint Norbert. In 1962 the American Geophysical Union established a medal in his honor to be awarded each year for special contributions to geophysical sciences by young scientists. The American Meteorological Society also presents annually a similar award (*Macelwane Award*) to three outstanding students of meteorology. With such awards these American scientific societies were acknowledging how great an educator Macelwane had been in the fields of geophysics and meteorology.

9.8 Pierre Teilhard de Chardin, Geologist, Philosopher and Mystic

Pierre Teilhard de Chardin (1881–1955) was born in Sarcenat, near Orcine, a short distance west of Clermont-Ferrand (France) (Fig. 9.4).[15] His family came from the French rural aristocracy. He studied at the Jesuit College of Notre Dame at Mongré and in 1899 joined the Jesuits at Aix-en-Provence. He began to take an interest in

Fig. 9.4 Pierre Teilhard de Chardin (1881–1955), geologist, philosopher and mystic (Courtesy Marie Bayon de la Tour)

[15] Literature about Teilhard de Chardin is very abundant, some examples are: Claude Cuénot, *Pierre Teilhard de Chardin. Les grands étapes de son évolution*. Paris: Librairie Plon (1958); Robert Speaight, *Teilhard de Chardin. A biograhy*. London: Collins (1967); Ursula King, *Spirit of fire. The life and vision of Teilhard de Chardin*. New York: Orbis Books (1998); Patrice Boudignon, *Pierre Teilhard de Chardin, sa vie, son oeuvre, sa reflexión*. Paris: Éditions du Cerf (2008).

geology during his studies of philosophy on the Island of Jersey, where the French Jesuits had established their house of studies after the expulsion due to the laws passed in France in 1901. Between 1905 and 1908 Teilhard was professor of physics and chemistry in the college run by the French Jesuits in Cairo. This was his first contact with the East, which would fascinate him later, and this is where he began his first work in field geology. He studied theology in Hastings (England), where he continued to develop his interest in geology and paleontology. He then became acquainted with Charles Dawson and was present at the "discovery" of the "Piltdown man", a supposed human fossil that turned out to be a forgery. Later, when Teilhard was already famous, some unjustly accused him of being responsible for this fraud. From 1912 to 1914 Teilhard studied geology in Paris under the direction of Marcellin Boule, director of the Institute of Human Paleontology, Museum of Natural History. During those years he became acquainted with Henri Breuil, professor at the Institute of Human Paleontology and a member of the *Collège de France*, a pioneer of human paleontology and archeology. With him Teilhard established a close friendship, which was to last during all his life. Together they visited caves having prehistoric paintings and human remains in south France, Normandie, northeast Spain, Switzerland and England. In 1915 he was called up to serve in the army as a stretcher-bearer and until 1919 he underwent the experience which he called a "baptism in reality", his immersion in the great human confrontation of the First World War. His dedication and bravery in attending to the wounded won him several decorations: the War Cross (1915), the Military Medal (1917), and a *Légion d'Honneur* (1920). Teilhard returned to Paris after the war and in 1919 he finished his licentiate in natural sciences and began teaching geology at the *Institut Catholique.* The following year he presented his doctoral thesis on the mammals of the lower Eocene in France, which won him an award from the French Geological Society.

In 1923, Teilhard made his first journey to China, invited by the Jesuit Émile Licent (1876–1952), who was organizing a museum of natural science in Tianjin. At this time Teilhard carried out field work in geology in the north of China and in Mongolia. As a result of this first journey his life remained bound to geological and paleontological work in China. Teilhard divided his time between China and France, where he continued to give his classes and work at the Museum of Natural History. In 1926 he began his study of the fossils deposit at Chukutien, which contain the human fossils of the Sinanthropus (*homo erectus pekinensis*). Thus the study of human fossils became the main line of Teilhard's research. In 1931 he took part as a geologist in the *Croisière Jaune* (Yellow Expedition), a car journey from Beijing to Turkestan crossing central Asia over the course of 9 months. This was organized by the automobile industry *Citroën* to demonstrate the endurance of its cars. Between 1932 and 1937 he worked at the Museum of Man in Beijing, while making frequent journeys to France. Teilhard also carried out geological field work outside China. In 1935 he joined the geologists George Barbour, Davidson Black and Helmut de Terra working in Kashmir, and in 1938 he worked in Java, invited by Gustav von Koenigswald, the discoverer of human fossils in this region. By 1939, Teilhard had become a recognized figure in the science of human paleontology. He traveled frequently, gave lectures in France and the United States, and collaborated in field work with other scientists, in addition to the work already mentioned in China, Kashmir, Java, Burma and South Africa. In 1947 he was appointed an Officer of the *Légion d'Honneur* and a corresponding member of the *Académie des Sciences*.

His ideas about human evolution and its effect on the formulation of the Christian faith were widely discussed in France. At that time the theory of evolution was still considered with hostile reserve in ecclesiastic circles and Teilhard's ideas aroused growing suspicion and rejection. He was forbidden to publish except for purely scientific work, but in spite of the prohibition his essays were copied and spread among his many enthusiastic followers. His presence in France soon became very controversial and he was kept away as long as possible. In 1948 Teilhard traveled to Rome and tried in vain to obtain permission for the publication of his main work *The phenomenon of man*. He was also denied permission to apply for a post at the prestigious *Collège de France*. In spite of all this, Teilhard's fidelity to the Church and to the Society of Jesus remained firm. From 1952 he spent longer periods in New York. There he continued to work on human paleontology in the Wenner Gren Foundation until his death in 1955.

As both a Jesuit priest and a passionate scientist, Teilhard's great concern was how to integrate Christian thought with the new world view presented by modern science of a universe in evolution, in particular, by geology and paleontology. This concern was evident already in his early writings during the war, and appears even in the last pages written a few days before his death. It sprang from the two forces on which his whole life rested: his scientific work and his mystical experience. For him the new evolutionary view of life and the universe, assimilated in his passionate dedication to science ever since his student days, had to affect deeply any religious thinking about the relation between God and the world. In the past such a relationship had been understood in terms of a static conception of the universe, but now it had to be adapted to the new evolutionary image discovered by modern science. A reformulation was required of many aspects of the Christian message, such as: God and creation; the problem of evil; the mystery of Christ in relation to the world. His writings on these topics were the source of serious concern for the ecclesiastical authorities, who forbade their publication. But for Teilhard this was not a theoretical problem but, rather, it affected the core itself of his life and spirituality. It is customary to see Teilhard either as a scientist specializing in geology and paleontology, which put him in contact with the early history of the human cosmos, or as a philosopher who elaborated a new evolutionary synthesis of man and the cosmos. However, he was also a Christian mystic, who discovered the cosmic Christ in the fibers of matter and as the ultimate end of the evolution of the universe.[16] The central theme of his mysticism was precisely how to understand the role of Christ in an evolutionary universe. This formed both the foundation of his Christian vision of evolution and the core of his Christology.

[16] Christopher F. Mooney, *Teilhard de Chardin and the mystery of Christ*. New York: Doubleday Image Book (1968); Édith de la Héronniere, *Teilhard de Chardin, une mystique de la traversée*. Paris: Albin Muchel (2003); Gustave Martelet, *Teilhard de Chardin, prophète d'un Christ toujours plus grande*. Brussels: Lessius (2005).

Teilhard kept his scientific work apart from his religious reflections, and in his more than 200 scientific articles there is no mention of the religious problem. He was a true scientist renowned for his work in geology and paleontology. However, for him the scientific work had in itself a religious value and constituted what he called some kind of "worship". He claimed that science and religion are two facets of the same approach to the knowledge of reality. Thus he found in his scientific work a form of spiritual nourishment: "I am convinced that there is no more powerful natural nourishment for the religious life than the contact with well understood scientific truths".[17] Thus, while active in scientific research, he was continuously producing works of philosophical and religious importance. This culminated in his two main works *The Phenomenon of Man*, written between 1938 and 1940, and *The Divine Milieu*, written between 1926 and 1927. The first contains his more philosophical thought, where Teilhard, taking as his starting point scientific data, gives a global view of the evolution of the universe, including man,. He finds that the evolutionary process must converge on what he calls the "Omega Point", which he identifies with God. At the end of this work he gives some considerations on what he calls the "Christian Phenomenon" and how it must be interpreted from these new perspectives. The second work has a more purely religious character. Here Teilhard presents a new view of Christian asceticism and mysticism in accordance with his evolutionary view of the world. In this and other writings Christ appears as the true Omega Point of the evolution of the universe. In many other religious and philosophical essays, Teilhard explains and develops his ideas about the evolutionary nature of the universe, the potentialities of matter, the future of humanity and the central role of Christ.

For Teilhard the central role of Christ in his evolutionary view of the world was not just a theoretical position but formed the center of his personal spirituality, his interior life and his mysticism. The notes he took during his "Spiritual Exercises" (the annual retreats made by Jesuits) allow us to follow the evolution of his personal meditation and prayer.[18] Year after year his "Exercises" turn around the same ideas. The term "Christ-Omega" is present already in his notes of 1922 and repeated in the following years. In 1940 appears the term "omegalize" to express the union of the universe with the "Total Christ". In later years he refers to his double task: to "universalize Christ" and "Christify the universe". Thus he conceives his life in fidelity to Christ-Omega and writes: "now more than ever it is Christ-Omega who sheds light and guides my life". However, he was aware that his image of Christ differed from that presented by many ecclesiastics. In his last "Exercises" he writes that nothing should enter into his life which is not "Christifiable". His great concern is to "end well", with total belief in Christ-Omega and the Cosmos. For him, to "end well", meant having had the time and ability to formulate the essentials of his message. On the last day of his last Exercises in 1954 he summarized his vision in a single word "Pan-Christism".

Another source on the extent to which Teilhard's conception of Christ's role in the world constituted the inspiration of his life is provided by the prayers found in

[17] Pierre Teilhard de Chardin, *Science et Christ, Oeuvres IX*, Paris: Éditions du Seuil (1965), 62.

[18] Pierre Teilhard de Chardin, *Notes de retraites, 1919–1954*. Paris: Éditions du Seuil (2003).

many of his works. His “Mass on the World”, written first in 1918 during the war and later in 1923 in the Ordos desert in Mongolia, deserves special mention. This very beautiful prayer follows the outline of the celebration of a Mass and presents the consecration of the world as an extension of that of the Eucharist: “mysteriously and in very truth, at the touch of the supersubstantial Word, the universe, an immense Host, becomes flesh. All matter is henceforth incarnate, O my God, through your Incarnation”. It is to this Jesus, incarnate in the world, that Teilhard addresses his prayer:

> To your Body, in all its extension, that is, to the World converted by your potency and by my faith in the wonderful and living crucible where everything disappears in order to be born again, I give myself up in order to live and die in it. Jesus … glorious Christ, divine influence secretly diffused and active in the depth of matter and the dazzling center where all the innumerable fibers of the manifold meet, … it is you to whom my being cries out with a desire as vast as the universe, in truth you are my Lord and my God.[19]

Such prayers clearly show that Teilhard’s Christology was not only the fruit of theological reflection, but sprang, above all, from a deep mystical experience. For him, Christ cannot be conceived as separated from the universe, nor can the universe be separated from Christ. He lived this presence of Christ in the world with an ardent passion and tried to share it with those with whom he came into contact, despite all the misunderstanding and suspicion he encountered.

9.9 From “Cosmogenesis” to “Christogenesis”

In order to understand Teilhard – as scientist, philosopher, theologian and mystic – it will be useful to present the key-lines of his thought. Such a summary will also serve to shed some light on the scientific work of Jesuits in general. Teilhard is a very special case because he has left in his writings a unique testimony of how he tried to integrate his passion for the world of science and his Christ-centered spirituality, which he based on the *Spiritual Exercises* of Saint Ignatius. He has made explicit elements of a spirituality that is to be found implicit in other Jesuit scientists.

He describes his vital attitude to life in the first of his essays, dated 1916, where many seeds of his thought are already expressed in words full of youthful enthusiasm:

> I write these lines from the exuberance of life and for the need to express a passionate view of the Earth and to seek a solution to the doubts about my activities; because I love the universe, its energies, its secrets and its hopes and at the same time I am devoted to God, the only Origin, the only End.[20]

---

[19] Pierre Teilhard de Chardin, La Messe sur le monde. *Le coeur de la matière. Oeuvres XIII*, Paris: Éditions du Seuil (1976), 141–156.

[20] Pierre Teilhard de Chardin, La vie cosmique. *Écrits du temps de la guerre (1916–1919), Oeuvres XII*, Paris: Éditions du Seuil (1976), 5.

Teilhard's passionate love of God and of the Earth is constant in his life. He often repeated that he felt himself to be at the same time a "child of heaven and of earth". In the same text he urges Christians to discover this "cosmic awakening" and the divine Ideal hidden in the interior of the most material and terrestrial objects. They are to grasp the sanctifying value and the eternal hope of what he calls "holy Evolution". For Teilhard, his main task in life was to integrate the fact of the evolutionary nature of the universe with Christian thought. Many of his insights in this direction are already to be seen in the 20 essays written during the war (1916–1919), but he will continue to rework them until a few days before his death.

He takes as his starting point the evolutionary nature of life and the universe as shown by science, but he proposes the "phenomenon of man" as the key to its understanding. A further consideration concerns what he calls the "Christian phenomenon" and how this is to be interpreted from these new perspectives.[21] Teilhard tries to understand the nature of matter not from its physical constitution of elementary particles, but from the evidence of consciousness. Instead of understanding human consciousness from its material components, its biology and chemistry, as do many neurobiologists today, he reverses the process and tries to understand matter from the presence of consciousness in the material being that is the human person. Thus Teilhard's fundamental work has as its title *The Phenomenon of Man*. For him, if man is a self-conscious material being, the quality of consciousness must be present is some way, if at a very primitive level, in all matter. This led him to suggest that in all material beings there is an "interior" in addition to the "exterior" which is the object of the experimental sciences. This "interior" is linked to the complexity of things, so that the greater their complexity the greater their "interiority". Interiority and complexity are both finally linked to consciousness and spirituality. Two types of energy, "tangential" and "radial", correspond to this double character of matter: tangential energy corresponds to physical energy, while radial energy is responsible for evolution always in the direction of greater complexity, consciousness, and spirituality. These two types of energy are but two aspects of a fundamental spiritual energy. Matter has, then, for Teilhard an internal dynamism which includes a spiritual dimension. Thus he dispenses with the traditional dualism of matter and spirit.

An important moment in evolution is the apparition of life. It forms a new envelope or layer on the Earth, which he calls the "biosphere". From the first single-cell living beings, animal life evolves into more and more complex systems. Increasing brain complexity becomes the preferred line of evolution, finally ending in primates and man. With greater brain complexity the level of consciousness increases and the spiritual dimension appears. In this process, the apparition of full consciousness in man constitutes a new beginning and forms a new envelope on the Earth, which

[21] On the religious aspects of Teilhard's thought: Henri de Lubac, *La pensée religieuse du Père Teilhard de Chardin*. Paris: Aubier (1962); Émile Rideau, *La pensée du Père Teilhard de Chardin*. Paris: Éditions du Seuil (1965). Georges Crespy, *La pensé théologique de Teilhard de Chardin*. Paris: Éditions Universitaires (1961).

Teilhard calls the "noosphere". The apparition of the biosphere and the noosphere on Earth are two cases of discontinuity in continuity in the evolving process which results in the apparition of life and consciousness.

The cosmic evolutionary process of matter toward spirit does not end with the apparition of man, but has to continue, led now by the human spirit. This whole evolutionary process must necessarily converge somewhere. This convergence of the evolutionary cosmic process is an essential element of Teilhard's thought. For him a non-convergent evolution would not have any meaning. He calls the convergent point of the whole cosmic evolutionary process the Omega Point. Thus, the evolution of the noosphere or human evolution must move toward its convergence in the Omega Point. In that convergence all elements of the cosmic evolution will achieve their final perfection. Thus matter, which had begun its evolutionary process at the origin of the universe, will not achieve its final completeness until its convergence in the Omega Point through the evolution of the spiritual dimension in the human being. This last stage in cosmic evolution takes place through what Teilhard calls human "collectivization" or "socialization". In this process, humanity can progress toward a convergent unity or a divergent multiplicity. The first implies continuous progress, the second a set-back in the evolutionary process.

At this level, given man's free will there is always a danger of falling into divergent plurality. To avoid this possibility an attraction from a Center, the Omega Point, is needed with sufficient attractive force to unify all the conscious elements. This attraction must be at the same time effective and consistent with the free condition of the human being. This can only be done if the attractive Center or Omega Point is at the same time transcendent and personal or "hyper-personal", so that it can realize without forcing it the unification of all that is human. Thus the Omega Point cannot be a passive center, but an always active one, attracting everything to itself. Teilhard will finally identify the Omega Point with God. The final convergent unity of the whole evolutionary process is, then, achieved at the level of the noosphere. There is a human evolution through a unifying process attracted by the personal and transcendent Omega Point. Teilhard has moved from the evidence of science, which showed the evolutionary nature of the universe and the role of man in this process, to the need for an ultimate convergent unity in God.

Teilhard takes a further step with his Christian interpretation of the whole evolutionary process when the Omega Point is finally identified with Christ. The universe is progressing at the human level toward a final unity, which according to the Christian faith can only be achieved by the union of all men and women in Christ. Thus at the same time Christ is both the Omega Point present in human history and is attracting everything to Himself as the unifying Pole of its final achievement. The unity in Christ of all that is human, and through this unity that of the whole universe, forms what he calls the "Total" or "Cosmic Christ". With this further consideration Teilhard transforms the "cosmogenesis" or evolution of the universe into what he calls a true "Christogenesis", that is, the process by which the universe itself through human evolution becomes the "total Christ". Thus, for Teilhard, Christ's role is central for the whole movement of cosmic evolution. These ideas find their most complete expression in one of his last essays, written only a month before his death, and that can be considered as his spiritual testament.[22] They show the connection in Teilhard between his scientific view of the world in evolution and his mystical experience of Christ's role in it.

This short presentation of Teilhard's ideas may seem out of place in a history of Jesuits in science. However, it has been added here, because, while Teilhard was a true Jesuit scientist, with a brilliant career in geology and paleontology, his religious views take into account the evolutionary character of the universe as presented by science. As a Jesuit his Christian faith and his Ignatian spirituality drove him to find a consistency between the way of science and the path of faith. He found this connection in the identification of the Omega Point of science with the Christ of faith, the Incarnated God who achieves in himself the whole evolutionary process. Moreover, Teilhard shows how a vision of the world, as discovered by science, can be seen to have a sacramental sense. In his terminology the world is then a "transparency" (*diaphanie*) of Christ. As will appear in more detail in the Epilogue, Teilhard explicitly formulates ideas often implicitly present in many Jesuit scientists, who try to integrate dedication to science with spirituality. One cannot write the history of Jesuit scientists and ignore the figure of Teilhard de Chardin.

[22] Teilhard de Chardin, Le Christique. *Le coeur de la matière. Oeuvres XIII*, Paris: Éditions du Seuil, 1976, 93–117.

Chapter 10
The Sciences in Colleges and Universities

10.1 In the Training of Jesuits

After their restoration in 1814, the Jesuits found (as was noted in Chap. 6) that the situation with respect to science had changed. This change affected the training of Jesuits themselves. In the old Society scientific studies, which included mathematics, astronomy and physics, were part of the philosophy program, the second stage in their training before the study of theology. During the eighteenth century, some modern topics in advanced mathematics and Newtonian physics and astronomy had been introduced, as part of the courses of scholastic natural philosophy (see Sect. 2.7). By the time the Society was restored in the nineteenth century, science had made great progress and was firmly established with branches in physics, chemistry, geology and biology. The situation was thus very different from what it had been before the suppression of the Society. In addition, there had been a radical change in education that would affect the new Jesuit schools. In general, education was divided into three levels: primary, secondary or middle, and higher or university. In most countries education began to be considered a public service and a responsibility of the state. Thus governments established the regulations and study programs that were to be followed at all three levels. There was a definite emphasis on science in secondary and university education. Thus, the traditional Jesuit program following the norms of the *Ratio Studiorum* – humanities, philosophy and theology, – did not fit these new educational structures. Jesuits found that they had to adapt to the new circumstances in their schools and universities. The training of Jesuits kept as much as possible to the old system, and now became separate from that given to lay students. For the Jesuits the teaching of science came during the course dedicated to philosophy. The level of scientific knowledge depended on what the young men brought with them when they entered the novitiate, generally after finishing their secondary education. To judge by the recommendation of the 21st General Congregation (1829) – "care must be taken that the study of mathematics

A. Udías, *Jesuit Contribution to Science: A History*,
DOI 10.1007/978-3-319-08365-0_10

and physics should not be in detriment of that of philosophy" (Decree 17)[1] – science must have acquired a great importance during the first years after the restoration, in accordance with the spirit of the times.

During these years, an interesting debate took place in the Roman College between on the one hand those who defended that modern science should be taught separately and on the other hand those who proposed a return to the teaching of natural philosophy to be taught now according to neo-scholastic Thomism or the neo-Thomist school.[2] Giovanni Battista Pianciani (1784–1862), professor of physics and chemistry between 1824 and 1848, and his successor Francesco Saverio Provenzali (1815–1891), were prominent among those who defended the first position. Secchi, who had installed the observatory and taught astronomy in the College, gave them his support. Pianciani was the author of the textbook *Istituzioni fisico-chimice* (Physico-chemical principles, 4 vols. 1833/1835) and insisted on teaching physics and chemistry. Salvatore Tongiorgi (1820–1865) and Domenico Palmieri (1829–1909) also supported this position and rejected the Aristotelian theory of hylomorphism defended by the neo-Thomist philosophers. However, Matteo Liberatori (1810–1892), author of *Institutiones Philosophicae* (Philosophical principles, 1860/1861), a very influential work published in several editions, and Giovanni Cornoldi (1822–1892) were among the defenders of the neo-Thomist doctrine. Cornoldi strongly criticized Secchi's *L'Unità* as being contrary to Aristotelian doctrine. One of the key issues in this controversy was the scholastic theory of the hylomorphic (matter-form) composition of material bodies as opposed to the theory of atomic composition, supported by the discoveries of modern physics and chemistry. Secchi, who defended the atomic theory, complained that the ideas of Boscovich had found so little acceptance among Jesuit professors of natural philosophy. Liberatori defended scholastic hylomorphism and strongly attacked Tongiorgi for accepting the atomism of modern chemistry. In 1863 the Superior General, Pieter Beckx (1795–1887), intervened in the controversy with a conciliatory letter asking for moderation on both sides. He requested that in doubtful areas freedom of opinion should be respected.

Later, Beckx changed his mind and in 1878 took the side of Liberatori defending scholastic philosophy. Finally, in his encyclical *Aeterni Patris* (1879), Pope Leo XIII clearly defended the views of neo-Thomism and demanded the restoration of Christian philosophy according to the doctrine of St Thomas Aquinas, claiming that it was necessary for theology. However, the pope was not opposed to science; he says at the end of the encyclical; "nothing is of greater use to the philosopher than diligently to search into the mysteries of nature and to be earnest and constant in the study of physical things" and "between certain and accepted conclusions of modern physics and the philosophical principles of the schools there is no conflict worthy of the name".[3]

---

[1] *Institutum Historicum Societatis Iesu*. Florence: Typ. SS. Conceptione, vol. 2, 480.

[2] Robert Jacquin, Une polémique romaine sur l'hylomorphisme il-y-a un siècle. *Revue des Sciences Religieuses* 32 (1958), 372–377.

[3] Vatican.va/holy_father_leo_xiii/encyclicals/documents/hf_l_xii_enc_04081879_aeternis_patris. No. 30. Acceded May 2013.

The part of natural philosophy which deals with the nature of material objects began to be known now with the term "cosmology", and finally the scholastic theses of neo-Thomism were adopted. In their 23rd General Congregation (1883) the Jesuits complied with the encyclical of Leo XIII and stated that nothing should be explained in experimental physics which contradicted the doctrine of St Thomas (Decree15), although the study of natural and physical sciences was to continue (Decree 17).[4] Thus, both approaches had to be respected in the study of nature: the philosophical, in accordance with neo-scholastic Thomism, and the scientific, acknowledging the modern positive sciences.

However, the Jesuits continued to teach science as part of the philosophy program. For example, between 1850 and 1873 there were courses in the Roman College of mathematics, mathematical physics, physical chemistry and astronomy. After the confiscation of the Roman College by the Italian government, in 1873, the Jesuit faculties of philosophy and theology became the Gregorian University, named in memory of the first patron, Gregory XIII, and there science continued to be taught until the 1930s. Among the professors of science at that time mention should be made of Giuseppe Gianfranceschi (1875–1934), who had a doctorate in physics and mathematics from the University of Rome. He carried out research in electrophysics and acoustics. His book, *La fisica dei corpuscoli* (Physics of corpuscles, 1916), was very popular and appeared in three editions and a translation into Spanish. In 1928 Gianfranceschi took part in an interesting expedition to the North Pole in an airship.

In Jesuit philosophy faculties cosmology was taught in accordance with neo-Thomism and it became one of the main subjects. The *Institutiones Philosophicae* (Philosophical Principles), published in eight volumes between 1890–1900 by Juan José Urráburu (1844–1904), a professor at the Gregorian University, became very popular. He combined the philosophy of Suárez with neo-Thomism, and his cosmology was widely used in many Jesuit faculties. Another popular textbook of cosmology in Jesuit circles was *Philosophia Naturalis* (Natural philosophy, 1894) by Heinrich Haan (1844–1909), who taught in the German Faculty of Philosophy at Valkenburg and propounded a neo-Thomist cosmology. Thus a clear separation was established between so-called "scholastic cosmology" and the modern sciences of physics, chemistry and biology. In those years attacks on the Church began to appear, presenting it as opposed to science and typical of the opposition between science and religion. Pianciani was one of the first Jesuits to deal with this problem in his *Cosmogonia naturale comparata col Genesi* (Natural cosmogony compared with Genesis, 1862). He defended the need to show the compatibility between Christian religion and science for apologetic reasons. This attitude became very common among Catholic authors when dealing with this subject.

In some faculties of philosophy, astronomical and meteorological observatories were set up to complement the teaching of science. One of them was installed in the German faculty of Valkenburg, and from 1872 to 1941 science was taught at a high standard, especially in physics. Among the professors Theodor Wulf (1868–1946)

[4] *Institutum Historicum,* vol. 2, 501–504.

is worth mentioning. He was the author of several textbooks of physics and of an early book (in 1921) about Einstein's theory of relativity (*Einstein Relativitäts Theorie*). His physics textbook (*Lehrbuch der Physik*, 1926) went into several editions and was translated into English, French and Spanish. Wulf built a sensitive electrometer, capable of detecting very low levels of radiation.[5] In 1910 he carried out an experiment on the Eiffel tower in Paris and discovered that radiation was stronger at the top than at the bottom, contrary to what was supposed to be the case. He concluded that radiations came not from the earth but from outside the atmosphere. Thus he was the first to discover the effect of cosmic rays and is considered a pioneer in the study of these.[6] Another professor of physics was Ludwig Dressel (1840–1917), author of a popular physics textbook, *Elementare Lehrbuch der Physik nach den neusten Anschauungen* (Elementary physics according to the new opinions, 1895), which ran to four editions. In 1870 Dressel and Wulf travelled to the Republic of Ecuador to take charge of the Polytechnic School founded by President García Moreno. They returned to Germany 6 years later after the death of the president and the closing of the school. Another scientific work undertaken by the Jesuits in Valkenburg was the study of the Babylonian astronomical tables. Johann Strasmaier (1846–1920), professor at the Faculty of Philosophy of Maria Laach in Germany before its move to Valkenburg, began this investigation. From 1878 to 1897 he patiently copied the Babylonian tables at the British Museum and began their interpretation. Joseph Epping (1835–1894) and Franz Kugler (1862–1929) continued these studies in Valkenburg using the copies made by Strasmaier.

In 1884, members of the Jesuit Belgian Province, aware of the need to maintain scientific studies as part of Jesuit training, presented a document to the Superior General so that it could be distributed for consideration by other provinces.[7] The document asked for physics, chemistry, geology, botany and zoology to be included in the programs of philosophical studies. The answers from the different provinces were very different, but they agreed that, although sciences had to be taught, it was not convenient to give general rules. Similar recommendations, like that from Wulf in 1921, were sent to Rome insisting on the need for scientific education. However, these proposals were not promulgated in any official documents. The 27th General Congregation (1931) repeated the obligation to follow St Thomas in theology and philosophy (Decree 99) and stressed the need for care to be taken with regard to dangerous doctrines (Decree 102).[8] Nothing was said about the study of science. With the reorganization of ecclesiastical studies proposed in 1931 (in the Apostolic Constitution *Deus scientiarum Dominus* by Pius XI), new programs of philosophy and theology were established. In 1934, the Jesuits adapted their programs to fit

[5] Theodor Wulf, *Die Faden-Elektrometer*. Berlin: Dümmlers (1933).

[6] P. Carlson, A century of cosmic rays. *Physics Today*, 65 (2012), 30–36.

[7] Examen postulati Provinciae Belgicae de studio scientiarum physicarum a patribus eiusdem Provinciae institutum (1884). Archivum Romanum S.I. (Rome), De scientiis, stud 4–21 (1884–1921).

[8] Statuta Congregationis Generalis XXVII, *Acta Romana S. I.* 4 (1924) 59–61.

these new norms.[9] Topics related to science – such as the structure of matter, the origin of the universe and the biological evolution – were included in a new additional subject called "Scientific questions related with philosophy". These topics were to be considered in relation to philosophical problems, so that there was no teaching of science in the strict sense. In this subject "scientific questions" were considered from the philosophical point of view.[10] In this way, the explicit teaching of mathematics and science practically disappeared from the Jesuit faculties of philosophy. What remained was "cosmology", seen from the point of view of neo-Thomist doctrine, and the problems discussed in the "scientific questions". In 1938 the 28th General Congregation approved this organization of philosophical studies.[11]

Among the professors of cosmology, there were some who kept a great interest in science. Pieter Hoenen (1880–1961) won a doctorate in science from the University of Leyden and became professor of physics and chemistry in several Jesuit colleges in Holland. His textbook *Cosmologia* (1931) ran to five editions. Filippo Selvaggi (1913–1995), taught cosmology and philosophy of science and he published a popular textbook *Cosmologia* (1959), on Thomistic natural philosophy, and several books on physics, such as *Orientamenti della fisica* (Overviews on physics, 1961) and *Problemi della fisica moderna* (Problems of modern physics, 1953). The Jesuit Superior General, Jean Baptiste Janssen (1889–1964), was concerned about the teaching of cosmology and its relation to science. In a letter in 1960 he made clear that cosmology, though a philosophical subject, had to be open to the problems raised by scientific progress in modern times. To profit from it, professors and students alike had to have sufficient scientific training. Thus Jesuit students needed sufficient knowledge of mathematics, physics, chemistry, biology, geography and astronomy before beginning their philosophical studies. Although he was aware that the scientific training acquired at the secondary school level was not sufficient, he gave no solution to this problem.[12]

The progress being made in science and the need for a more advanced training meant that after finishing their philosophical studies some Jesuit students with a special interest were sent to study science at universities. This special training was reserved to those appointed to become professors or to undertake research at observatories. The time for this was the period between philosophy and theology, when young Jesuits were usually sent to teach at schools. For those selected this was a period of so called "special studies". Another possibility was to undertake scientific studies after finishing theology. The custom of specialized studies began soon after the restoration of the Society of Jesus. For example, between 1858 and 1860 Perry

[9] Approbantur statuta facultatum theologiae et philosophiae in collegiis Societatis Iesu erectarum, *Acta Romana S.* I., 7 (1934), 606–633.

[10] F. M. Palmés, J. Torra Almenara y J. M. Caballería. Las "cuestiones científicas" en las Facultades de Filosofía. *Estudios Eclesiásticos* 55 (1935), 1–32.

[11] Decreta Congregationis Generalis 28, 4. De studiis, *Acta Romana S. I.*, 9 (1938–1940), 37–40.

[12] Jean Baptiste Janssen, Instructio de cosmologia in nostris scholasticatibus docenda. *Acta Romana S. I.*, 13 (1960), 809–814.

studied mathematics and science in the universities of London and Paris before beginning his theological studies; and Macelwane in 1923, after finishing his theology, obtained his doctorate in science from the University of California, Berkeley. In a letter (1947) addressed to Jesuits about their work, Janssen put academic studies in first place: "this applies to sacred disciplines but also to *profane sciences*, which should not be considered as foreign to our vocation in the Church or the Society". Among the examples he mentions are archeology, paleontology, and physics. He insisted on the need for serious scientific work in these fields for the prestige of the Church: "Today people do not esteem the theologian or the Catholic philosopher, but would have a great appreciation for the historian, mathematician and astronomer".[13]

At an official level the 31st General Congregation (1965) was the last to refer explicitly to the work of Jesuits in the natural sciences (Decree 29 on scientific research). The decree begins: "Jesuits must greatly value scientific work and especially authentic research and must consider it as one of the most important works of the Society".

Although the decree refers in the first place to the sacred sciences, it includes also "the mathematical-natural sciences and the technology that is derived from them".[14] The decree justifies such dedication given the mentality of the time so deeply penetrated by science. In 1976 the Superior General, Pedro Arrupe (1907–1991) put forward the same argument in a letter on the intellectual apostolate:

> How can we make the Church present, and keep the necessary personal contacts in a social context of so vital importance as the scientific and technological, without giving to science the importance it deserves? How can we make a theological reflection that is intelligible without a profound knowledge of the scientific roots of this mentality?[15]

These questions remain relevant today, when the Jesuit presence in scientific fields has radically diminished, as will be seen. In this decrease, among other factors, the 32nd General Congregation (1974) has had an influence, especially with Decree 4 (On the service of faith and promotion of justice). This decree marked a change in the priority given by Jesuits to the social apostolate, and although not intended to do so, it supposed in practice a diminution in the involvement of Jesuits with the natural sciences. The General Congregations since then, the 33rd (1983), 34th (1995), and 35th (2008), barely made any mention of scientific work as part of the intellectual apostolate.

Jesuits working in the natural sciences felt the need to form associations; the first was the American Association of Jesuit Scientists, established in 1922 in the United States. It began only with Jesuits from the northeastern States, but later included others from all the States. In 1954 there were 254 members; it met once a

---

[13] Jean Baptiste Janssen, Epistola de ministeriis nostris (22 June 1947). *Acta Romana S. I.*, 11 (1946–1950), 299–336.

[14] Decreta Congregationis Generalis 31, De labore et investigatione scientifica, *Acta Romana S. I*, 14 (1961–1966), 958–960.

[15] Pedro Arrupe, El apostolado intelectual en la misión de la Compañía hoy. *Acta Romana S. I.*, 16 (1973–1976), 1015.

year and had a quarterly Bulletin from 1922 to 1966.[16] This Bulletin is an important source of information about Jesuit scientific activity in the United States during those years. In 1968, Pedro Arrupe, concerned with the problems of Jesuit scientists, appointed François Russo (1909–1998), a philosopher and historian of science, to be the "International Coordinator of Jesuits Working in the Positive Sciences". In 1970 he launched the publication of the *Bulletin de Liaison des Jesuites Scientifiques* (six issues until 1975). In 1976 Charles Courtoy succeeded Russo, and in 1982 Alberto Dou (1915–2009), a professor of mathematics at the *Universidad Complutense* of Madrid became coordinator. In 1985 Dou began the publication of a new bulletin, *Jesuits in Science* (15 issues up to 1999). These Bulletins contain interesting information about the problems of modern Jesuit scientists. In Europe a group named "European Jesuit Scientists" met in 1989 at Aix-en-Provence (France). Meetings continued every 2 years until 2009, with 20–30 participants.

10.2 Science Departments in Universities

The suppression of the Society of Jesus brought with it the loss of the universities the Jesuits had managed in Europe and America from the sixteenth century. When the Society was restored in 1814, the situation of universities in Europe had radically changed. The medieval institutions, where theology had played a central role, had suffered a steep decline. With the nineteenth century a process of reform began whereby theology either lost its predominant position or in many cases totally disappeared, while the natural sciences took the leading place. Moreover, the German universities, beginning with Berlin and Göttingen, were pioneers in giving a major role to research. This double change of orientation was adopted by all modern universities. In France a radical reform of all the universities took place during the Napoleonic period and they became directly dependent on the state. A similar process of university dependence on the state spread to other countries in Europe, notably Italy and Spain. Thus, it became practically impossible for Jesuits to take back the universities they once had in Europe, except for those devoted exclusively to ecclesiastical studies. In other countries, such as the United States of America, the situation was different, and there the Jesuits could establish institutions of higher education (colleges and universities) from the nineteenth century. India was also different, and following the English system Jesuits could establish colleges associated with state universities. Over the years the European situation has changed, and Jesuits have been able to set up institutions of higher education in many countries.

[16] *Bulletin of the American Association of Jesuit Scientists (The Jesuit Science Bulletin)*, vol. 1–43 (1922–1966).

Fig. 10.1 Georgetown University (Washington, D.C.) (Georgetown University Archives)

Today there are 133 Jesuit universities in the world established from the end of the nineteenth century.[17] In the United States, where there is a strong tradition of private universities, the first colleges and universities were established shortly after the restoration. The first was Georgetown University in Washington D.C., established by John Carroll, a former Jesuit and first Catholic Bishop of Baltimore, in 1789 while the Jesuits were still suppressed (Fig. 10.1). After the restoration, the university management was entrusted to the Jesuits. Nineteen of the twenty-eight Jesuit American universities existing today were founded before 1900. Those considered most important are: Georgetown (1789), Saint Louis (1818), Fordham in New York (1841), Boston (1863), Loyola in Chicago (1870), San Francisco (1875), Regis in Denver (1877), and Marquette in Milwaukee (1881). Between 1950 and 1960, all Jesuit universities in the United States experienced a rapid growth in students, faculty and resources.[18]

[17] *Wikipedia*: List of Jesuit institutions – List of Jesuit universities (Wikipedia.org/wiki/List_of_Jesuit_Institutions) acceded June, 2013.

[18] Raymond A. Schroth, *The American Jesuits. A history*. New York: New York University Press (2007), 174–196; Gerald McKevitt, Jesuits schools in the USA, 1814-c.1970. In Thomas Worcester (ed.), *The Cambridge companion to the Jesuits*. Cambridge: Cambridge University Press (2008), 278–293.

In India, there are today 16 Jesuit colleges, among them St Xavier's in Kolkata (1860), St Xavier's in Mumbai (1869), and Loyola in Chennai (1925); in Europe, 8; in Central and South America 24, among them: the *Pontificia Universidad Javeriana* in Bogota (1623 and 1937), the *Pontificia Universidade Catolica de Rio de Janeiro* in Brazil (1940), and the *Universidad Iberoamericana* in Mexico (1943); in Asia, 11, among them the *Ateneo de Manila* (1859), Sofia in Tokyo (1913), and Sogang in Seoul (1960); in Africa and the Middle East, there are 5 Jesuit universities.

Not all these universities have science programs and the teaching of science varies greatly as does the quality of the science departments. Some Jesuits teach in the science departments, but not in all. In most cases, the Jesuits are mainly involved in teaching, but some in research. A few did research in astronomy, geophysics and meteorology in those universities which had observatories, as we have already seen. Jesuits also carried out important research in the natural sciences, especially botany and entomology, as will be seen later. At present, in the Jesuit colleges in India there are important programs, mainly in mathematics, physics, botany and environmental science: e.g. in St Xavier's College (Mumbai), St Xavier's College (Kolkata), Loyola College (Chennai) and St Joseph's College (Bangalore). To have an idea of how many Jesuits were involved in scientific work, some numbers for the universities of the U.S.A and India may help.[19] When their numbers were at their highest in 1975, there were some 135 Jesuit professors of science in the United States (where most of the 28 Jesuit universities had science programs at different levels), approximately 3–6 in each institution; the highest number was at Boston College (14), then Fordham (8) and Saint Louis (7), and these were divided between mathematics 40, physics 39, chemistry 22, biology 27, geology 4, and engineering 2. In the same year, in the 16 colleges in India there were a total of 68 Jesuit professors of science: mathematics 15, physics 20, chemistry 15, and biology 18.

In Europe there are few Jesuit institutions of higher education with science programs. In Belgium one finds the *Facultés Universitaires Notre Dame de la Paix* (today University of Namur) in Namur, founded in 1831. In France the *Institut Catholique d'Arts et Métiers* (ICAM) in Lille was founded in 1898 and offers higher studies in engineering from 1911; this is more technological than scientific in character. In Spain, Jesuits established in 1905 the *Laboratorio Químico del Ebro* and in 1908 the *Laboratorio Biológico del Ebro*, for studies in chemistry and biology respectively. In 1916 they moved to Barcelona and merged together under the name *Instituto Químico de Sarriá* (IQS). In 1990 the IQS amalgamated with other institutions to form the *Universidad Ramón Llull.* In 1908, the Jesuits established in Madrid the *Instituto Católico de Artes e Industrias* (ICAI), an engineering school inspired by the ICAM in Lille. In 1978 the ICAI became part of the Jesuit *Universidad Pontificia Comillas* in Madrid.

[19] Information obtained from the survey of the *International Coordinator of Jesuits Working in the Positive Sciences*, made by F. Russo and A. Dou between 1975 and 1985 (Biblioteca Comunidad ICAI, Madrid).

The shifts in the number of Jesuit science professors in Jesuit colleges have varied from place to place.[20] Thus, in Boston College (U.S.A.), while in 1940 there were 6 science professors out of 79 Jesuit professors, in 1960 the number increased to 10 out of 117 (physics 4, chemistry 2, biology 2 and mathematics 2). In 2000 they had decreased to 3 with 3 emeriti. In Georgetown University, recent changes have been more abrupt: in 1940 there were 5 out of a total of 58 Jesuit professors; in 1960, 7 out of 90, and in 2000, none out of 68. In India, in St Xavier's College, Kolkata, out of 30 Jesuit professors there were 4 teaching science in 1960, but only 2 in 2000. In St. Xavier's College, Mumbai, in 1960 there were 7 in science out of a total of 19, but in 2000 only 2 out of 20. In the *Facultés Universitaires* of Namur in 1940 there were 7 Jesuit science professors (chemistry 2, biology 2, mathematics 2, physics 1) out of 43 Jesuit professors: This number remained stable and in 1960 there were still 7. In 2000 there were only 2 in biology and another 2 in what was called "scientific research". In the ICAI Madrid, between 1930 and 1980 the number of Jesuit professors of science and engineering was very stable with about 10. But the number fell abruptly to 4 in 1990 and to only 1 in 2000. In the two Jesuit universities in Brazil, the *Pontificia Universidade Catolica* in Rio de Janeiro and the *Unisinos* (*Universidade do Vale do Rio dos Sinos*) in San Leopoldo, in 1975 there were 4 professors of science in each, but in 2005 there was only 1 for both.

These examples show that the highest number of Jesuit professors of science in Jesuit universities began to fall between 1960 and 1970, which was the time when there was the highest overall number of Jesuits. They also show that the number of Jesuit professors of science in Jesuit universities was fairly stable until about 1980, in some cases reaching up to 10 % of a Jesuit faculty, but fell rapidly thereafter. While many factors influenced this rapid decrease, an important one was the fall in number of those joining the Society, especially in Europe and the United States. The more general problem of the decrease of Jesuit involvement in science in recent times will be discussed later.

Following in the tradition of scientists in the early Society, modern Jesuits have also written a number of science textbooks. Some have already been mentioned: Secchi, *Lezioni di Fisica Terrestri* (1877); Wulf, *Lehrbuch der Physick* (1926); Macelwane, *Introduction to theoretical seismology* (1936). But more examples of science textbooks at university level written by Jesuits deserve to be mentioned: August Bellynck and Égide Pâque (Université N.D. de la Paix, Namur, Belgium), *Cours de Botanique* (1899); Joseph Bayma (Santa Clara University, California), *Elements of Infinitesimal Calculus* (1889); Adolf Müller (Gregorian University, Rome), *Elementi di astronomía* (1904); Eduardo Vitoria (Instituto Químico de Sarriá). *Manual de química moderna,* (1910 which has had 15 editions, the last in 1955); José A. Pérez del Pulgar, *Electrodinámica Industrial* (4 vols. 1915–1919); Edmund J. Burke (Fordham University, New York), *Lectures in biology: zoology* (1928); B. Rajam (St. Joseph College, Trichinopolis, India),

[20] Data from the corresponding *Catalogues of the Provinces of the Society of Jesus* for the years 1940, 1960 and 2000 (Archivum Romanum S. I., Rome).

Atomic Physics (1950), with a preface by Louis de Broglie; Claudio Mataix (ICAI, Madrid), *Mecánica de fluidos y máquinas hidráulicas* (1970).

10.3 Science Teaching in Secondary Schools

The reform of education in the nineteenth century established three levels: the primary and secondary schools and the universities. Most of the Jesuit schools founded after the restoration were at the secondary level. They spread rapidly throughout the entire world. Their development is usually divided into three periods. During the first, 1812–1917, there was much political instability and the Jesuits were expelled from several countries with as a consequence the opening and closing of several schools. The second period, 1920–1965, was one of rapid expansion and saw the opening of many schools. The third, 1970 to today, has witnessed the closing of many schools, due to the fall in the number of Jesuits, and radical changes in the organization of schools. Thus there has been a massive intake of lay teachers with a drop in the number of Jesuits teaching. Between 1965 and 1998 the number of Jesuits teachers decreased from about 10,000 to 4,500 and the proportion of lay teachers increased in many cases above 95 %. At the present day there are about 315 Jesuit secondary schools in the world, 120 in India and 19 in the rest of Asia, 62 in Europe, 54 in Central and South America, 52 in the United States, and 8 in Africa and the Middle East.[21] There are also other educational ventures apart from the traditional schools, such as the networks of *Fe y Alegría* schools in Central and South America with more than 400,000 students, the SAFA schools in south Spain, and the school network of *Cristo Rey* in the United States. In all these schools science teaching varies enormously with regard to subjects and levels.

Most Jesuits teaching science at these schools had to restrict themselves to teaching duties, given the elementary and introductory character of science education at this level, and many devoted themselves to this hidden work, laying the foundations of scientific knowledge in their young students. Little more can be said about them, except that these selfless teachers laid the seeds for what in some cases would be brilliant scientific careers. Observatories were installed in some of these schools, such as those at Stonyhurst, Kalocsa, Manila and Belén, reaching a high level in astronomy, geophysics and meteorology. They also established good modern laboratories of physics and chemistry and set up museums of natural science. For a while they carried out significant research in the fields of botany and entomology. As was the case with university professors, Jesuit teachers in secondary schools also published a number of good scientific textbooks: for example, Bonifacio Valladares, *Física elemental* (1900, 1913, 1921); Ferdinando Trassarelli, *Corso de Biologia* (1948); Ignacio Puig, *Curso general de química* (1927); Leon Talpain, *Mathématiques. Classe de première D* (1966).

[21] *Wikipedia*: List of Jesuit institutions – List of Jesuit secondary schools. (Wikipedia.org/wiki/List_of_Jesuit_Institutions) acceded March 2013.

10.4 The New Naturalists and Biologists

As explained in Chap. 5, during the seventeenth and eighteenth centuries many Jesuit naturalists did important work producing the first studies of the flora and fauna of America and Asia. This tradition continued in the modern Society of Jesus thanks to Jesuits teaching in universities and secondary schools. Biology is a relatively modern science: in the mid-eighteenth century Carl von Linné (Linnaeus) established with biological taxonomy a basis for the classification of living beings and only at the beginning of the nineteenth century Gottfried Reinhold Treviranus and Jean Baptiste Lamarck first used the term "biology" and are considered its pioneers. Thus, several Jesuits contributed to the beginning of this science especially in the fields of botany and entomology.

The Austrian Erich Wasmann (1859–1931) devoted his life to the study of ants, termites, and their parasites.[22] He was on the staff of the Jesuit journal *Stimmen der Zeit*, in which he published many of his articles on scientific problems. These included a long list of articles on the study of ants and termites. He also published two books, where he describes their system of communication and social organization. He collected and classified some 1,000 species of ants, 200 of termites and 2,000 of myrmercophiles, 933 of them being new species. Wasmann was an early defender among Catholic authors, of the theory of evolution, though in a moderate version which rejected natural selection as the only mechanism involved and the descent of man. His *Die moderne Biologie und die Entwicklungtheorie* (Modern biology and the theory of evolution, 1906) had a wide circulation (three editions and several translations). Wasmann claimed that the theory of evolution was not opposed to Christian faith and that divine Providence guides the evolutionary process according to the internal laws of nature. Opposition from conservative sectors in the Church forced him to stop further editions of his book. He took part in a long debate with Ernst Haeckel and rejected his philosophical monism in two series of conferences held in Berlin in 1905 and 1907. In 1921 the University of Freiburg (Switzerland) awarded him an honorary doctorate. Karl Frank (1875–1950) was a follower of Wasmann. He combined his work as a biologist with his role as professor of natural philosophy in the Jesuit Faculty of Philosophy at Pullach (Bavaria). Adolf Haas (1914–1982) succeeded Frank in Pullach; his doctoral thesis was on the flight of bumblebees, and he worked with the ethologist Konrad Lorenz on the behavior of certain fish and also on the concept of instinct in animals. Hass contributed to the spread of Teilhard's thought in Germany and to the development of the theory of divine creation through evolution.

Pelegrín Franganillo (1873–1955) began his work with a study of spiders in Galicia and Asturias (northern Spain), published as *Las arañas. Manual de aracnología* (Spiders. Handbook on arachnology, 1917). In 1918 he moved to the Belén

[22] Charles J. Widerman, *Erich Wasmann S. J. 1859–1931*. San Francisco: Wasmann Biological Society, (1942); Christian Kummer, Forschung und Weltanschaung: Die Rolle del deutschen Jesuiten in der Biologie. *Sammelblatt des Historischen Vereins Ingolstadt* 109 (2000) 105–120.

School (Havana, Cuba), where he continued his work on spiders and published *Arácnidos de Cuba* (Spiders in Cuba, 1923). Longinos Navás (1858–1938), a professor in the Jesuit school in Zaragoza, was an entomologist specializing in neuroptera. He classified up to 3,316 new forms (388 genera, 2,684 species, and 244 varieties).[23] A tireless worker, he published 588 articles in scientific journals and proceedings of congresses between 1898 and 1935. An important publication was his *Entomologia de Catalunya. Neuròpters* (Entomology of Catalunya. Neuroptera. Vols. 1–3, 1923, 1926, 1934). Jaime Pujiula (1869–1958), founder of the Biological Institute of Sarriá (Barcelona), was a pioneer in embryological, histological, and cytological research in Spain. He had studied in Innsbruck and Trieste and obtained his doctoral degree in Vienna. He was a member of the Academy of Medicine of Barcelona and published *Citología* (Cytology, 1928) and *Histología, embriología y anatomía humana y animal* (Human and animal histology, embryology and anatomy, 1934).

Biology was especially developed in the two Portuguese Jesuit colleges of Campolide (Lisbon) and São Fiel (Louriçal do Campo). Joaquim da Silva Tavares (1866–1931), Cândido Azevedo Mendes (1874–1943), and Carlos Zimmermann (1871–1950), all teachers at the Jesuit college of São Fiel, were the founders of the journal *Brotéria* (named after the Portuguese botanist Felix Avelar Brotéro) dedicated originally (between 1902 and 2002) to the natural sciences, especially, botany, zoology and genetics. Today it is a cultural journal.[24] Tavares was an entomologist and a member of the Academy of Science of Lisbon; he worked on the cecidia and zoocecidia of Portugal, Brazil, Argentina, and Madeira. Mendes devoted his efforts to the study of butterflies. Luis Jorge Archer (1926–2011), a professor of molecular genetics at the *Universidade Nova* in Lisbon and a research fellow at the Gulbenkian Institute of Sciences, was director of *Brotéria* from 1962 to 1979 and kept up the tradition of Portuguese Jesuit biologists, so much so that the Portuguese Society of Genetics has established an award in his honor.[25] Xavier Ferenc Speiser (1854–1933), taught at the school at Kalocsa (Hungary) and was an entomologist. He studied 9500 species of coleopterons, most of them from Africa, which Laszlo Menyarth (1849–1897), also a professor at Kalocsa, brought from his journeys to the Zambezi. Balduino Rambo (1905–1961), professor at the *Colegio Anchieta* in São Leopoldo (Brazil) followed in the tradition of naturalists of the old Society in South America. He founded the Museum of Natural Sciences of Rio Grande do Sul and organized the *Herbario Anchieta* which contained more than 60,000 plants. The English botanist, Gerald Wilson Browne (1911–1975), studied the flora of Kanuku in Guyana.

At the start of the twentieth century some Jesuits in India began a brilliant series of botanical studies in relation to agriculture, environmental studies and ecology

[23] Juan Jesús Bastero Monserrat, *Longinos Navás, científico jesuita*. Zaragoza: Universidad de Zaragoza (1989).

[24] Francisco Malta Romeiras, The journal Brotéria (1902–2002). Jesuit science in the 20th century. *Journal of History of Science and Technology*, 6 (2012), 100–109.

[25] Francisco Malta Romeiras, The emerging of molecular genetics in Portugal: the enterprise of Luís Archer. *Archivum Historicum S. I.* 82 (2013), 501–512.

Fig. 10.2 Ethelbert Blatter (1877–1934), botanist at India (Xavier's College, Mumbai)

that have been continued to the present day.[26] A Swiss Jesuit, Ethelbert Blatter (1877–1934), who arrived in India in 1903, was among the first (Fig. 10.2). He taught at St. Xavier College, Mumbai, and became a member of the Bombay Natural History Society. He established in the college a collection of nearly three million plants known today as the *Blatter Herbarium*. Among his publications one finds *The palms of British India and Ceylan* (1926), published by the Oxford University Press, *Beautiful flowers of Kasmir* (2 vols. 1927–1928), and in collaboration with Walter S. Millard, *Some beautiful Indian trees* (1937, 1997). The Asian Society of Calcutta acknowledged his work on botany and awarded him the first Bruehl Memorial Award. Another important professor at St Xavier's College was Hermenegildo Santapau (1903–1970), from Tarragona (Spain); he had a doctorate from London University and was an accomplished botanist. In 1955 he became president of the Indian Botanical Society and between 1961 and 1967 was director of the Botanical Survey of India. He published some 216 scientific articles and reports, and several books: e.g. *The Flora of Purandhar* (1958). In 1967 the Government of India awarded him the Padma Shri Award.

Other names worth mentioning are the following: K.M. Mathew (1930–2004) had been born in Kerala and was professor of botany at St. Joseph College, Tiruchirappalli. He worked with great energy for the study and conservation of the *shola* forests of the subtropical regions of south India. Well known as a taxonomist, he published among other books, *The flora of the Palni Hills* (1999) and *The flora of the Tamilnadu Carnatic* (5 vols., 1983). Mathew organized the *Rapinat Herbarium*, a collection of more than 54,000 plants collected and catalogued by Jesuits since 1895, and he encouraged the local people to take an interest in their plants. In 2002 the Government of India awarded him the Indira Gandhi Paryavaran Puraskar Award for his environmental studies. Cecil Saldanha (1926–2002), professor of botany at St. Joseph's College, Bangalore, studied the flora and the environment of the Karnataka State. He published *Flora of Karnataka* (2 vols. 1978), with

[26] Leo D'Souza, Jesuit contribution to environmental protection. Sjweb-info/sjs/networks/ecology/Indian (2012) acceded March 2012.

descriptions of 1,700 species. Botany, applied botany, environmental and ecological studies are taught today in many of the Jesuit Indian colleges, for example, in those of St. Xavier's (Mumbai), St. Xavier's (Ahmedabad), Loyola (Chennai), St. Joseph's (Tiruchirappalli), St. Aloysius (Mangalore), St. Xavier's (Kolkata) and St. Xavier (Palayamkottai).

10.5 Jesuit Scientists in Other Institutions

During these decades, the Jesuits active in different fields of science were not confined to their own universities and schools, and several Jesuit scientists were, and still are, working at non-Jesuit universities and research centers, both private and public. Some examples have been already mentioned: Teilhard de Chardin worked mainly in non-Jesuit institutions, such as the Museum of Natural History in Paris, the Museum of Man in Beijing, and the Wenner Gren Foundation of New York. Lejay was director of the Observatory of Zikawei and after his return to France worked at the *Laboratoire National de Radioélectricité* of the *Centre National de la Recherche Scientifique* (CNRS). Mayaud, who collaborated with the Observatory of Addis Ababa, had a post as *Maître de Recherches* at the CNRS in the *Institut de Physique du Globe* in Paris. From 1978 he wrote on the history of science, especially on the conflict between the new astronomy and the Church with his monumental *Le conflit entre l'astronomie nouvelle et l'Écriture sainte aux XVIe et XVIIe siècles* (The conflict between the new astronomy and the Sacred Scriptures in the sixteenth and seventeenth centuries, 6 vols., 2005). Archer worked at the *Universidade Nova* in Lisbon and at the Gulbenkian Institute of Sciences. Some Jesuits working at Jesuit observatories were also professors at other centers. Drake besides being Director of Riverview Observatory was professor at the Macquarie State University. Ingram, Director of the Observatory of Rathfarnham Castle, also taught mathematics at University College Dublin. Wulf and a group of German Jesuits managed during a few years the new Polytechnic School in Quito, Ecuador. Sarasola and Ramírez were for some time in charge of the National Meteorological Service of Colombia. Macelwane organized studies and research in seismology at the University of California, Berkeley for a couple of years.

During the 1970s a few Jesuits worked, mainly in biology, for the CNRS in France and at the *Institut Catholique* in Paris and Toulouse. In the same years there were Jesuit professors of science at State Universities in other countries in Europe, such as the University of Padua (Italy), Warsaw (Poland) and the National University (Ireland). In countries like the United States and India, where there are many Jesuit colleges and universities, there are few examples of scientists working in other institutions. But an interesting exception is Timothy Toohig (1928–2001), a nuclear physicist associated with Boston College. He did his research in Brookhaven National Laboratory between 1965 and 1970 and afterwards was at the Fermi National Accelerator Laboratory (Fermilab), a center dependent on the U.S. Atomic Energy Commission. Here Toohig collaborated on the design of the *Tevatron,* a particle accelerator. From 1988 he worked with the Project for the new particle accelerator, the Superconducting Super Colliding (SSC), which was never finished. Later he was one of the U.S.A. representatives in the *Conseil Européen pour la Recherche Nucléaire* (CERN) in Geneva.

Between 1970 and 1980, there was an interesting development in Spain in connection with Jesuit scientists working at State institutions. Given the many vocations in the 1950s and an increase of Jesuits who studied science in State universities, many remained in them as professors while others held research posts in the *Consejo Superior de Investigaciones Científicas* (Higher Council of Scientific Research, CSIC). This may have been partly due to a certain detachment vis-à-vis Jesuit institutions. In those years some Jesuits thought it was better to work in non-Jesuit institutions, and there were even some who thought that all Jesuit institutions should be abandoned. For those who had studied science there was the additional factor that the only existing Jesuit higher educational institutions with programs in science in Spain were the IQS in Barcelona and the ICAI in Madrid. The number of Jesuits in State institutions in 1973 reached 75, i.e. professors and research specialists in different fields; among them there were 27 in science. That year the Provincial of Spain called a meeting in Madrid for such Jesuits and in 1977 he created a group under the title "University Mission in non-Jesuit Institutions" (*Misión Universitaria en Instituciones que No son de la Compañía de Jesús*: MUINSI). He appointed Julián Rubio (1925–1996), a Professor of Genetics in the University of Oviedo (North Spain), to be the coordinator of the group and it met every year. An evaluation in 1980 of the apostolic value of this experiment recognized the positive value of having members of the Society of Jesus and of the Church in scientific fields and in particular in State institutions. In those years among the Jesuit scientists occupying posts in State Universities there were 11 top-level professors (*catedráticos*) in the fields of mathematics, theoretical physics, atomic physics, optics, geophysics, physical chemistry, biochemistry, biology, genetics and paleontology at the universities of Madrid-Complutense, Barcelona, Barcelona-*Autónoma*, Barcelona-Politécnica, Seville, Oviedo, Malaga and Cordoba. Meetings of the MUINSI continued until 1988, when they stopped because no new Jesuits had joined any of the State Universities. No continuation was possible due to the sharp decrease in the number of new Jesuits in Spain and the departure from the Society of some young Jesuits dedicated to science; but also due to changes in the choice of apostolic priorities, preference now being given to social and pastoral work.

10.6 Modern Jesuits and Science

From about 1980 there has been a downward trend in the number of Jesuits involved in scientific research in both observatories and universities. With the exception of the Vatican Observatory, which is really a pontifical institution entrusted to the Society, and the Manila Observatory, all the observatories have been closed, or entrusted to other hands. As pointed out above, the number of Jesuit science professors in colleges and universities has also diminished radically. The new situation deserves further analysis in greater detail.

A recent development associated with the presence of Jesuits in science is the activity of Jesuits on the topic of the relation between science and religion. We have seen already (Sect. 6.5) the work carried out on this subject by the Vatican Observatory in collaboration with the CTNS since 1988, especially by George Coyne and William Stoeger. Robert A. Brungs (1931–2006), a professor of physics at Saint Louis University, co-founded in 1968 the Institute for Theological Encounter with Science and Technology (ITEST) and was its director for 38 years. This is one of the oldest institutes active on the relation between science and religion. Since 2000 the Gregorian University collaborates with six other catholic universities in the project "Science Theology and the Ontological Quest" (STOQ) to further the connections between science, philosophy and theology. The *Universidad Pontificia Comillas* (Madrid, Spain) began in 2003 the chair of *Ciencia, Tecnología y Religión* (Science, Technology and Religion, CTR) with a variety of activities.

The period from 1950 to 1980 saw the highest number of Jesuit scientists in modern time. A fairly complete overview can be gained from a survey compiled by the coordinators of Jesuits in science, Russo and Dou, compiled between 1970 and 1980[27]; this can be conveniently dated to 1975. The survey covered professors in Jesuit universities and colleges, Jesuits working in observatories, some (not all) professors in secondary schools, and those working in non-Jesuit institutions. In the four areas – mathematics, physics, chemistry, and biology – and excluding engineering and medical sciences, the total number came to 547: mathematics 152, physics 177, chemistry 89 and biology 129. Thus most were in physics, which included astronomy, geophysics and meteorology, followed by mathematics and biology. The lowest number was in chemistry. The two countries with highest numbers were the U.S.A. 181, and India 136. In Europe there were 127 (mostly in Spain, 44). In other countries: Central and South America 58, Canada 15, Africa 13 and Asia-Australia (without India) 17. The distribution for each country by scientific field was as follows: mathematics – U.S.A./Canada 64, India 39, Europe 30, Asia-Australia 4, Central and South America 13; giving a total of 152; physics – U.S.A/Canada 65, India 40, Europe 43, Central and South America 18, Asia-Australia 7, Africa 4: with a total of 177; chemistry – U.S.A/Canada 31, India 27, Europe 17, Central and South America 10, Asia-Australia 2, Africa 2: total 89; biology – U.S.A/Canada 36, India 30, Europe 37, Central and South America 17, Africa 5, Asia-Australia 4: total 129. Given 28,856 as the total number of Jesuits in 1975, the Jesuits involved in science came to 2 %, which is not a small number as work in the natural sciences is not a main occupation of Jesuits. This survey was made when the number of Jesuit scientists was at its highest, and the total number of Jesuits was still high (the highest number of Jesuits world-wide ever reached was 36,038 in 1965). However, the numbers in this survey are only approximate, since the survey was not complete, especially for teachers in secondary schools.

[27] Information about the number of Jesuit professors of science in the different scientific fields at 1975 have been obtained from the source of note 19.

The contrast with the present situation becomes obvious when one considers how many Jesuits are active in scientific fields in the year 2005.[28] In that year there were 143: mathematics (including computer science) 47, physics 42, chemistry 19 and biology 35. Clearly the total number is much smaller than in 1975. The distribution is similar, but now there is an increase in the number of mathematicians vis-à-vis physicists because the number includes specialists in computer science. The geographic distribution is the following: Europe 40 (10 emeriti) of whom 11 at the Vatican Observatory; U.S.A. and Canada 60 (5 emeriti); India 18; Asia-Australia 14; Africa 11. The two countries with the highest numbers are, as in 1975, the U.S.A. 57, with 22 mathematicians, and India 18, with the highest number in biology, 9. Between 1975 and 2005 there has a notable diminution: in 2005 only a fourth of those in 1975. The largest decrease is in the U.S.A, India, and Europe. However, numbers have increased in Africa and Asia-Australia, where numbers in 1975 were low. The closure or transfer of observatories (about 38 between 1960 and 1980) affected Jesuit involvement in scientific work. Today the only observatories with Jesuit presence in their staff are those of the Vatican and Manila. In the 1970s many young Jesuits left the Society in Europe and USA and this had an effect on those studying science. Those who remained in the Society and had begun their scientific careers in the 1970s were active until about 2000. On retiring they were often not replaced. Some young Jesuits were still studying science in 2005, but very few of them, so that while there will continue to be a few Jesuits engaged in science, they will be much less in number than in the 1970s and not in as many countries.

These numbers we have given must be compared with the total number of Jesuits at each time. The overall number of Jesuits rose rapidly in the twentieth century: from about 16,000 in 1910 to a maximum of 36,038 in 1965, then decreased steadily to 18,247 in 2010: in Europe the decrease has been most dramatic: from 11,000 in 1982 to 5,700 in 2010; in the U.S.A. from 5,600 in 1982 to 2,700 in 2010.[29] In the years given for the two surveys we have considered, 1975 and 2005, the total numbers of Jesuits dropped from 28,856 to 19,844 respectively, and the number of Jesuits in science dropped from 547 to 143. This represents 2 and 0.7 % of the total number of Jesuits. Thus, the overall decrease of Jesuits between 2005 and 1975 has been 31 %, but the decrease of Jesuit scientists in the same period comes to 74 %, much larger than that of the total number of Jesuits and is not proportional to the general decrease. Thus while the decrease in vocations has been an important factor, it does not totally explain what has happened. Other factors besides the demographic have to be considered to explain this decrease. Among them must be the influence of a change in apostolic priorities which began with the 32nd General Congregation in 1974. There a new emphasis was put on the promotion of faith and justice. This was not intended to decrease the work of Jesuits in science, but in fact it has contributed, even if is not the only factor. The priority given by that Congregation to work

[28] Data are from the *Catalogues of the Provinces of the Society of Jesus* for the year 2005 (Archivum Romanum S. I., Rome). Students are not included.

[29] Thomas Gaunt, (2011, February 1). The Changing Jesuit Geography. [Blog posting]. http://nineteensixty-four.blogspot.com/2011/02/changing-jesuit-geography.html.acceded June 2012.

with the poorer and less favored social groups had in fact an unintended negative influence in the involvement of Jesuits with science.

Today Jesuit superiors no longer see the natural sciences as an urgent need for Jesuits. The apologetic aspect that was so important in the Jesuit choice of scientific investigation at the beginning of the twentieth century, when the Church was being accused of opposing science, is no longer dominant. Dedication to science is not seen as a priority in the Church. In addition, especially in North America and Europe, sociological and cultural currents among the young have become imbued with a distrust and suspicion toward science and technology. Movements such as those of environmentalists and ecologists, may also have had an influence. However, these sociological currents contrast paradoxically with the increasing influence and dependence in the modern world on science and technology. A recent trend among young Jesuits is their interest in spirituality and their lack of interest in scientific research. Moreover, given the present phenomenon of men joining the Jesuits at an older age it becomes more difficult to begin scientific studies, unless they have been pursued beforehand. But even then, many are reluctant to continue in a scientific field and prefer to undertake direct pastoral work or studies in theology or spirituality. Such factors influence today the Jesuits and their attitude to science. However, in spite of all this, Jesuits continue to be active in science, even if less than in the past. Thus, the scientific tradition remains a characteristic of the Jesuit order as of no other group in the Church.

Chapter 11
Epilogue: The Jesuit Scientific Tradition

11.1 Science and Spirituality

One may well ask, having seen the long history of Jesuit scientists, how this constant involvement of Jesuits in science is to be explained. Obviously other religious orders have also had important scientists: for example, the Augustinian, Gregor Mendel, was a pioneer in the understanding of genetics. But among Jesuits, there is a continuous, almost institutional, presence from the very early period up to the present day. An indication of this continuity can be seen in the fact that Pope Pius X entrusted the Jesuits with the direction of the Vatican Observatory only 18 years after its foundation and they have been in charge ever since. This dedication to science was present during the first period of the history of the Society of Jesus until its suppression in 1773 and it continued after its restoration in 1814. What then is special about Jesuits which would explain this phenomenon?[1]

Steven J. Harris was one of the first who tried to find an answer to this question.[2] He restricted himself to Jesuits of the seventeenth and eighteenth centuries, but his ideas are relevant to modern times. In a detailed and well founded analysis, Harris puts what he calls "apostolic spirituality" as the basis of Jesuit scientific interest and sees this as the core of "Jesuit ideology". The term "Jesuit ideology" was first used by Ritka Feldhay to denote the Jesuit way of viewing knowledge as a path to salvation.[3] For Harris the main elements of this spirituality are the following: firstly, an emphasis placed on Christian *service*, which channels religious dedication outward into worldly activities not usually associated with religious life; but to this he adds an active engagement with the world, which leads to a respect for experience, or a "testing against experience" and "proof by trial" which provide standards of evaluation

[1] Daniel O'Connell, Jesuit men of science. *Studies*, 45 (1956), 307–318. Agustín Udías, Jesuit scientific tradition and Ignatian spirituality. *Lo Sguardo-Rivista di Filosofia*, 10 (2012) 207–219.

[2] Steven J. Harris, Transposing Merton's thesis: Apostolic spirituality and the establishment of the Jesuit scientific tradition. *Science in Context*, 3 (1989), 29–65.

[3] Rivka Feldhay, Knowledge as salvation in Jesuit culture. *Science in Context*, 1 (1987), 195–213.

A. Udías, *Jesuit Contribution to Science: A History*,
DOI 10.1007/978-3-319-08365-0_11

for ideas and projects. In the scientific field the empirical aspects of observation and experiment are of key importance. Thus, for Harris this particular apostolic spirituality – the basic force behind the activity of every Jesuit – automatically includes the motivation for scientific work. Another element is a special "esteem for learning", leading to what he calls the "sanctification of learning". This is responsible for the enormous Jesuit effort ever since their foundation in the field of education with the establishment of colleges and universities. According to Harris, the sanctification of learning drove them to accept commitment to secular activities at the service of an ultimately spiritual goal: the salvation of souls. He sees this being carried out in three main areas: education; the European royal courts; and the foreign missions.

Historical factors, also mentioned by Harris, played an important role. In the first place, there was the coincidence of the foundation of the Society of Jesus and its choice of education in schools and universities with the beginning of modern science. When the Jesuits established their schools they did not have an established academic tradition which might have conditioned them, as was the case, for example, with the Dominicans and Franciscans. Jesuits could incorporate more easily in their teaching the new currents of modern science and thus respond to the social demands of their times, in particular with reference to the advances in mathematical and experimental sciences. Clavius, from his influential position, made a great effort to introduce the teaching of mathematical science in Jesuit colleges, and this was finally accepted among the norms established for all schools by the *Ratio Studiorum*. He insisted, contrary to the opinions predominant in the traditional circles of his time, that mathematics is a true science, totally necessary for the study of natural philosophy. The introduction of mathematical sciences as an important element in the programs of Jesuit colleges was also in answer to the growing social importance that modern science was acquiring. Christopher Scheiner was aware of this social aspect and of its apostolic importance. In a letter to Johannes Rader he observed that the study of mathematics should be more highly valued among Jesuits. To Paul Gudin, professor of mathematics in the college of Graz, he wrote: "It is evident that mathematics is the net with which one can catch the elites and nobles and bring them to God's service".[4] The novelty of this interest in mathematics is manifest in the letter sent to Philip IV by the professors of the universities of Salamanca and Alcalá de Henares, when it was suggested that the Jesuit college in Madrid, the *Colegio Imperial*, should be raised to university rank and have the category of *Reales Estudios* (Royal Studies). These professors were opposed to the royal plan and argued that it was neither convenient nor "decent" for religious to teach profane matters, such as mathematics, astrology, navigation and military science. The Jesuits replied that they would be justified in teaching these subjects as they were part of the curriculum of an ecclesiastical and religious school, and were necessary for philosophy; also they would serve to prevent "any abusive use by the general public in their forecasts and superstitions".[5]

[4] Steven J. Harris, Les chaires de mathématiques. In Luce Giard (ed.), *Les jésuites a la renaissance*. Paris: Presses Universitaires de France, (1995), 239–262, (253–254).

[5] José Simón Díaz, *Historia del Colegio Imperial de Madrid*. Madrid: Consejo Superior de Investigaciones Científicas (1952), 72–84.

After the restoration of the Society of Jesus in the nineteenth century, the role of Jesuits in the sciences was partly justified with an apologetic argument, i.e., they could defend the Church against those who attacked it at that time as an enemy of science. Jesuit scientists and the existence of their scientific institutions were a clear argument against these false accusations and were an example of the compatibility between science and Christian faith. Aloysius Cortie, the Director of the Stonyhurst Observatory, wrote in 1923 about Angelo Secchi:

> The enemies of Holy Church have made such unwarranted use of science as a weapon of attack against her most fundamental truths, that an impression has sometimes been introduced among many of her children that the pursuit of science is damaging and dangerous to faith…Father Secchi is a striking example of one who knew how to unite religion and science.[6]

However, in addition to these historical circumstances, connected with the apostolic needs of educational work in colleges and universities, some more specific intrinsic elements of Jesuit spirituality need to be considered. After all, some elements of *apostolic spirituality* are to be found in other modern religious orders (founded after the seventeenth century), some with an apostolic character similar to that of the Jesuits. But they have not had a similar impact in the scientific world. How is one to explain this peculiar tradition?.[7] The fundamental lines of so-called "Ignatian spirituality", based on the ascetic and mystical experiences of Saint Ignatius, are to be found in his two great works: the *Spiritual Exercises* and the *Constitutions*.[8] The former outlines a month-long program of prayers and meditations. This is followed at the beginning and the end of Jesuit training and repeated annually for a period of eight days. It forms the core of this spirituality, and it is there that one should search for the motivation which can explain the especial devotion of Jesuits to scientific research. As mentioned earlier, the most important characteristic of Ignatian spirituality is that it is a mystique of *service*, understood as a service of God through the apostolic ministry to souls.[9] The *Constitutions* define the aim of the Society of Jesus with the words:

> not only to assist the salvation and perfection of its own members by divine grace, but with the same grace intensely to assist the salvation and perfection of their fellow men.[10]

For St. Ignatius this service is understood as a service to the Church. He often repeated in his writings the phrase that became the Jesuit motto: "all for the greater glory of God" (*Ad maiorem Dei Gloriam*). Saint Paul urges the faithful to "do all for the glory of God" (1Cor 10,31), but St. Ignatius added the comparative "greater" to

---

[6] A. L. Cortie, Angelo Secchi, S. J. (1818–1878). In B. Windle (ed.) *Twelve catholic men of science*. Londres: Catholic Truth Society, (1923), 149–164.

[7] Udías, *Searching*, 7–12.

[8] St. Ignatius, *Obras*, 221–238, 432–646.

[9] Jean-Claude Dhôtel, *La spiritualité ignatienne. Points de repère*. Paris: Vie Chrétienne (1991); Rosano Zas Friz, Espiritualidad ignaciana. In José García de Castro (ed.) *Diccionario de espiritualidad ignaciana*. Bilbao-Santander: Mensajero-Sal Terrae (2007), 811–820.

[10] St. Ignatius, *Obras*, (Examen Primero y General, Chap. 1. 3), 466.

indicate that Jesuits should seek not just the glory, but the greater glory of God. In the first meditation of the Spiritual Exercises, entitled the "Principle and Foundation", he exhorts that one should always act "desiring and choosing only what is "most" conducive for us to the end for which we were created".[11] The term "more" (in Latin *magis*) indicates an important feature of this spirituality. In all times and circumstances Jesuits are inspired to seek this "more" or "greater" in their work. For many Jesuits this was to be found in scientific work.

Another phrase that recurs frequently in the writings of St. Ignatius is "finding God in all things". This points to a growing awareness that God can be found in every person, place and thing, and that a Jesuit needs to have an attitude of prayer even in an active life. This is an important aspect of the last meditation of the *Spiritual Exercises*, the "Contemplation to attain love". Here one is asked:

> to look how God dwells in creatures, in the elements, in the plants, in the animals, in men [in me, myself]... and to consider how God works and labors for me in all things created on the face of the earth.[12]

Thus all things, people and circumstances are occasions for finding God. Moreover, by his stress on the humanity of Christ, St Ignatius invites one to see that everything human, and there one can include science, can serve for our encounter with God.

Jerónimo Nadal, a companion of St. Ignatius, refers to what has been called the "Jesuit way" with the expression "contemplatives in action". This implies a union between prayer and action; we can find God in the silence of prayer and in the activity of work. There is no activity, no matter how profane it may look, that cannot be transformed into prayer. Teaching mathematics or physics in a university, observing the light from a distant galaxy, or drawing a map of an unknown region are activities that a Jesuit finds perfectly compatible with his vocation. Through them he can try to find God in his life. In other religious groups, this may become a personal attitude, but for Jesuits it stems from the core of their spirituality. Angelo Secchi, the renowned astrophysicist, acknowledged this saying: "The contemplation of God's works is one of the noblest works of the spirit; this is the principal aim of the study of nature".[13] So Jesuits involve themselves in activities, among them scientific research, which others consider incompatible or at least not suitable for religious. We find them active in social work and as parish pastors, but also carrying out research in biology or astrophysics, or writing poetry, all motivated by the same spirituality of "seeking God in all things"... for the greater glory of God.

Another aspect of Ignatian apostolic spirituality is its preference for situations and activities that may be called "frontier work". By this is meant that instead of devoting themselves to traditional pastoral activities inside the Church, Jesuits try to get into contact with those who are outside the Church, in activities sometimes considered to be non-conventional. Pope Benedict XVI recognized this when he said to the Jesuits:

---

[11] St. Ignatius, *Obras*, 228–229.

[12] St. Ignatius, *Obras*, 272–273.

[13] Angelo Secchi, *Le soleil*, Paris: Gauthier-Villar, 1875, (Prologue), 6.

> As my Predecessors have said to you on various occasions, the Church needs you, relies on you, and continues to turn to you with trust, particularly to reach those physical and spiritual places which others do not reach or have difficulty in reaching.[14]

This explains how a Jesuit may be praying in a Buddhist monastery or carrying out physical research in a particles accelerator. Jesuits are always searching for frontiers, for places and situations where the Christian message is not yet known. For example, this spirit drove Mateo Ricci and his companions in the seventeenth century to present themselves in the imperial Court of China as astronomers of the West, adopting the dress and manners of Chinese scholars. In the nineteenth century and beginning of the twentieth century, science was considered to be a field that was alien, if not hostile, to religion. Jesuits felt the need to show by their scientific work that there is no incompatibility between science and faith, and that scientific work itself can also be a vehicle to finding God. With this motive there emerged the foundation of a network of Jesuit observatories spread around the world. A modern Jesuit scientist considers the scientific community to be his parish and overcomes by his presence any barriers of estrangement or lack of understanding. Linehan, a tireless seismologist and explorer, aware of his function as a priest and a scientist, engraved on the base of the chalice that he used: "First Mass at Magnetic North Pole 1954" and "First Mass at South Pole 1958". He made very clear what his priorities were when he told a journalist from *Life* magazine: "I would give up all my seismology, to celebrate one such Mass as you came to this morning".[15]

11.2 Testimony from Jesuit Scientists

Jesuits of the seventeenth to eighteenth centuries did not feel the need to justify their scientific research, considered to be perfectly normal. However, some modern Jesuit scientists did feel it necessary to write on the subject. Secchi was among the first to do so. He presented his work as a demonstration of the compatibility between science and religion, a controversial stance at that time: "True faith is not hostile to science, but both are like two rays coming from the same Sun which should illuminate our mind by the way of truth".[16] Somewhat later, Teilhard de Chardin, commenting to a group of young Jesuits on the Superior General Fr Janssen's 1947 letter on the intellectual apostolate, showed how the involvement of Jesuits with science entered into the context of his thought.[17] He began by stressing the enormous importance of science in the modern world, pointing out that it has become what he calls the *Grande Affaire du Monde*, a "human function as vital as nutrition and reproduction". For him scientific research constitutes the arrow-head of human evolution. He asked himself why it

[14] *Congregación General 35 de la Compañía de Jesús*. Bilbao: Mensajero (2008), 264.

[15] Charlotte B. Harvey, The voyage of the Monte Carlo. *Boston College Magazine*, (Fall 2000).

[16] Quoted by Van Tricht, *Secchi*, 370.

[17] Pierre Teilhard de Chardin, Sur la valeur religieuse de la recherche. *Science et Christ. Oeuvres 9*, Paris: Éditions du Seuil, 1965, 255–263.

was so important for Jesuits to have a share in such work, and answered: "this is the form under which the creative power of God is hidden and operates more intensively in nature around us". Through it, he adds, a new increase of consciousness appears in the world. He is convinced of the need to reconcile the "problem of the two faiths", faith in God and faith in man. This reconciliation is needed in order to advance the Kingdom of God in the world. For him, these two faiths are in fact the two essential components of a complete human-Christian mystique. Their reconciliation and synthesis cannot be achieved in the world unless it is concretely lived. Hence the importance science has for Jesuits. He finishes by saying:

> We Jesuit priests should not only be interested in scientific research, but we must believe in it, since only there the human-Christian mystique can develop by which a human unanimity can be established in the future.[18]

Thus, under different form, the same Ignatian spirituality appears which finds in scientific research the best means to the expansion of the Kingdom of God with the reconciliation of the two faiths. In his well-known essay, *Mass on the world,* Teilhard is expressing in different words the Ignatian principle of finding God in all things.[19] He consecrates in this prayer all human efforts and sufferings offered to God on the altar of the world as a universal host. Teilhard had also very clear the apostolic character of his work: "My gospel and my mission are to bring Christ to the heart of those realities considered to be the most dangerous, the most naturalistic, the most pagan".[20]

In the 1960s there was a large number of Jesuit scientists in the United States, some 300. This led some of them to explicitly ask about the compatibility of the vocation of the priest and the Jesuit as a scientist. The problem was one particular aspect of a more general problem, which was called at the time the status of the "hyphenated" priest or the priest with other professional activities. Frank Haig, a professor of physics, claimed that the priest-scientist could be explained only in two ways: in terms of personal interests or with a view to an apologetic role in defense of the Church.[21] For him these are necessary but insufficient reasons, and he suggested that the priest-scientist had primarily a symbolic role, illustrating how a life of science and a life of grace could be integrated. He justified such integration with the need in the Church today for a rethinking of its message, using the language of the scientific world. Only thus could it be relevant at the frontiers of knowledge. According to him the priest-scientist "attempts to make the world of faith and the world of science transparent to each other". These thoughts were applied by Ernest Spittler, a professor of chemistry, to the concrete case of Jesuits; he puts at the

[18] Ibid., 263.

[19] Teilhard de Chardin, La Messe sur le monde. *Le cœur de la Matière Oeuvres 13*, Paris: Éditions du Seuil (1976), 139–156.

[20] Teilhard de Chardin, Le pêtre. Ecrits de temp de la guerre, 1916–1919. *Oeuvres 12*, Paris: Éditions du Seuil (1965), 329.

[21] Frank. R. Haig, vocation of the priest-scientist. *Bull. American Association of Jesuit Scientists,* 36 (1961), 3–21.

center their role as priests, mediators between God and men.[22] This role must be realized inside the Church and is "mediatorial" in relation to the scientific community, to the Catholic laity, and to non-Catholics. Such a role constitutes a true apostolate, but often a hidden one, and for Jesuits it will be realized most often in educational work. The psycho-analyst, William Meissner, presents the Jesuit-scientist as one who adopts two roles, those of the priest and of the scientist. He says that this places him in a situation which has its conflicts, frustration, insecurity and anxiety.[23] To resolve them there is a need to integrate the values of science with the values of Jesuit life. The priest-scientist is primarily a priest and exercises a specifically religious function, which is at the same time both apostolic and symbolic: it is *apostolic* insofar as he comes across as one who bears testimony to truth, it is *symbolic* insofar as he embodies the fullness of the Christian religious life. Meissner states:

> He symbolizes, then, the unification in truth of religious and scientific knowledge, of faith and reason, of Catholic wisdom and secular learning, of the Church and modern science.

Meissner presents what is specifically Jesuit as what he calls the "Jesuit proposition", a way of referring to Jesuit spirituality. He first considers the means by which the aims of the Society of Jesus are to be pursued, as proposed by St. Ignatius in the *Constitutions*. Some of these are supernatural and others natural.[24] This means in the case of the Jesuit-scientist the complementarity of his religious ideal of personal sanctity and his devotion to science. For a Jesuit, scientific work must be always apostolic and Meissner takes this to mean "incarnational" – sharing in the presence of the incarnate Word of God among men for the sanctification of all human activity. He recognizes the difficulty there is in carrying out this dynamic synthesis between being truly a man of God and a man of science. Meissner admits that this is a complex role, and consequently there will always be a certain tension in the life of a Jesuit-scientist.

Timothy Toohig, who worked on particle physics at the Fermilab, also tries to find the connection between his work as a physicist and his spirituality.[25] He starts with the thought that physics seeks to understand the ultimate structure of matter, its elementary constituents, and the forces that bind them together – as well as the origin and final destiny of the universe. For this reason, he says, we should not be surprised that physicists like Albert Einstein and Werner Heisenberg and more recently Stephen Hawking and Steven Weinberg discuss in some of their writings problems concerning the ultimate questions of existence and end up talking, in one way or another, about God. Toohig tries to show that there is a certain analogy

[22] Ernest G. Spittler, The priest-scientist in the Church. *Bull. American Association of Jesuit Scientists,* 39 (1962), 30–37.

[23] William W. Meissner, The Jesuit as a priest-scientist. *Bull. American Association of Jesuit Scientists,* 40 (1963), 25–33.

[24] San Ignacio, *Obras,* (Constituciones X, 3, 813–814), 641–642.

[25] Timothy E. Toohig, Physics research, a search for God. *Studies in the Spirituality of Jesuits*, 21/2 (1999), 1–26.

between the work of physicists and the search for God. He examines how Einstein and Niels Bohr recognize the presence of a hidden mystery behind the structure of the universe, which always remains open to new intuitions and inspirations. Both these thinkers, according to Toohig, exercised an intuitive approach to physical problems and proceeded in their work with a deep faith in the existence of an underlying simplicity in the structure of matter. He asks himself if these attitudes are not indicative of an underlying spirituality and a certain implicit experience of God, and he evokes in his analysis the concept of a "transcendental experience" developed by the German Jesuit theologian Karl Rahner. By this is meant an experience of the transcendence present in every person, which constitutes in itself an "unthematic and anonymous knowledge of God". Toohig claims that there is a similarity between this transcendental experience and the experience undergone by physicists brought face to face with the mystery of the universe. He affirms that, although not explicitly so, research in the physical sciences can be considered as a search for God. Even if the similarity may be open to question, Toohig's attempt to find a certain "religious" sense in the scientific experience may be compared with Teilhard's ideas and, indeed, with the Ignatian proposal of "finding God in all things". More links with other aspects of Ignatian spirituality, such as the experience of personal redemption at very deep levels of life (as presented in the *Spiritual* Exercises), and the hunger for an ever deeper penetration of the mystery of the universe, are also suggested by Toohig. He even dares to suggest that Einstein's experience of the discovery of new physical theories was not dissimilar to St. Ignatius' mystical experience (beside the River Cardoner) about the dogma of creation.

A more specific and practical approach is that of José Gabriel Funes, now Director of the Vatican Observatory, who also begins by stating that the Jesuit scientific tradition has its roots in Jesuit spirituality.[26] He discusses the challenges to the Christian faith that science poses today and the need for the "scientific apostle" to be present at the frontiers between the Church and the world of science. He offers his own testimony that there is no real conflict between them. To the practical objection that there are other more urgent problems for a priest today, Funes answers with a reference to the incident of the anointing of Jesus in Bethany (Mc 14, 3–9). What can be considered as a life "wasted" in the study of galaxies, has its purpose in the service of the Lord. He stresses the idea, already mentioned by others that the priest-scientist serves as a bridge between the Church and the scientific community.

Another member of the Vatican Observatory, Paul Gabor, asks the question: is Jesuit involvement in science an expression of Jesuit identity, part of the Order's charisma, or is it the result of external influences?[27] He starts with a review of scientific work in the Society, from Clavius in the sixteenth century to the 31st General Congregation in 1965. Analyzing the motives behind this work, Gabor presents first some external reasons, such as the apologetic and utilitarian which he considers

[26] José G. Funes, Una imagen evangélica para el apostolado científico (Mc 14, 3–9). *Revista de Espiritualidad Ignaciana*, 37, (2007) 87–95.

[27] Paul Gabor, Jesuit identity and science. An inside look. In Petra Čemus (ed.) *Proceedings of the Bohemia Jesuitica Conference* (Prague 25–27 April 2006) Prague: Carolinum (2010), 647–660.

insufficient. Then he examines the internal motivation and connects scientific research with Jesuit spirituality as expressed in St. Ignatius' *Constitutions* and *Spiritual Exercises*. Gabor states that "science is done in the Society for its own sake as a true spiritual quest". Consequently he claims that the extraneous reasons – pragmatic, pastoral, pedagogical – are not essential, and that the fundamental motive for Jesuits is scientific research in itself. He also asserts that one may find many parallels between the life of scientists and the spiritual life. With reference to physics in particular, Gabor finds that there are several aspects of work in this area, such as playfulness, wisdom, and ethical and ascetic dimensions that have spiritual aspects. He ends by saying, "there is an unmistakable affinity between science and Ignatian spirituality".

With these few testimonies it is clear that Jesuit scientists themselves see their activity not only as compatible with, but as a consequence of Ignatian spirituality. It can, then, be justified in itself and not only for apostolic or utilitarian reasons. It is also clear that efforts have been made to show the affinity between scientific work and such a spirituality, although it may not be completely clear in what this affinity consists. Gabor recognizes that these are testimonies rather than in-depth analyses, and he finds frustrating the lack of systematic studies into the spirituality of scientific work. Funes suggests that there are different ways among Jesuits to live spiritually the scientific apostolate, as there are many different ways to live spiritually the Jesuit vocation itself. As an example, he refers to the differences between Matteo Ricci and Teilhard de Chardin, and he identifies himself more with the former.[28] All testimonies start with the premise that for much of its history some Jesuits have distinguished themselves in the Church as belonging to the institution which has had and still has a special interest and involvement in science. They conclude that, in some way, this involvement must be related to Ignatian spirituality, the foundation of Jesuit life, even if this relationship is expressed from different points of view and calls for further study.

11.3 The Practice of Scientific Research and Jesuit Life

A question, still remaining, concerns how the involvement in scientific research was actually lived by Jesuits in the context of their religious order, and how it was seen by other Jesuits, especially, by their superiors. In the first place, one has to bear in mind that scientists are, and always have been, a small minority among Jesuits. Their training somehow separates them from the rest of their brethren, mainly involved in theology, philosophy and the humanities, or in directly pastoral work. Although they have followed a similar training in theology and philosophy, their scientific training and work is not shared by the rest. This explains that sometimes the Jesuit scientist may feel a little misunderstood or isolated in his work, like an "outsider" in his own community, as José Oriol Cardus, director of Ebro Observatory,

[28] Funes, "Una Imagen", 92.

expressed it. Superiors do not always understand their work and sometimes regret that they cannot appoint scientists to other works considered more "useful".

Mordechai Feingold examines in detail the situation of Jesuit scientists in the seventeenth and eighteenth centuries and why there were no top-notch scientists among them, like Galileo, Descartes or Newton?[29] The relevance of the question may be questioned, since scientists at that level are so rare that it is difficult for any institution to produce even one of them. But in answer to the question Feingold examines a series of circumstances that raised obstacles to the advance of Jesuit scientists. He even wonders whether there is not a real incompatibility between the Jesuit vocation and dedication to a secular science, since science is always for Jesuits a means to an end and not an end in itself. One has to admit that scientific work for a Jesuit, like any other activity, is finally oriented to an apostolic end, the salvation of souls. However, this does not invalidate their dedication to science, since even the greatest scientists had a variety of motives in undertaking their work. The supposed purity of intention in the pursuit of science "for its own sake" is actually a myth, as has been shown by Robert K. Merton, a sociologist of science.[30] Often scientists are moved, among other things, by the sake of peer recognition, social prestige and power. Feingold also examines a series of practical problems which might place obstacles to progress in the scientific career of a Jesuit. Among them he includes the little interest of superiors for this type of study and their need to attend to other matters considered more important. A concrete instance would be the priority given by superiors to teaching over research in the colleges and universities, exemplified in the publication of textbooks rather than those of research monographs. Feingold shows how Jesuits excelled more as trainers of others in research rather than by their own work. So, among noted scientists who were students in Jesuit schools one finds Torricelli, Descartes, Laplace, Volta, Buffon and Lalande. Another obstacle to the scientific work of Jesuits, according to Feingold, was the superiors' objection to novelties. And added to that were the rules for unanimity of doctrine, the interior censorship of the order, and the auto-censorship of the authors themselves. As was seen earlier, a concrete problem that affected Jesuit scientists in the seventeenth and eighteenth centuries concerned acceptance of both the Copernican system in astronomy and atomism in physics. These problems, recognized by Jesuit scientists themselves, certainly existed, but they were not insurmountable obstacles. Also, one cannot separate so drastically teaching and research; Jesuits were good science teachers because they combined teaching and research.

In the restored Society after 1814 Jesuit scientific work was largely centered in observatories and universities. By then there was no longer prejudice against scientific theories as such, since scientific questions had been separated from philosophy and theology. Jesuit scientists had complete freedom in their work in scientific fields. For example, Teilhard de Chardin never had any problems with the publica-

[29] Mordechai Feingold, Jesuits: Savants. In M. Feingold (ed.), *Jesuit science and the republic of letters*. Cambridge (Mass.): The MIT Press (2002), 1–45.

[30] Robert K. Merton, *The sociology of science. Theoretical and empirical investigations.* Chicago. The University of Chicago Press, 1973, chap. 15.

tion of his scientific work; indeed, superiors advised him to stay in the scientific field and not to enter the theological. Work in the new observatories, founded after 1824, was highly esteemed by Jesuits and fostered and encouraged by superiors until the 1970s, when the interest for them began to decline. In 1893 the Superior General, Luis Martín, praised the work done in the Observatory of Manila (Philippines) saying that it contributed not only to the "splendor of science and the good name of our Society, but also to the benefit of scientists and the help of souls, which is the end of all our studies and work according to our institute".[31] Pierre Lejay, Director of the Observatory of Zikawei (China), complained in 1933 that the scientific work of the observatory was not sufficiently appreciated in the Jesuit Chinese Mission and that some missionaries considered it a waste of time. Wlodmir Ledochowski, then Superior General, wrote to him that he "personally always protested against this unjust opinion" and referred to "the high importance of intellectual work and the appreciation it has always enjoyed in the Church and the missionary traditions".[32] One may conclude that scientific work is still held in high esteem among Jesuits, except for some who do not understand its importance.

In modern times the emphasis on teaching, found in the previous period, is still present, owing to the increased number of universities managed by Jesuits and the need for Jesuit professors in the science departments. Ernesto Gherzi, the last director of Zikawei Observatory, who worked afterwards at the Jesuit Geophysical Observatory in Montreal, complained in 1970 about the little support given to him by other Jesuits in his geophysics research. Louis Larendeau, Provincial of the Jesuit French Canadian Province, acknowledged that often superiors are not familiar with the world of science so that the vocation of Jesuit scientists is often unrewarding, especially for those involved in research, while teaching is more recognized and accepted.[33] Carl-Henry Geschwind, writing with reference to modern Jesuit work in seismology in the United States, also asks, like Feingold, whether Jesuits engage in science for its own sake or are seeking other ends. He claims that in the nineteenth century the Jesuit scientific tradition faded and in North America "an ideological distaste for unfettered research" developed.[34] This generalization ignores Jesuits like Angelo Secchi, Stephen Perry and Benito Viñes, who in the nineteenth century all made significant and lasting contributions to astrophysics, geophysics and meteorology. Geschwind contrasts the attitudes to seismological research of Frederik Odenbach and James Macelwane, claiming that only the latter had a true scientific mentality. Similar criticism is to be found in Lewis Pyenson's account of French Jesuit scientists in China, whom he calls "functionary missioners", for whom science was a mere adjunct and not the work of real researchers, whom he

[31] Luis Martín's letter to Fr. Miguel Saderra (3 Nov 1893), *Archivum Romanum Societatis Iesu* (*ARSI*), Letters of Fr. Luis Martín, 1892–1906,158.

[32] Wladimir Ledokowski's letter to Fr. Pierre Lejay (17 April 1933). *ARSI*, Letters of Fr. Wladimir Ledokowski, 1915–1942, 482–483.

[33] Letter of the Provincial of the French Canadian Jesuit Province Louis Laurendeau to Fr. Ernesto Gherzi (Montreal, 3 Feb.1970) *Saint-Jérôme, ASJCF*, Bo-214–35.

[34] Geschwind, "Embracing science", 28 and 49.

calls "seekers". However, Aitor Anduaga has shown that the Spanish Jesuits in the Philippines were real scientists.[35] Mention has already been made of the complexity of motivation in scientific work and an ultimate religious end does not invalidate the work of Jesuits as scientists.

In this book it has been shown how from their foundation in the sixteenth century Jesuits have shown a continuous and institutional commitment to the natural sciences, which is unparalleled by any other religious order in the Catholic Church. This scientific work, carried out largely as part of their educational commitments in a network of schools, colleges and universities, was marked by different characteristics in the two periods of Jesuit history, the first between the sixteenth and eighteenth centuries and the second from the nineteenth century until today. But the presence of Jesuits in science has continued throughout their long history. In addition to a very pragmatic motivation, the basic foundation for such work is to be found in Ignatian spirituality. The core of this spirituality lies in the emphasis on finding God in all things, the union of prayer and work, the search for what leads to the greater glory of God, and the preference for work "on the frontiers". This has often involved Jesuits in unconventional activities and situations, including scientific research. Jesuit scientists, who have reflected on their work, acknowledge this special affinity between the scientific vocation and their spirituality and are aware of the difficulty in combining this vocation with that of a Jesuit, of being at the same time priests and scientists. To conclude, the Jesuit scientific tradition, in spite of all the problems encountered during its long history, is still alive and serves as a special characteristic in the Catholic Church.

[35] Anduaga, "Spanish Jesuits" 2–3 and 24–25; Lewis Pyenson, Pure research, Jesuit institutions and metropolitan ambitions: The evolution of French Policies overseas, 1880–1940. *Chine et Europe: Évolution et particularités des rapports est-ouest du XVIe au XXe siècle. Actes du IVe Colloque International de Sinologie de Chantilly,* Paris, (1991), 249–271.

Appendices

Appendix 1

List of Jesuit Scientists

Mathematicians

Bayma, Joseph
Benavente, Miguel
Billy, Jacques de
Bourdin, Pierre
Bouvet, Joachim
Castel, Louis Bertrand
Cerdá, Tomás
Ceva, Tommaso
Chales, Claude François Milliet de
Clavius, Christopher
Dou, Alberto
Eximeno, Antonio
Faille, Jean Karen della
Figluicci, Vicenzo
Gerbillon, Jean François
Gouye, Thomas
Grassi, Giovanni Antonio
Guldin, Paul
Halde, Jean Baptiste du
König, Johannes
Kresa, Jakub
Lembo, Giovanni Paolo
Loubere, Antoine de la

A. Udías, *Jesuit Contribution to Science: A History*,
DOI 10.1007/978-3-319-08365-0

Maelcote, Odo van
Monteiro, Inácio
Nicolas, Pierre
Pardies, Ignace Gaston
Pfriem, Joseph
Pons, Jean François
Rafael, Enrique de
Riccati, Vicenzo
Ricci, Matteo
Richard, Claude
Rieger, Christian
Saccheri, Giovanni Girolamo
Saint Vincent, Grégoire de
Sagner, Gaspar
Schwab, Johann
Semple, Hugh
Souciet, Etienne
Tacquet, André
Talpain, Leon
Torres, Baltasar
Trentel, Franz Xavier
Villalpando, Juan Bautista
Vydra, Stanislas
Wedlingen, Johann
Zaragoza, José
Zebrwoski, Tomasz
Zucchi, Nicola

Physicists

Aguilón, François de
Ango, Pierre,
Benvenutti, Carlo
Biwald, Leopold Gottlieb
Bono, Michele de
Boscovich, Roger
Brungs, Robert A.
Cabeo, Nicolò
Cornaeus, Melchior
Dressel, Ludwig
Fabri, Honoré
Fournier, George
Garzoni, Leonardo
Gautruche, Pierre
Gianfranceschi, Giuseppe

Grandami, Jacques
Grimaldi, Francesco Maria
Haig, Frank
Hoenen, Pieter
Horváth, Ker János
Kircher, Athanasius
Lana-Terzi, Francesco
Line, Francis
Lozano, Pedro
Lucas, Anthony
Mangold, Maximus
Mataix, Claudio
Mayr, Anton
Mesa, José de
Palmieri, Domenico
Paulien, Aimé-Henri
Pereyra, Antonio
Pérez del Pulgar, José Agustín
Pianciani, Giovanni Battista
Rajam, B.
Rapin, René
Regnault, Noël
Russo, François
Sarrabat, Nicolas
Schott, Kaspar
Selvaggi, Filippo
Tongiorgi, Salvatore
Toohig, Timothy
Valladares, Bonifacio
Wulf, Theodor
Zallinger, Jacob von

Astronomers

Almeida, José Bernardo de
Amiot, Jean Joseph Marie
Asclepi, Giuseppe
Bauer, Alfred
Beati, Gabriele
Benoist, Michel
Béraud, Laurent
Biancani, Giuseppe
Bonfa, Jean
Borgondio, Orazio

Borri, Cristoforo
Bouchet, Jean Venant
Boudier, Claude Stanislas
Braun, Carl
Bussolini, Juan Antonio
Cairon de Merville, Jean Nicolas
Carbone, Giovanni Battista
Chevalier, Stanislas
Comellas, Juan
Cortie, Aloysius
Costa, Francisco da
Coyne, George V.
Curley, James
Cysat, Johann Baptista
Dias, Manuel
Duchamp, Pierre François Xavier
Dumouchel, Dominique
Esch, Michael
Espinha, José da
Fenyi, Gyula
Franz, Josef
Funes, José Gabriel
Gabelsberger, Anton
Gabor, Paul
Gatterer, Alois
Gaubil, Antoine
Goetz, Edmund
Gottignie, Giles François de
Gramatici, Nicasius
Grassi, Orazio
Grienberger, Christoph,
Grimaldi, Filippo
Gruber, Johann
Hagen, Johann Georg
Hallerstein, Augustin von
Hell, Maximilian
Heyden, Francis J.
Hisgen, Joseph
Huberti, Franz
Kastner, Kaspar
Kirwitzer, Wenzel
Kögler, Ignaz
Lafont, Eugene
Lagrange, Louis
Laval, Antoine de
Liesganig, Joseph

Longobardi, Nicolò
Luskina, Stefan
Maire, Christopher
Marchetti, Francesco
Martínez, Ramón
Mascardi, Nicoló
Mayer, Christian
McGeary, James L.
McNally, Paul A.
Mier y Terán, José
Morand, Jean Claude
Müller, Adolf
O'Connell, Daniel
Pantoja, Diego de
Penaranda, Alphonse de
Pereira, Tomas
Pezenas, Esprit
Pilgram, Anton
Poczobut, Martin Odlanicki
Rho, Giacomo
Riccioli, Giovanni Battista
Richaud, Jean
Rigge, Joseph F.
Rigge, William F.
Rocha, Felix da
Rodrigues, André
Rubino, Antonio
Saint-Bonnet, Jean de
Schall von Bell, Johann Adam
Scheiner, Christoph
Scherffer, Karl
Schönberger, Georg
Schreck, Johann (Terrentius or Terrenz)
Schreier, Joseph
Secchi, Pietro Angelo
Sestini, Benedetto
Smogulecki, Mikolaj
Spina, Pedro
Stansel, Valentin
Stein, Johan
Stepling, Josef
Strobl, Andre
Stumpf, Kilian
Suárez, Buenaventura
Tachard, Guy
Thaillandier, Pierre

Thomas, Antoine
Treanor, Patrick
Trisnecker, Franz
Ursis, Sabatino de
Veiga, Eusebio da
Verbiest, Ferdinand
Vico, Francesco de
Weld, Alfred
Ximenes, Leonardo

Geophysicists, Geologists and Meteorologists

Antinori, Luigi
Algué, José,
Belgrado, Giacobo
Berloty, Bonaventure
Birkenhauer, Henry
Bradley, Edward
Brunner, George
Buist, Maurice
Burgaud, Maurice
Burke-Gaffney, Thomas N.
Cabré, Ramón
Cardús, José Oriol
Cirera, Ricardo
Colin, Elie
Combier, Charles
Dechevrens, Marc
Deppermann, Charles
Descotes, Pierre M.
Drake, Lawrence A.
Due Rojo, Antonio
Egidi, Giovanni
Eisele, Louis J.
Faura, Federico
Francotte, Edward
Froc, Louis
Gambara, Bernardino
Gangoiti, Lorenzo
Gherzi, Ernesto
Goberna, Rafael
Gouin, Pierre
Gutierrez Lanza, Manuel
Hennesey, James J.
Heredia, Gustavo

Ingram, Richard E.
Joliat, Joseph
Juan, Martín
Lejay, Pierre
Linehan, Daniel
Lynch, Joseph
Macelwane, James B.
Mayaud, Pierre Noël
Menyhart, Laszo
Merino, Baltasar
Moidrey, Joseph de
O'Connor, Edward
Odenbach, Frederick L.
O'Leary, William J.
Perry, Stephen J.
Pigot, Edward F.
Plassard, Jacques
Poisson, Charles
Ramírez, Jesús Emilio
Reppeti, William C.
Rey, Charles
Rey, Jean
Ricard, Jerome S.
Roblet, Désiré
Rodés, Luís
Romañá, Antonio
Rowland, James P.
Saa, German
Saderra Masó, Miguel
Sánchez Navarro-Neumann, Manuel
Sarasola, Simón
Selga, Miguel
Sidgreaves, Walter
Stauder, William
Teilhard de Chardin, Pierre
Tondorf, Francis A.
Viñes, Benito

Chemists

Puig, Ignacio
Saz, Eugenio
Spittler, Ernest
Vitoria, Eduardo

Biologists

Archer, Jorge Luis
Azevedo Mendes, Candido de
Bellynck, August
Blatter, Ethelbert
Browne, Gerald Wilson
Burke, Edmund J.
Franganillo, Pelegrín
Frank, Karl
Haas, Adolf
Mathew, K. M.
Navás, Longinos
Pâque, Egide
Pujiula, Jaime
Rambo, Balduino
Rubio, Julián
Saldanha, Cecil
Santapau, Hermenegildo
Silva Tavares, Joaquim da
Speiser, Xaver Ferenc
Trassarelli, Ferdinando
Wasmann, Erich

Naturalists, Geographers and Cartographers

Acosta, José de
Alieni, Giulio,
Bonnécamps, Joseph Pierre de
Bono, Michele de
Boym, Michael Pyotr
Briet, Philippe
Capassi, Domenico
Cardim, Fernão,
Clavijero, Francisco Javier
Cobo, Bernabé
Falkner, Thomas
François, Jean
Fritz, Samuel
Gilg, Adam
Giulio, Giuseppe
González Chaparro, Juan
Grossinger, Johann Baptista
Gumilla, José

Gusta, Francisco
Jolis, José
Kamel, Georg Josef
Licent, Emile
Loureiro, João de
Martínez, Carlos
Martini, Martino
Molina, Juan Ignacio
Murillo Velarde, Pedro
Ovalle, Alonso de
Ratkay, Juan María
Recupito, Julio Cesar
Rosales, Diego
Sajnovics, Johann
Sánchez Labrador, José
Soares, Diego
Strobel, Matías
Termeyer, Ramón María
Tieffenthaler, Joseph
Vasconcellos, Simão
Vega y Terán, Claudio de la
Wendel, Francis Xavier

Explorers

Acevedo, Francisco
Acuña, Cristóbal de
Albanel, Charles
Allouez, Claude,
Andrade, Antonio de
Arce, José Francisco de
Artieda, Andrés de,
Blende, Bartolomé de
Cabral, João
Cacela, Estevâo
Camaño Bazán, Joaquín
Cardiel, José
Dorville, Albert
Ezpeleta y Goñi, Jerónimo de
Goes, Bento de
Jiménez, Bartolomé
Kino, Eusebio Francisco
Lobo, Jerónimo
Marqués, Manuel

Marquette, Jacques
Montserrat, Antonio de
Páez, Pedro
Quiroga, José
Roth, Heinrich

Appendix 2

Jesuit Scientists in Gillespie's Dictionary

Jesuit scientists included in Charles Gillespie (ed.), *Dictionary of Scientific Biography*, vols. 1–16. New York: Charles Scribner, 1970–1980.

Aguilón, François de, 1, 81
Billy, Jacques de, 2, 131
Boscovich, Roger, 2, 326–332
Cabeo, Nicolò, 3, 3
Ceva, Tommaso, 3, 183
Clavius, Christopher, 3, 311–312
Cysat, Johann Baptista, 3, 528
Fabri, Honoré, 4, 505–506
Grimaldi, Francesco Maria, 5, 542–544
Guldin, Paul, 5, 588–589
Hell, Maximilian, 6, 233–235
Kircher, Athanasius, 7, 374–378
Liesganig, Joseph, 8, 358–361
Mayer, Christian, 9, 231–232
Pezenas, Esprit, 10, 571–572
Pardies, Ignace Gaston, 10, 314–315
Ricci, Matteo, 11, 402–403
Riccioli, Giovanni Battista, 11, 411–412
Saccheri, Giovanni Girolamo, 12, 55–57
Saint Vincent, Grégoire de, 12, 74–76
Scheiner, Christoph, 12, 151–152
Schott, Kaspar, 12, 210–211
Secchi, Pietro Angelo, 12, 266–210
Stein, Johan, 13, 12
Stepling, Josef, 13, 39–40
Tacquet, André, 13, 235–236
Teilhard de Chardin, Pierre, 13, 274–277
Villalpando, Juan Bautista, 14, 29
Zucchi, Nicola, 14, 636

Bibliography

Acosta, José de. 1954. *Obras. Historia natural y moral de las Indias*. (Preliminary study and edition by Francisco Mateos). Biblioteca de Autores Españoles 73. Madrid: Rivadeneira.

Adams, Frank D. 1954. *The birth and development of the geological sciences*. New York: Dover.

Alay, Josep Lluis. 2006. (Edition and introduction) *Antoni de Montserrat, Embajador en la corte del Gran Mogol. Viajes de un jesuita catalán del siglo XVI por la India, Pakistan, Afganistán y el Himalaya*. Lerida: Milenio.

Altamore, Aldo, and Sabino Maffeo (eds.). 2012. *Angelo Secchi. L'avventura scientifica del Collegio Romano*. Foligno: Quater.

Anduaga, Aitor. 2013. Spanish Jesuits in the Philippines: Geophysical research and synergies between science, education and trade, 1865–1898. *Annals of Science* 1–25. doi: 10.1080/00033790.2013.841289.

Arrupe, Pedro. 1973–1976. El apostolado intelectual en la misión de la Compañía hoy. *Acta Romana S. I.* 16: 1015–1027.

Ashbrook, Joseph. 1966. Father Perry's expedition to Kerguelen Islands. *Sky and Telescope*, (June 1966), 340–341.

Aspaas, Per Pippin. 2012. *Maximilian Hell (1720–1792) and the eighteenth century transits of Venus: A study of Jesuit science in Nordic and Central European context*. Doctoral Thesis, University of Tromsø, Tromsø.

Baldini, Ugo. 1992. *Legem impone subactis. Studi su filosofia e scienza dei gesuiti in Italia, 1540–1632*. Rome: Bulzoni.

Baldini, Ugo. 2003. The Academy of Mathematics of the Collegio Romano from 1553 to 1612. In *Jesuit science and the republic of letters*, ed. Mordechai Feingold. Cambridge, MA: The MIT Press.

Bastero Monserrat, Juan Jesús. 1989. *Longinos Navás, científico jesuita*. Zaragoza: Universidad de Zaragoza.

Bellino Carrara. 1914. I gesuiti e Galileo. *Revista de Apologia Cristiana* (Julio-Agosto 1914), 1–62.

Berloty, Bonaventure. 1912. L'Observatoire de Ksara (Liban). *Ciel et Terre* 23: 103–108.

Birkenhauer, Henry. 1956. Father Macelwane and the Jesuit Seismological Association. *Earthquakes Notes* 27: 12–13.

Bishop, George D. 1977. *Stephen Joseph Perry (1833–1889) priest, scientist, educator*. Master of Science Thesis, University of Manchester.

Bishop, George D. 1979. Stephen Perry (1833–1889); forgotten Jesuit scientist and educator. *Journal British Astronomical Association* 89: 473–484.

Bishop, George D. 1998. *A Lion to Judah. The travels and adventures of Pedro Páez S. J.* Anand: Gujarat Sahitya Prakash.

A. Udías, *Jesuit Contribution to Science: A History*,
DOI 10.1007/978-3-319-08365-0

Bishop, George D. 2005. *Jesuit pioneers of modern science and mathematics*. Anand: Gujarat Sahitya Prakash.

Blum, Victor. 1956. Sketch of the life of James Bernard Macelwane, S.J. *Earthquakes Notes* 27: 9–11.

Boistel, Guy. 2003. Inventaire chronologique des oeuvres imprimées et manuscrites du père Esprit Pezenas (1692–1776), jésuite, astronome et hydrographe marseillaise. *Revue d'Historie des Sciences* 56: 221–245.

Borgato, M.T. (ed.). 2002. *Giambattista Riccioli e il merito scientifico dei gesuiti nell'età barroca*. Florence: L.S. Olschki.

Boscovich, Roger. 1755. *De litteraria expeditione per pontificiam dictionem ad dimidiendos duos meridiani gradus et corrigendam mappam geographicam, iussu et auspicii Benedicto XIV, suscepto a Patribus Societatis Iesu Christophoro Maire et Rogerio Josepho Boscovich*. Rome: Nicolaus e Marcus Pelearini.

Boscovich, Roger. 1763. *Theoria philosophiae naturalis redacta ad unicam legm virium in natura existentium, nunc ab ipso perpolita et aucta, ipso auctore praesente et corrigente*. Venice: Typographia Remondiniana.

Bosmans, Henri. 1912. Ferdinand Verbiest, directeur de l'Observatoire de Peking (1623–1688). *Revue des Questions Scientifiques* 71: 195–273, 375–464.

Boudignon, Patrice. 2008. *Pierre Teilhard de Chardin, sa vie, son oeuvre, sa reflexión*. Paris: Édition du Cerf.

Boyle, Henry A. 1939. Girolamo Saccheri, S. J. and Euclid's parallel postulate. *Bulletin American Association of Jesuit Scientists* 17: 93–99.

Brannagan, David. 2010. Earth, sky and prayer in harmony. Aspects of the interesting life of Father Edward Pigot, S.J., B.A., M.B., B.C.H. (1858–1929). A Jesuit seismologist. Part I. *Earth Science History* 29(2010): 69–99.

Brockey, Liam M. 2007. *Journey to the East. The Jesuit mission to China, 1572–1724*. Cambridge, MA: Harvard University Press.

Buisseret, David. 1997. Jesuit cartography in Central and South America. In *Jesuit encounters in the World. Jesuit chroniclers, geographers, educators and missionaries in the Americas, 1549–1767*, ed. J.A. Galiano and C.E. Ronan. Rome: Institutum Historicum S.I.

Burgaleta, Claudio M. 1999. *José de Acosta (1540–1600) his life and thought*. Chicago: Loyola Press.

Burrus, Ernest J. 1967. *La obra cartográfica de la Provincia Mexicana de la Compañía de Jesús*. Madrid: José Porrua Turanzas.

Burrus, Ernest J. 1971. *Kino and Monje. Explorers of Sonora and Arizona. Their vision of the future. A study of their expeditions and Plans*. Saint Louis: Jesuit Historical Institute.

Byerly, Perry, and William Stauder. 1958. James B. Macelwane, S. J., September 28, 1883 – February 15, 1956. *Memoirs of the National Academy of Science* 31: 254–281.

Cabré, Ramón. 1988. 75 Años en la vanguardia de la sismología. *Revista Geofísica* 29: 47–56.

Carlson, P. 2012. A century of cosmic rays. *Physics Today* 65: 30–36.

Casanovas, Juan. 1984. Il P. C. Clavio professore di matematica del P. M. Ricci nel Collegio Romano. *Atti del Primo Convegno Internazionale di Studi Ricciani*, 229–239. Macerata.

Casanovas, Juan. 1988. Per il secondo centenario della morte del P. Rugiero Boscovich. *La Civiltà Cattolica* IV: 531–544.

Casanovas, Juan. 2005. The discovery of sunspots. En: W. Schröder (ed.) *Historical events and people in aeronomy, geomagnetism and solar-terrestrial physics. Beitr. Gesch. Geophysik und Kosmischen Physik* 7, 243–260.

Cervera, José Antonio. 2011. *Las varillas de Napier. Giacomo Rho S. J. (1592–1638) y su trabajo como matemático y astrónomo en Beijing*. Mexico: El Colegio de México.

Chen Minsun. 1994. Ferdinand Verbiest and the geographical works by Jesuits, 1584–1674. In *Ferdinand Verbiest (1623–1688) Jesuit missionary, scientist, engineer and diplomat*, ed. J.W. Witek, 123–133. Nettetal: Steyler Verlag.

Chinnici, Ileana. 1996. An "Italian observatory" in India: The history of the Calcutta Observatory. *Studies in History of Medicine and Science* 14: 1–21.

Chinnici, Ileana. 2012. Il profilo scientifico e umano di Angelo Secchi. In *Angelo Secchi. L'avventura scientifica del Collegio Romano*, ed. A. Altamore and S. Maffeo, 43–64. Foligno: Quater.

Cobo, Bernabé. 1959. *Obras. Historia del nuevo mundo.* (Preliminary study and edition by Francisco Mateos). Biblioteca de Autores Españoles 91 and 92. Madrid: Rivadeneira.

Coenraads, Robert R. 1993. The San Calixto Observatory in La Paz, Bolivia. Eighty years of operation. Director Dr. Lawrrence A. Drake S. J. *Journal and Proceedings. Royal Society of New South Wales* 126: 191–198.

Colin, Elie. 1905. *Les cyclones à Madagascar.* Tananarive: Imprimerie Officielle.

Colubi, Mariano. 1999. *Boskovich y la visión mecánica de la naturaleza (1740–1785).* Madrid: Universidad Autónoma de Madrid.

Congregación General 35 de la Compañía de Jesús. 2008. Bilbao: Mensajero.

Cortie, Aloysius L. 1890. *Father Perry, the Jesuit astronomer*, 2nd ed. London: Catholic Truth Society.

Cortie, Aloysius L. 1923. Angelo Secchi, S. J. In *Twelve Catholic men of science*, ed. B. Windle, 149–164. London: Catholic Truth Society.

Coyne, George V., M.A. Hoskin, and O. Pedersen (eds.). 1983. *Gregorian Reform of the Calendar. Proceedings of the Vatican conference to commemorate its 400th anniversary, 1582–1982.* Vatican City: Specola Vaticana.

Crespy, Georges. 1961. *La pensé théologique de Teilhard de Chardin.* Paris: Éditions Universitaires.

Cronin, Vincent. 1959. *The wise man from the west.* London: Rupert Hart-Davies.

Cuénot, Claude. 1958. *Pierre Teilhard de Chardin. Les grands étapes de son évolution.* Paris: Librairie Plon.

Daxecker, Franz. 1996. *Das Hauptwerk des Astronomes P. Christoph Scheiner SJ "Rosa Ursina sive Sol" eine Zusammenfassung.* Berichte nat-med Verein Insbruck. Supplement 13.

Daxecker, Franz. 2004. *The physicist and astronomer Christopher Scheiner. Biography, letters, works.* Innsbruck: Innsbruck University.

de Dainville, François. 1940. *La géographie des humanistes.* Paris: Beauchesnes.

de Dainville, François. 1954. L'enseignement des mathématiques dans les collèges jésuites de France du XVI au XVIII siècle. *Revue d'Histoire des Science* 7: 6–21, 109–123.

de Fonseca, G. 1906. Rerum naturalium in Lusitania cultores. P. Johannes de Loureiro. *Broteria* 5: 98–114.

de Kort, Jules, and Jacques Marie Adrian. 1952. Obituary: In memoriam Father Johan Stein, S. J. (1871–1951). *Richerche Astronomiche* 2: 372–374.

de la Héronniere, Édith. 2003. *Teilhard de Chardin, une mystique de la traversée.* Paris: Albin Muchel.

de Lubac, Henri. 1962. *La pensée religieuse du Père Teilhard de Chardin.* Paris: Aubier.

de Vrégille, Pierre. 1905. Les jésuites et l'étude du magnétisme terrestre. *Études* 104: 495–511.

de Vréguile, Pierre. 1905. *L'observatoire du Collège de la Trinité à Lyon.* Paris: Victor Retaux.

de Vrégille, Pierre. 1906. Les observatoires de la Compagnie de Jésus au début du XXe siècle. *Revue des Questions Scientifiques* 59: 10–72, 493–579.

de Vrégille, Pierre. 1907. Galilée et les jésuites. *Études* 110: 460–478, 584–600.

Dear, Peter. 1987. Jesuit mathematical science and the reconstruction of experience in the early seventeenth century. *Studies in History and Philosophy of Science* 18: 133–175.

Débarbat, Suzane and Simone Dumont. 1990. Antoine-François Laval (hydrographe du Roi, jésuite et astronome). *Histoire des Sciences et des Techniques. 115 Congress Nat. Soc. sav. Avignon*, 17–26.

Dechevrens, Marc. 1887. *Sur la variation de température observée dans les cyclones.* Shanghai: Observatoire de Zikawei.

Delzel, Franz, and Joseph Stepling. 1782. *Abbildungen Böhmischer und Mährischer Gelehrten und Künstler* (Prague), vol. 4, 164–172.

Devane, John F., and James P. McCaffrey. 1988. Reverend Daniel Linehan (1904–1987). *Bulletin Seismological Society of America* 78: 1618–1620.

Dhôtel, Jean-Claude. 1991. *La spiritualité ignatienne. Points de repère*. Paris: Vie Chrétienne.
Didier, Hugues. 2002. *Les portugais au Tibet. Les premières relations jésuites (1624–1635)* (translation and presentation). Paris: Chandeigne.
Dopp, H. 1931. Johann Georg Hagen, S. J., 1847–1930. *Revue des Questions Scientifiques* 99: 5–37.
Dou, Alberto. 1970. Logical and historical remarks on Saccheri's geometry *Notre Dame Journal of Formal Logic* 11: 385–415.
Drake, Stillman. 1957. *Discoveries and opinions of Galileo*, 237–238. Garden City: Doubleday Anchor Books (The Assayer).
Drum, Walter M. 1905. *The pioneer forecasters of hurricanes*. Washington, DC: Styormont and Jackson.
D'Souza, Leo. 2012. *Jesuit contribution to environmental protection*. Sjweb-info/sjs/networks/ecology/Indian.
Due Rojo, Antonio. 1940. *Labor científica del R. P. Manuel Sánchez Navarro, S. J.* Granada: Observatorio de Cartuja.
Due Rojo, Antonio. 1941. El R. P. Manuel M. Sánchez Navarro, S. J. *Boletín Real Sociedad de Historia Natural* 39: 117–120.
Due Rojo, Antonio. 1950. El fundador de la heliofísica. *Urania* 223–224: 121–137.
Due Rojo, Antonio. 1953. El cincuentenario del Observatorio de Cartuja. *Urania* 234: 67–81.
Dunn, Gordon E. 1951. Tropical cyclones. In *Compendium of meteorology*, ed. T.F. Malone, 887–901. Boston: American Meteorological Society.
Dunn, Gordon E., and Banner I. Miller. 1960. *Atlantic hurricanes*. Baton Rouge: Louisiana State University Press.
Dunne, George H. 1962. *Generation of giants. The story of the Jesuits in China in the last decades of the Ming dynasty*. London: Burns and Oates.
Ebel, John. 2009. The Jesuit science. *Company* (Summer 2009) 28–31.
Engelfriet, Peter. 1998. *Euclides in China. The genesis of the first translation of Euclid's Elements in 1607 and its reception up to 1723*. Leiden: Brill.
Epistolae selectae Praepositorum Generalium ad Superiores Societatis. 1911. Rome: Polyglottis Vaticanis.
Euvé, François. 2012. *Mathématique, astronomie, biologie et soin des âmes. Le jésuites et les sciences*. Brussels: Lessius.
Fantoli, Annibale. 1996. *Galileo for Copernicanism and for the church*. Vatican City: Vatican Observatory Publications.
Feingold, Mordechai (ed.). 2003. *The new science and Jesuit science: Seventeenth century perspectives*. Dordrecht: Kluwer Academic Press.
Feingold, Mordechai (ed.). 2005a. *Jesuit science and the republic of letters*. Cambridge, MA: The MIT Press.
Feingold, Mordechai. 2005b. Jesuits: Savants. In *Jesuit science and the republic of letters*, ed. M. Feingold, 1–45. Cambridge, MA: The MIT Press.
Feldhay, Rivka. 1987. Knowledge as salvation in Jesuit culture. *Science in Context* 1: 195–213.
Fényi, Julius. 1906. Die Sternwarte in Kalocsa, das Haynald Observatorium. *Nachr. der öster-ungarische Provinze* 3: 14–23.
Fernández, Jaime, and Eduardo di Vita. 2004. *Presença dos jesuitas no mundo científico*. São Paulo: Edições Loyola.
Findlen, Paula (ed.). 2004a. *Athanasius Kircher, the last man who knew everything*. London: Routledge.
Findlen, Paula. 2004b. Introduction. The last man who knew everything… or did he? Athanasius Kircher S. J. (1602–1680) and his world. In *Athanasius Kircher, the last man who knew everything*, ed. P. Findlen, 1–48. London: Routledge.
Fischer, Karl A.F. 1978. Jesuiten-Mathematiker in der Deutschen Assistenz bis 1773. *Archivum Historicum S. I.* 47: 159–224.
Fischer, Karl A.F. 1983. Jesuiten-Mathematiker in der Französischen und Italienischen Assistenz bis 1762 bis 1773. *Archivum Historicum S. I.* 52: 52–92.

Fletcher, John (ed.). 1988a. *Athanasius Kircher und seiner Beziehungen zum gelehrten Europa seiner Zeit*. Wiesbaden: Otto Harrassowitz.

Fletcher, John E. 1988b. Athanasius Kircher: A man under pressure, Kircher's works and bibliography. In *Athanasius Kircher und seiner Beziehungen zum gelehrten Europa seiner Zeit*, ed. J. Fletcher. Wiesbaden: Otto Harrassowitz, 1–15 and 179–195.

Fletcher, John E. 1988c. Athanasius Kircher and his correspondence. In *Athanasius Kircher und seiner Beziehungen zum gelehrten Europa seiner Zeit*, ed. J. Fletcher, 170–178. Wiesbaden: Otto Harrassowitz.

Fletcher, John E. 2011. *A study of the life and Works of Athanasius Kircher 'Germanus Incredibilis'* (Edited for publication by Elizabeth Fletcher). Leiden: Brill.

Fontana, Michela. 2005. *Matteo Ricci. Un gesuita alla corte di Ming*. Milan: Mondadori.

Francisco Javier. 1979. *Cartas y escritos de San Francisco Javier*. (Annotated by Felix Zubilaga, 3rd ed.). Madrid: Biblioteca de Autores Cristianos.

Froc, Louis. 1920. *Atlas of the tracks of 620 typhoons*. Shanghai: Observatoire de Zikawei.

Funes, José G. 2007. Una imagen evangélica para el apostolado científico (Mc 14, 3–9). *Rev. de Espiritualidad Ignaciana* 37: 87–95.

Furlong Cardiff, Guillermo. 1929. *Glorias Santafesinas*. Buenos Aires: Surgo.

Furlong Cardiff, Guillermo. 1936. *Cartografía jesuítica del Río de la Plata*. Buenos Aires: Jacobo Peuser.

Gabor, Paul. 2010. Jesuit identity and science. An inside look. In *Proceedings of the Bohemia Jesuitica Conference*, ed. Petra Čemus, 647–660 (Prague 25–27 April 2006). Prague: Carolinum.

García Doncel, Manuel, and Antonio Roca. 2007. *El Observatorio del Ebro. Un siglo de historia (1904–2004)*. Roquetas: Observatorio del Ebro.

García Villoslada, Ricardo. 1954. *Storia del Colegio Romano, dal suo inizio alla soppressione della Compagnia de Gesù (1773)*, Analecta Gregoriana, vol. LXVI. Rome: Universitàt Gregoriana.

Garzoni, Leonardo. 2005. *Trattati della Calamita a cura di Monica Ugalia* (Introduction by M. Ugalia: Introduzione 7–83) Milan: FrancoAngeli.

Gauthier, Henri. 1924. Un maître en physique du globe. *Études* 178: 272–291.

Gauvin, Jean-François. 2003. La physique du globe tous azimuts: La carrière scientifique d'Ernesto Gherzi, S. J., 1910–1973. *Archivum Historicum S. I.* 72: 45–80.

Gehlert, Oswald. 1990. Jesuiten als Naturwissenschaftler. *Kalksburger Korrespondence* 218: 28–36.

Geschwind, Carl-Henry. 1998. Embracing science and research. Early twenty-century Jesuits and seismology in the United States. *Isis* 89: 27–49.

Gherzi, Ernesto. 1924. Étude sur les microséismes. *Observatoire de Zikawei: Notes séismologiques* 5: 1–16.

Gherzi, Ernesto. 1939. On the constitution of typhoons. *Bulletin American Meteorological Society* 19: 59–66.

Giard, Luce (ed.). 1995. *Les jésuites à la Renaissance. Système éducatif et production du savoir*. Paris: Presses Universitaires de France.

Gill, Henry V. 1941. *Roger Boscovich S. J. (1711–1787). Forerunner of modern physical theories*. Dublin: Gill and Sons.

Gillispie, Charles C. (ed.). 1970–1980. *Dictionary of scientific biography*, vol. 1–16. New York: Charles Scribner and Sons.

Godwin, Joscelyn. 1986. *Athanasius Kircher, la búsqueda del saber de la antigüedad*. El Escorial: Swan, Avantos y Hakeldama.

Godwin, Joscelyn. 2009. *Athanasius Kircher's theatre of the world. The life and work of the last man to search for universal knowledge*. Rochester: Inner Traditions.

Golvers, Noël. 1993. *The Astronomia Europaea of Ferdinand Verbiest, S.J. (Dillingen, 1687). Text, translation notes and commentaries*. Nettetal: Steyler.

Golvers, Noël. 2003. *Ferdinand Verbiest, S. J. (1623–1688) and the Chinese Heaven*. Louvain: Leuven University Press.

Gómez de Liaño, Ignacio. 2001. *Athanasius Kircher, itinerario del éxtasis o las imagines de un saber universal*. Madrid: Siruela.
Graney, Christopher M. 2012. Anatomy of a fall: Giovanni Battista Riccioli and the story of g. *Physics Today* 65: 36–40.
Gray, Richard O., and Christopher J. Corbally. 2009. *Stellar spectral classification*. Princeton: Princeton University Press.
Gumilla, José. 1946. *El Orinoco ilustrado*. (Introduction and notes by Constantino Bayle). Madrid: Aguilar.
Gutenberg, Beno, and Charles F. Richter. 1949. *Seismicity of the Earth and associated phenomena*. Princeton: Princeton University Press.
Gutierrez Lanza, Manuel. 1904. *Apuntes históricos acerca del Observatorio del Colegio de Belén, Habana*. Havana: Avisador Comercial.
Haig, Frank R. 1961. Vocation of the priest-scientist. *Bulletin American Association of Jesuit Scientists* 36: 3–9.
Hansen, T.L., and P.P. Aspaas. 2005. *Maximilian Hell's geomagnetic observations in Norway 1769*. Tronsø: University of Tronsø.
Harris, Steven J. 1989. Transposing Merton's thesis: Apostolic spirituality and the establishment of the Jesuit scientific tradition. *Science in Context* 3: 29–65.
Harris, Steven J. 1995. Les chaires de mathématiques. In *Les jésuites a la renaissance*, ed. Luce Giard, 239–262. Paris: Presses Universitaires de France.
Harris, Steven J. 2005. Jesuit scientific activity in the overseas missions 1540–1773. *Isis* 96: 71–79.
Harvey, Charlotte B. 2000. The voyage of the Monte Carlo. *Boston College Magazine*, (Fall 2000), 1–4.
Heck, N.H. 1944. The Jesuit contribution to seismology in the U.S.A. *Thought* 19: 221–228.
Heilbron, John L. 1982. *The elements of early modern physics*. Berkeley: University of California Press.
Heilbron, John L. 1999. *Electricity in the 17th and 18th centuries. A study in early modern physics*. Mineola: Dover.
Hellyer, Marcus. 2005. *Catholic physics. Jesuit natural philosophy in early modern Germany*. Notre Dame: University of Notre Dame Press.
Hennessey, James J. 1957. Charles Deppermann S. J., Philippine scientist. *Philippine Studies* 5: 311–335.
Hennessey, James J. 1960. The Manila observatory. *Philippine Studies* 8: 99–120.
Heyden, Francis J. 1975. *The beginning and end of a Jesuit observatory (1841–1972)*. Manila: Publications of Manila Observatory.
Hill, Elizabeth. 1961. Roger Boscovich: A biographical essay. In *Roger Joseph Boscovich, S. J., F. R. S., 1711–1787: Studies of his life and work on the 20th anniversary of his birth*, ed. Lancelot L. Whyte, 16–101. London: George Allen and Unwin.
Hine, William. 1988. Athanasius Kircher and magnetism. In *Athanasius Kircher und seiner Beziehungen zum gelehrten Europa seiner Zeit*, ed. J. Fletcher, 79–97. Wiesbaden: Otto Harrassowitz.
Hingle, Peter D. 2005. The priest and the stuffed penguin; Father Stephen Perry S.J. and the transit of Venus expeditions to Kerguelen Island, 1874 and Madagascar, 1882. *Journal British Astronomical Association* 115(2005): 150–158, 168–170.
Ho Peng Yoke. 1985. Li Qi and Shu: *An introduction to science and civilization in China*. Hong Kong: Hong Kong University Press.
Hsia, Florence C. 1999. Jesuits, Jupiter's satellites, and the Académie Royale des Sciences. In *The Jesuits. Cultures, science and the arts*, ed. J.W. O'Malley, G.A. Bailey, S.T. Harris, and F. Kennedy. Toronto: University of Toronto Press.
Hsia, Florence C. 2009. *Sojourners in a strange land. Jesuits and their scientific missions in late imperial China*. Chicago: University of Chicago Press.

Ignatius of Loyola. 1991. *Obras Completas. Edición Manual*. Madrid: Biblioteca de Autores Cristianos.
Institutum Societatis Iesu. 1893. (vol. 1–3), Florence: SS. Conceptione.
Jacquin, Robert. 1958. Une polémique romaine sur l'hylomorphisme il-y-a un siècle. *Revue de Sciences Religieuse* 32: 372–377.
Janssen, Jean Baptiste. 1946–1950. Epistola de ministeriis nostris (22 June 1947). *Acta Romana S. I.*, 11: 299–336.
Janssen, Jean Baptiste. 1960. Instructio de cosmologia in nostris scholasticatibus docenda. *Acta Romana S. I.* 13: 809–814.
Jones, Elisabeth. 1992. *Gentlemen and Jesuits: Quest for glory and adventure in the early days of New France*. Toronto: University of Toronto Press.
King, Ursula. 1998. *Spirit of fire. The life and vision of Teilhard de Chardin*. New York: Orbis Books.
Kircher, Athanasius. 1641. *Magnes sive de arte magnetica, Opus Tripartitum*. Rome: Ludovici Grignani.
Kircher, Athanasius, *Itinerarium exstaticum*. Rome: Mascardi, (1656); *Iter exstaticum II*. Rome: Mascardi (1657). Edited by Kaspar Schott, *Iter exstaticum coeleste*. Würzburg: Endter (1660 and 1671).
Kircher, Athanasius. 1664. *Mundus subterraneus in XII libros digestus*. Amsterdam: Joannem Janssonium Waesberghe.
Knobloch, Eberhard. 1989. Sur la vie et l'oeuvre de Christophore Clavius (1538–1612). *Revue d' Histoire des Sciences* 41: 331–356.
Kochhar, R.K. 1994. Secondary tools of empire: Jesuit men of science in India. In *Discoveries, missionary expansion and Asian cultures*, ed. T.R. de Souza, 175–183. New Delhi: Concept Pub.
Kummer, Christian. 2000. Forschung und Weltanschaung: Die Rolle der deutschen Jesuiten in der Biologie. *Sammelblatt des Historischen Vereins Ingolstadt* 109: 105–120.
Labrador, Carmen, M. Bertrán-Quera, A. Diez Escanciano, and J. Martínez de la Escalera. 1986. *La Ratio Studiorum de los jesuitas*. Madrid: Universidad Pontificia Comillas.
Lattis, James M. 1994. *Between Copernicus and Galileo, Christopher Clavius and the collapse of Ptolemaic cosmology*. Chicago: The University of Chicago Press.
Latusseck, Arndt. 2009. *Via Nubila am Grund des Himmels. Johan Georg Hagen und die kosmichen Wolken*. Doctoral Thesis, Hamburg University.
Leitao, Henrique. 2007. *A ciência na "Aula da esfera" no colégio de Santo Antão, 1590–1759*. Lisbon: Ministério da Cultura.
Lejay, Pierre. 1925. Trois observatoires de Mission. *Études* 182: 594–602.
Lerner, Michel-Pierre. 1995. L'entrée de Tycho Brahe chez les jésuites ou le chant du cygne de Clavius. In *Les jésuites à la Renaissance*, ed. Giard Luce, 145–185. Paris: Presses Universitaires de France.
Libois, Charles. 2001. Pedro Páez, descubridor de las fuentes del Nilo. In *Exploradores españoles olvidados de África*, 89–112. Madrid: Ed. Pedro Páramo.
Linehan, Daniel. 1984. Jesuits in seismology. *Earthquake Information Bulletin* 16: 156–165.
L'Observatoiore de Zi-ka-wei. Cinquante ans de travail scientifique. 1930. Paris: Imp. D'Art G. Boüan.
López de Santa Anna, Antonio. 1957. *Contribución a una biografía completa del P. Benito Viñes Martorell, S.I., célebre meteorólogo de las Antillas*. Santander: Hermanos Bedia.
Lukács, Ladislaus. 1986. *Monumenta Paedagogica Societatis Iesu*, vol. I–VII. Rome: Institutum Historicum Societatis Iesu.
Lynch, Joseph. 1970. *Watching our trembling Earth for 50 years*. New York: Dodd, Mead and Co.
Macelwane, James B. 1939. Evidence of the interior of the Earth derived from seismic sources. In *Internal constitution of the Earth*, ed. Beno Gutenberg, 219–290. New York: McGraw Hill.
Macelwane, James B. (ed.). 1950a. *Jesuit Seismological Association, 1925–1950. Twenty-fifth anniversary commemorative volume*. St. Louis: Saint Louis University.

Macelwane, James B. 1950b. The forerunner: The Jesuit Seismological Service. In *Jesuit Seismological Association, 1925–1950. Twenty-fifth anniversary commemorative volume*, ed. J.B. Macelwane, 3–16. St. Louis: Saint Louis University.

Macelwane, James B. 1950c. The Jesuit Seismological Association. In *Jesuit Seismological Association, 1925–1950. Twenty-fifth anniversary commemorative volume*, ed. J.B. Macelwane, 17–23. St. Louis: Saint Louis University.

MacDonnell, Joseph. 1989. *Jesuit geometers*. St. Louis: Institute of Jesuit Sources.

Machado Mota, Bernardo. 2010. A cultura matemática no contexto Jesuita anterior ao século XVIII. *Broteria* 170: 27–42.

Maffeo, Sabino. 1991. *In the service of nine Popes. 100 years of the Vatican Observatory*. Vatican City: Vatican Observatory.

Maffeo, Sabino. 2001. *La Specola Vaticana. Nove Pappi, una missione*. Vatican City: Pulicazioni della Specola Vaticana.

Maffeo, Sabino. 2006. *J.G. Hagen S.I. Astronomo e direttore spirituale della beata E. Hesselblad*. Rome: Edizioni AdP.

Maffeo, Sabino. 2012. Il Collegio Romano e l'insegnamento de la scienze. In *Angelo Secchi. L'avventura scientifica del Collegio Romano*, ed. Aldo Altamore and Sabino Maffeo, 15–41. Foligno: Quater.

Malone, T.F. (ed.). 1951. *Compendium of meteorology*. Boston: American Meteorological Society.

Marcović, Zeijko. 1970–1980. Bošcović, Rudjer J. In *Dictionary of scientific biography*, ed. C.C. Gillespie, vol. 1, 226–332. Nueva York: Scribner.

Martelet, Gustave. 2005. *Teilhard de Chardin, prophète d'un Christ toujours plus grande*. Brussels: Lessius.

McKevitt, Gerald. 2008. Jesuits schools in the USA, 1814–c. 1970. In *The Cambridge companion to the Jesuits*, ed. Thomas Worcester, 278–293. Cambridge: Cambridge University Press.

McMullin, Ernan (ed.). 2005. *The Church and Galileo*. Notre Dame: University of Notre Dame Press.

Meissner, William W. 1963. The Jesuit as a priest-scientist. *Bulletin American Association of Jesuit Scientists* 40: 25–33.

Merton, Robert K. 1973. *The sociology of science. Theoretical and empirical investigations*. Chicago: The University of Chicago Press.

Millones Figueroa, Luis, and Domingo Ledezma (eds.). 2005. *El saber de los jesuitas, historias naturales del Nuevo Mundo*. Madrid: Iberoaméricana.

Minghui Hu. 2002. Provenance in contest: Searching for the origins of Jesuit astronomy in early Qing China. *International History Review* 24: 1–36.

Mitchell, Brian J., and Agustín Udías. 2003. William Stauder (1922–2002). *Seismological Research Letters* 74: 373.

Mohr, Paul (ed.). 2007. *Geophysical observatory of Addis Ababa University: 50th anniversary symposium proceedings*. Addis Ababa: Addis Ababa University.

Moigno, François. 1879. *Le Révérend Père Secchi, sa vie*. Paris: Gauthiers-Villars.

Mooney, Christopher F. 1968. *Teilhard de Chardin and the mystery of Christ*. New York: Doubleday Image Book.

Moraes, George M. 1951–1952. Astronomical missions to the court of Jaipur, 1730–1743. *Journal Bombay Branch of Royal Asiatic Society* 27: 61–65, 85.

Murphy, Thomas. 1968. The Rev. R. E. Ingram, S. J. *Quarterly Journal of Royal Astronomical Society* 9: 226–227.

Naux, Charles. 1983. Le père Christophore Clavius (1537–1612). Sa vie et son oeuvre. *Revue des Questions Scientifiques* 154: 55–67, 181–193, 325–347.

Naux, Charles. 1994. Gregoire de Sant-Vicent: A pioneer of the calculus. *The Mathematical Gazette* 78: 315–319.

Needham, Joseph. 1959. *Science and civilization in China*, Mathematics and the sciences of heavens and earth. The time of the Jesuits, vol. III. Cambridge: Cambridge University Press.

Núñez de Castro, Ignacio. 2005. De la amistad y desencuentro de Galileo con los jesuitas. *Archivo Teológico Granadino* 68: 69–109.

Observatorio de Cartuja, 1902–2002. 2002. Granada: Caja de Granada.
O'Connell, Daniel. 1956. Jesuit men of science. *Studies* 45: 307–318.
O'Malley, John W. 1993. *The first Jesuits*. Harvard: Harvard University Press.
Oñate Guillén, Carmelo. 1988. Un sabio del siglo XVIII: R. J. Boscovich. *Letras de Deusto* 18: 199–209.
Oñate Guillén, Carmelo. 1992. Rodrigo de Arriaga, filósofo del barroco ¿también científico? *Letras de Deusto* 22: 77–94.
Oñate Guillén, Carmelo. 1993. Primer texto de la difracción de la luz. *Revista Española de Física* 7: 54–57.
Oñate Guillén, Carmelo. 1997. Galileo científico y procesado, su relación científica y procesal con los jesuitas. *Letras de Deusto* 27: 105–129.
Oñate Guillén, Carmelo. 2000. Christopher Clavius y el calendario gregoriano. *Letras de Deusto* 89: 55–70.
Oñate, Carmelo. 2000. El gran astrónomo jesuita G. B. Riccioli (1598–1671) (manuscript, 2000).
O'Neill, Charles E., and Joaquín M. Domínguez (eds.). 2001. *Diccionario histórico de la Compañía de Jesús*, vol. 1–4. Madrid: Universidad Pontificia Comillas.
Orella, José Luis. 1967. Un elenco jesuítico de proposiciones filosóficas. *Pensamiento* 23: 273–306.
Oswald, Julius. 2008. Leben und Werk des würzburger mathematiker Kaspar Schott S.J. *Archivum Historicum S. I.* 78: 417–441.
Ott, Richard W. 1997. Rev. Ramón Cabré Roigé, S. J., May 2, 1922-February 24, 1997. *Seismological Research Letters* 68: 901.
Palmés, F.M., J. Torra Almenara, and J.M. Caballería. 1935. Las "cuestiones científicas" en las Facultades de Filosofía. *Estudios Eclesiásticos* 55: 1–32.
Pärr, Nora. 2013. *Maximilian Hell und sein wissenschaftliches Umfeld in Wien des 18. Jahrhunderts*, Religiongeschichte der frühen Neuzeit, vol. 14. Vienna: Traugott Bautz.
Perry, Stephen J. 1876. *Notes on a voyage to Kerguelen island to observe the transit of Venus, December 8, 1874*. London: Henry S. King.
Perry, Stephen J. 1880. L'Observatoire de Stonyhurst. *Annales de la Soc. Scient. de Bruxelles* 4: 147; 281–302.
Po-Chia Hsia, R. 2010. *A Jesuit in the forbidden city: Matteo Ricci, 1552–1610*. Oxford: Oxford University Press.
Poisson, Charles. 1939. *Un cinquantenaire, l'Observatoire d'Abohidempona, Tananarive, Madagascar, 1889–1939*. Paris: Editions Dillen.
Polzer, Charles. 1998. *Kino, a legacy: His life, his works, his missions, his monuments (A Kino Guide II)*. Tucson: Jesuit Fathers of Southern Arizona.
Prieto, Andrés I. 2011. *Missionary scientists. Jesuit science in Spanish South America, 1570–1810*. Nashville: Vanderbilt University Press.
Puig, Ignacio. 1927. *El observatorio del Ebro. Idea general sobre el mismo*. Tortosa: Imprenta Moderna del Ebro de Algeró y Baiges.
Puig, Ignacio. 1930. El P. José Algué, S. J. *Ibérica* 833: 392–398.
Puig, Ignacio. 1935. *El Observatorio de San Miguel, República Argentina*. San Miguel: Observatorio de San Miguel.
Pyenson, Lewis. 1991. Pure research, Jesuit institutions and metropolitan ambitions: The evolution of French policies overseas, 1880–1940. *Chine et Europe: Évolution et particularités des rapports est-ouest du XVIe au XXe siècle. Actes du IVe Colloque International de Sinologie de Chantilly,* Paris, 249–271.
Ramírez, Jesús Emilio. 1940. An experimental investigation on the nature and origin of microseisms at St. Louis, Missouri. *Bulletin Seismological Society of America* 30: 34–84; 139–178.
Ramírez, Jesús Emilio. 1977. *Historia del Instituto Geofísico al conmemorar sus 35 años (Universidad Javeriana) I. Antecedentes y Fundación; II. Los años de la guerra 1944–1945; III. Consolidación del Instituto Geofísico*. Bogota: Editora Guadalupe.
Ramos Guadalupe, Luis E. 1996. *Benito Viñes, s.j. Estudio biográfico*. La Habana: Editorial Academia.

Randles, W.G.L. 1999. *The unmaking of the medieval Christian Cosmos, 1500–1760. From solid heavens to boundless aether*. Aldershot: Ashgate.

Reilly, Conor. 1974. *Athanasius Kircher S. J., Master of a hundred arts (1602–1680)*. Rome: Edizioni del Mondo.

Repetti, William C. 1948. *The Manila observatory*, Washington, DC.

Reverte, Javier. 2001. *Dios, el Diablo y la aventura*. Barcelona: Plaza Janes.

Rideau, Émile. 1965. *La pensé du Père Teilhard de Chardin*. Paris: Le Seuil.

Riehl, H. 1951. Aerology of tropical storms. In *Compendium of meteorology*, ed. T.F. Malone, 902–916. Boston: American Meteorological Society.

Rodes, Luis, Eduardo Fontseré and Eduardo Alcobé. 1930. Sesión necrológica en homenaje del. Rvdo. P. Federico Faura, S. J. Discursos. *Memorias de la Real Academia de Ciencias y Artes de Barcelona*, Barcelona, Vol. 22: 69–92.

Rodríguez Carracido, José. 1899. *El P. José de Acosta y su importancia en la literatura científica española*. Madrid: Sucesores de Rivadeneira.

Romano, Antonella. 1999. *La contre-réforme mathématique. Constitution et diffusion d'une culture mathématique jésuite à la Renaissance*. Rome: École Français de Rome.

Romano, Antonella. 2005. Les jésuites entre apostolat missionnaire et activité scientifique (XVIe–XVIIIe siécles). *Archivum Historicum S.I.* 74: 213–236.

Romeiras, Francisco Malta. 2012. The journal Brotéria (1902–2002). Jesuit science in the 20th century. *Journal of History of Science and Technology* 6: 100–109.

Romeiras, Francisco Malta. 2013. The emerging of molecular genetics in Portugal: The enterprise of Luís Archer. *Archivum Historicum S. I.* 82: 501–512.

Rowbotham, Arnold H. 1942. *Missionary and mandarin. The Jesuits at the court of China*. Berkeley: University of California Press.

Rowland, Ingrid D. 2004. Athanasius Kircher, Giordano Bruno and the Panspermia of the infinite universe. In *Athanasius Kircher, the last man who knew everything*, ed. En.P. Findlen, 191–205. London: Routledge.

Saderra-Masó, Manuel. 1915. *Historia del Observatorio de Manila*. Manila: E. C. McCullough.

Sainz Ollero, Héctor, Helios Sainz Olleros, Francisco Suárez Cardona, and Miguel Vázquez de Castro. 1989. *José Sánchez Labrador y los naturalistas jesuitas del Río de la Plata*. Madrid: Ministerio de Obras Públicas.

Sánchez Navarro-Neumann, Manuel M. 1937a. Jesuitas astrónomos. *Broteria* 24: 423–436.

Sánchez Navarro-Neumann, Manuel M. 1937b. Os Jesuitas e a sismologia. *Broteria* 24: 145–151.

Sarton, George. 1944. Vindication of Father Hell. *Isis* 35: 97–105.

Schreiber, Johann. 1903. Die Jesuiten des 17 und 18 Jahrhunderts und ihre Verhältnis zur Astronomie. *Natur und Offenbarung* 49: 129–143; 208–221.

Schroth, Raymond A. 2007. *The American Jesuits. A history*. New York: New York University Press.

Schumacher, John N. 1965. One hundred years of Jesuit scientists: The Manila observatory. *Philippine Studies* 13: 258–286.

Secchi, Angelo. 1856. *Descrizione del nuovo Osservatorio del Collegio Romano*. Rome: Tipografia delle Belle Arti.

Secchi, Angelo. 1864. *L'unita delle forze fisiche*. Rome: Tipografia Forense.

Secchi, Angelo. 1875–1877. *Le Soleil*, Paris: Gauthier-Villar.

Semaine d'étude sur le problème des microséismes. 1952. Vatican City: Pontificae Academiae Scientiarum (Scripta Varia 12).

Sharma, Virendra N. 1982. The impact of the eighteenth century Jesuits astronomers on the astronomy of India and China. *Indian Journal of History of Science* 17: 345–352.

Sharma, Virendra N., and Lila Hubbert. 1984. Jesuit astronomers in eighteenth century India. *Archives Internationales d'Histoire des Sciences* 34: 99–107.

Shea, William R. 1970. Galileo, Scheiner and the interpretation of sunspots. *Isis* 61: 498–519.

Sidgreaves, Walter. 1883. The last scientific voyage of H.M.S. Fawn. *The Month* 47: 457–482.

Sierra, Eduardo. 1981. El geocosmos de Kircher. Una cosmovisión científica del siglo XVII. *Geocrítica*, 33–34: 5–19.

Sommervogel, Carlos. 1890–1909. *Bibliothèque de la Compagnie Jésus. Bibliographie*, vols. 1–10. Brussels: Alphonse Picard.

Speaight, Robert. 1967. *Teilhard de Chardin. A biography*. London: Collins.

Spittler, Ernest G. 1962. The priest-scientist in the Church. *Bulletin American Association of Jesuit Scientists* 39: 30–37.

Tannehill, Ivan R. 1945. *Hurricanes, their nature and history*, 6th ed. Princeton: Princeton University Press.

Teilhard de Chardin, Pierre. 1962–1976. *Oeuvres I-XIII*, Paris: Éditions du Seuil.

Teilhard de Chardin, Pierre. 2003. *Pierre, Notes de retraites, 1919–1954*. Paris: Édition du Seuil.

Thirion, J. 1890. Le R. P. Perry. *Revue des Questions Scientifiques* 27: 201–208.

Toohig, Timothy E. 1999. Physics research, a search for God. *Studies in the Spirituality of Jesuits* 21(2): 1–26.

Udías, Agustín. 1986. José de Acosta (1539–1600): A pioneer of geophysics. *Eos, Transactions of the American Geophysical Union* 67: 461–462.

Udías, Agustín. 1994. Jesuit astronomers in Beijing, 1601–1805. *Quarterly Journal Royal Astronomical Society* 34: 463–478.

Udías, Agustín. 1996. Jesuits' contribution to meteorology. *Bulletin American Meteorological Society* 77: 2307–2315.

Udías, Agustín. 2001. Serving god and science. *Astronomy and Geophysics* 42: 23–24.

Udías, Agustín. 2003. *Searching the heavens and the earth. The history of Jesuit observatories*. Dordrecht: Kluwer.

Udías, Agustín. 2005. Los libros y manuscritos de los profesores de matemáticas del Colegio Imperial de Madrid, 1627–1767. *Archivum Historicum S. I.* 74: 369–448.

Udías, Agustín. 2006. El P. José Zaragoza y su obra. In *Esphera en común, celeste y terraquea*(1675), ed. José Zaragoza. (Facsimil edition), 11–46. Madrid: Real Sociedad Geográfica and Escuela T. S. Ingenieros de Topografía, Geodesia y Cartografía.

Udías, Agustín. 2007a. Jesuits, Role in Geomagnetism. In *Encyclopaedia of Geomagnetism and Paleomagnetism*, ed. D. Gubbins and E. Herrero-Berrera, 460–462. Dordrecht: Springer.

Udías, Agustín. 2007b. Lawrence Drake, S. J. (1931–2007). *Seismological Research Letters* 78: 420.

Udías, Agustín. 2009. Jesuits' studies of earthquakes and seismological stations. In *Geology and religion: A history of harmony and hostility*, The Geological Society S.P. 310, ed. M. Kölbl-Ebert, 135–143. London: Geological Society.

Udías, Agustín. 2010. Profesores de matemáticas en los colegios de la Compañía en España. *Archivum historicum Societatis Iesu S. I.* 79: 3–27.

Udías, Agustín. 2012. Jesuit scientific tradition and ignatian spirituality. *Lo Sguardo-Rivista di Filosofia* 10: 207–219.

Udías, Agustín, and William Stauder. 1996. The Jesuit contribution to seismology. *Seismological Research Letters* 67: 10–19.

van Tricht, V. 1878. Le Père Secchi. *Revue des Questions Scientifiques* 4: 353–402.

Väth, Alfons. 1991. *Johann Adam von Bell S.J. Missionar in China, kaiserlicher Astronom und Ratgeber am Hofe von Peking 1592–1666. Ein Lebens und Zeitbild.* (new edition) Netettal: Steyler.

Vigano, Mario. 1978. Cento anni di astrofisica. Nel centenario della morte del padre Angelo Secchi. *Civiltà Cattolica* 129: 538–549.

Vollrath, Hans-Joachim. 2007. *Wundebar berechenbar: die Welt des Würzburger Mathematiker Kaspar Schott, 1608–1666*. Würzburg: Echter Verlag.

Wallace, William A. 1989. The problem of apodictic proof in early seventeenth century mechanics. Galileo, Guevara and the Jesuits. *Science in Context* 3: 67–87.

Wessels, C. 1997. *Early Jesuit travellers in Central Asia, 1603–1721*. Delhi: Asian Educational Service.

Whyte, Lancelot L. (ed.). 1961. *Roger Joseph Boscovich, S. J., F. R. S., 1711–1787: Studies of his life and work on the 20th anniversary of his birth*. London: George Allen and Unwin.

Widerman, Charles J. 1942. *Erich Wasmann S. J. 1859–1931*. San Francisco: Wasmann Biological Society.
Witek, John W. (ed.). 1994. *Ferdinand Verbiest (1623–1688) Jesuit missionary, scientist, engineer and diplomat*. Nettetal: Steyler.
Woolf, Harry. 1959. *The transits of Venus. A study of eighteenth century science*. Princeton: Princeton University Press.
Worcester, Thomas (ed.). 2008. *The Cambridge companion to the Jesuits*. Cambridge: Cambridge University Press.
Wulf, Theodor. 1933. *Die Faden-Elektrometer*. Berlin: Dümmlers.
Zas Friz, Rosano. 2007. Espiritualidad ignaciana. In *Diccionario de Espiritualidad Ignaciana*, ed. Grupo de Espiritualidad Ignaciana, 811–820. Bilbao: Mensajero and Santander: Sal Terrae.
Ziggelaar, August. 1971. *Le physicien Ignace-Gaston Pardies S. J. (1636–1673)*. Copenhagen: Odense University Press.
Ziggelaar, August. 1983. *François de Aguilón S.J. (1567–1617) Scientist and Architect*. Rome: Institutum Historicum S.I.

Index

A. Udías, *Jesuit Contribution to Science: A History*,
DOI 10.1007/978-3-319-08365-0